INTERNATIONAL LOGISTICS: GLOBAL SUPPLY CHAIN MANAGEMENT

International Logistics:
Global Supply Chain Management

By Douglas Long

Distributors for North, Central and South America:
Kluwer Academic Publishers
101 Philip Drive
Assinippi Park
Norwell, Massachusetts 02061 USA
Telephone (781) 871-6600
Fax (781) 681-9045
E-Mail <kluwer@wkap.com>

Distributors for all other countries:
Kluwer Academic Publishers Group
Post Office Box 322
3300 AH Dordrecht, THE NETHERLANDS
Telephone 31 786 576 000
Fax 31 786 576 254
E-Mail <services@wkap.nl>

 Electronic Services <http://www.wkap.nl>

Library of Congress Cataloging-in-Publication Data

Long, Douglas (Douglas C.)
International logistics : global supply chain management / by Douglas Long.
   p. cm.
   Includes bibliographical references and indexes.
   ISBN 1-4020-7453-0 (alk. paper)
    1. Business logistics 2. Physical distribution of goods--Management. 3. Freight and
    freightage. 4. Shipping. 5. International trade. I. Title

HD38.5.L683 2003
658.7--dc22   2003058808

This Book is Dedicated to the Memory of my father, Albert Long

Previously published as "International Logistics and Transportation"

CONTENTS IN BRIEF

TABLE OF CONTENTS

SECTION I
GLOBAL OPERATIONS MANAGEMENT

List of Illustrations

BY PETER M. TIRSCHWELL
Editor in Chief
The Journal of Commerce

When we think of international logistics, what comes to mind might be a profession or an area of academic study. But though entirely true, international logistics is perhaps best described as a process—specifically the process by which international trade in merchandise (as opposed to services) is carried out in a real world, operational sense. For nations to generate trade surpluses or deficits, for companies to seek out and exploit new markets overseas, for foreign direct investment to be successful, trade on the ground must ensue.

And that means putting into motion the complicated, multifaceted process of getting goods from here to there, with as much foresight and as few delays and hiccups as possible, because that means added, unanticipated cost. Choreographing the movement of goods through often congested seaports and airports, onto railcars, ships, trucks, or planes and sent across borders—making them automatic objects of suspicion by governments the same way people are screened by customs authorities as they disembark weary-eyed from airplanes, passports in hand—is the essence of international logistics. It's a field that's been long scoffed at as unglamorous, but many intelligent, highly capable people have made careers doing it, and say they wouldn't have traded it for anything else.

International logistics, it should be stated clearly, involves far more than merely transportation, though that is of course an indispensable component. Because we're talking specifically about those goods moving across national borders, this brand of logistics is in a category all by itself. It's a process that integrates trade finance, transportation, insurance, customs, export controls, trade agreements and cultural nuance.

The system of payments and insurance for goods moving internationally, for example, is treated differently from purely domestic transactions because of the complexity and the often higher risk associated with such transactions. Borders, unlike state lines, create an entirely different and vastly more intricate regulatory environment that does not affect goods moving domestically. Nations are sovereign entities that can, and do, demand payment of duties on imports (though generally in lower amounts than in the past) and restrict imports when they are perceived to threaten domestic industries or to tread upon social values such as an abhorrence of child or prison labor, or prop up totalitarian regimes.

Companies that take a cavalier attitude to these laws risk prosecution like one who evades taxes, or at the very least can acquire a bad reputation with customs authorities resulting in all of its shipments being suspect and thereby more likely to be stopped for timely and costly inspections at the border. The intersection of government regulation and international logistics goes well beyond that, however. Occasionally nations also ban or restrict exports, as the United States does in cases where specific types of products could be used for a military purpose if in the wrong hands. And since Sept. 11, 2001, the possibility that a terrorist could use the international logistics system to smuggle weapons of mass destruction into a country has added an entirely new layer of government interest in international trade, one with enormous significance for how international logistics will be carried out in coming years.

As goods move through the international logistics system, they come into contact with all of these elements, and the alert logistics manger must constantly be aware of how they come into play and interact with each other. It is therefore little surprise that for the growing number of companies that depend on international markets, the individuals watching over this process have become quite important. So much so that the logistics director at a large company can today draw a salary of hundreds of thousands of dollars.

That was not always the case.

International logistics as a term, in fact, is a relatively new construct, though the fact it has come into common use today reveals much about how this "process" has evolved and changed in recent years into the dynamic field it is today. Before international logistics there was simply transportation, and the individual at the company contracting to move goods was usually known as the traffic manager, a suitably unglamorous status that reflected how transportation was seen within the corporate environment- as a cost of doing business, and little more.

Vendors to the international traffic manager, in addition to truckers, steamship companies and airlines handling cargo, included customs brokers and freight forwarders. The customs broker represented the import side, darting in and out of the local customs house to get the clients goods' cleared, and the forwarder on the export side booked the freight and assisted the client to ensure its goods weren't lost or confiscated overseas. Within the world of international trade, these were largely discrete professions. Customs brokerage and freight forwarding services usually resided within the same firm, but steamship lines never handled customs brokerage, and terms like Third-Party Logistics or 3PLs, had yet to be coined.

But beginning slowly in the 1980s, and picking up steam in the 1990s, the world of logistics began to undergo significant change. U.S. deregulation of trucking and railroads in the early 1980s brought freight rates down substantially, giving companies a taste of what was possible by way of profitability and operational enhancements emanating from the gritty and overlooked loading dock. The ocean shipping container, invented in the 1950s but by the 1980s ubiquitous in seaports around the world, enabled international shipments to move from inland point to inland point without having to be unpacked and reloaded into trucks near the dock.

Meanwhile, academic theorists began thinking expansively about how transportation fits into the overall scheme of corporate operations. The "supply chain" became a common phrase, representing a series of interconnected events beginning with sales forecasts, progressing through the ordering, manufacturing, transportation, warehousing and distribution, and finally ending up with the final sale to the customer. A well coordinated supply chain could significantly reduce the amount of inventory—and thus expensive capital—that a company has committed at any given time, savings that reflect directly on the bottom line.

When seen from this perspective, transportation is hardly the isolated cost of doing business that it was for years, but rather a key enabler of corporate profitability to the extent that the opportunities it presents are recognized and maximized. Automobile companies were among the first to exploit the opportunities inherent in logistics through "just in time" manufacturing, in which components arrive at an assembly plant precisely at the time they will be used, avoiding unnecessary inventory and obviating the need for expensive warehouse space. By the time the Millennium arrived logistics planning had permeated a good portion of the Fortune 500 and many mid-market companies as well.

This trend in corporate thinking did not go unnoticed by those companies that provide international transportation and trade services. But it was not until the early 1990s that companies heretofore specializing in various segments of the international logistics process - freight forwarding, customs brokerage, steamship, air cargo, and trucking or information technology - began to see opportunities in providing expertise beyond their traditional business to companies looking to logistics to improve their bottom line.

Thus was born the 3PL. Unburdened by capital intensive ships, plans, or other transportation assets, freight forwarders and customs brokers, such as Fritz Companies of San Francisco and Expeditors International of Seattle, Washington, initially moved the fastest in expanding their business beyond their traditional core competencies, often to take advantage of the surge in exports from Asia, especially China. Their revenues and earnings grew apace, and the sector began attracting more attention from Wall Street. Rapid consolidation in the field led to the emergence of other traditional forwarders as global 3PL players such as European companies Exel, Schenker and Panalpina.

The 1990s saw this trend taken to yet another level, as companies began demanding that their logistics providers be global in scope and on the cutting edge in Internet-based information technology. Armed with billions in capital from a record initial public offering, and needing new avenues for growth UPS, traditionally a domestic package delivery company, has quickly emerged as a global leader in logistics, thanks in part to its acquisition of Fritz. But it is not alone on the global stage. Entering the field from a different direction has been Deutsche Post, the privatized German postal carrier, which has absorbed companies as diverse as U.S. air cargo forwarder AEI, the Swiss forwarder Danzas, and the package delivery company DHL, under which it is now branding its worldwide logistics services.

Even traditional steamship lines, not content to rely on the unpredictable and often unprofitable business of ocean container transportation, have increasingly been looking to augment their services with logistics offerings, with Maersk Logistics, APL Logistics and OOCL Logistics among the leaders among ocean carriers. Ocean carrier companies have been assisted by the 1998 partial deregulation of the ocean shipping business in the U.S., which allowed contracts between carriers and their customers to be confidential. That is a marked change from the days when all contracts had to be placed on file for public inspection with the Federal Maritime Commission. Beyond that, some firms have entered the "4PL" business in which they become lead contractors, managing other logistics companies a company might have hired.

But as far as logistics has progressed—as a subject of academic study, as a phenomenon driving higher profits, and as a business that has turned traditional transportation and forwarding on its head—it remains very much a work in progress. The progress of logistics as a process that ultimately increases profits for companies has been one of stop and start progress, with many false starts, setbacks, and diminished expectations. The dot-com boom of the late 1990s gave birth to dozens of logistics technology startups that sold their wares but then couldn't deliver the promised benefits, leaving many buyers disillusioned and the tech sector in a prolonged slump. But there are more fundamental issues that hold back logistics. One is the result of a paradox. For a company to truly take advance of the logistics expertise a company like UPS might offer, the company must do what makes few CEOs comfortable –open the very heart of their operations to an outsider. It's one thing to pay a shipping company to move your goods from point A to point B, or to pay a customs broker to get your goods cleared into the country. It's something else entirely to let that company know when, if you are Wal-Mart, for example, you plan to hold a sale of lawn furniture, information that competitors would love to get their hands on. One of the main reasons for the failure of relationships between companies and their logistics providers is a lack of information provided to the other side to allow them to the job they said they can do.

It is one of the reasons many companies still consider logistics to be a core internal competency and have no plans to outsource. Some companies, in fact, have felt themselves to be so proficient at logistics that in an effort to generate revenues they started to make available their logistics departments' services to other companies.

Overall, some studies have indicated that logistics has stalled in terms of its ability to be able to deliver consistently lower transportation and inventory carrying costs for the U.S. economy.

But international logistics is hardly a discipline that's sitting still. In fact, yet another revolution in logistics thinking is taking shape, driven by a consideration few had ever thought about, much less done anything about prior to Sept. 11—security. It was clear the moment the dust settled from the terror attacks on New York and Washington that that an international logistics system full of security holes is no longer tolerable from a national security perspective. In the U.S., Customs, the Coast Guard and other agencies with direct involvement in imports such as the Agriculture Department's Animal and Plant Health Inspection Service have been moved into the new Department of Homeland Security, indicating the vested interest the U.S. government has in supply chain security.

Shortly after Sept. 11, the Customs Service began the long process of securing the supply chain by creating a voluntary program called the Customs-Trade Partnership Against Terrorism, or C-TPAT, in which companies agree to take steps to ensure security in their own supply chains in return for fewer inspections and holdups for their goods at the border. Customs also began putting its own agents at key foreign seaports and, in a significant and controversial move, began demanding that information on the contents and key parties involved in all ocean shipments be transmitted to Customs 24 hours before it is loaded on a ship at the foreign port. The message is clear: the government is suggesting that it wants to know as much about the whereabouts and circumstances of inbound containers as the private sector knows itself.

But herein many observers see the next big step forward for international logistics. Here is the reason: information provided in advance of loading at the foreign port is just the start of a so-called "total information awareness" the government will eventually want with regard to all cargo shipments coming into the country. Foreign governments, similarly concerned about the scrooge of terrorism, are likely to demand similar information. The government, in essence, is demanding information about the current whereabouts and exact contents of shipments for security purposes. For everyone involved, providing this information is costly, and many predict that smaller freight forwarders and logistics firms that don't have the resources to comply will eventually falter and disappear.

But interestingly, this is the exact information that logistics experts have long argued is needed in order to maximize the value a company can extract from its logistics operations. If you know where a shipment is at any given time, you can divert that shipment en route if it's needed elsewhere. If you know precisely what is inside a shipping container, you can be that much more accurate when alerting other parties about what exactly is arriving, and when. This is a simplified description, of course, but many have long said that logistics, in its most useful form, is nothing more than information made available. As companies comply with government mandates, they could in the process be vaulting themselves into the next era of international logistics.

This book is the result of several years working in logistics, teaching international business, and talking to professionals and academics in this field in every continent of the world. It is also the result of generous contributions from working professionals who reviewed the manuscript and wrote case studies. Ten translated versions are currently being prepared, which will make this the first logistics textbook every published in so many languages. In doing this, I have been able to gain insights into how logistics is done in those other regions, especially those parts of the economy that are not inherently international.

This book deals with **theory, management, and practice**. Writing a book with three different audiences on a subject that is constantly changing has proven quite a challenge. However, there are some general principles to the subject that have stayed the same. Theory and research helps the reader understand not just what is happening, but why and how. It is also important for those who want to generalize on the lessons learned to other fields.

The managerial approach is for business students and professionals who need to take these concepts and put them into practice. In fact, the academic audience is also interested in seeing how their theories end up when put into practice. The managerial approach is more about strategic decisions, but there is also the third approach, which is practical and technical. No matter how much management experience one has, there are some nuts and bolts things about the industry that one must know. I have taken great care to identify these tactical issues without burdening the reader with too much detail and turning the book into a how-to manual.

It should be noted, however, that I am beholden to no company or product/service. While some companies are gaining exposure in this book, I have made every effort to steer clear of bias and favoritism. It would be impossible and undesirable to discuss an industry without mention of examples or key players. There is no geographic emphasis in this book. While some areas of the world have much more economic and trade activity than others, that is not an excuse to ignore the rest of the world. In fact, fortunes are to be made identifying those opportunities that will arise in the future. What may now be a poor and undeveloped region only means that it is waiting for people with energy and ideas to develop it.

This book is not designed as a reference book for those who need specific rules, dimensions, etc. I recommend for that purpose that one refer to Edward Hinkelman's "Dictionary of International Trade".

An Instructor's Manual is available, with 1,000 essay, multiple choice and true/false questions for each chapter. Instructors should contact the author directly.

Acknowledgements

First and foremost, Professor Don Wood provided me the opportunity to teach at San Francisco State University, and introduced me to the academic world of logistics. He has been an inspiration and mentor, as well as an accomplished author and expert in the field. Also at San Francisco State are the fellow faculty members who have provided their support, time and advice, particularly Edwin Duerr, Joel Nicholson, Dan Wardlow, Yim Yu Wong and Tai Furuse.

The first edition was completed with the generous assistance of Christian Perlee, and the second edition with Barbara Duhon, both of McGraw-Hill.

A few individuals distinguished themselves for their generous assistance and have earned my eternal gratitude, including Allison Acosta, Eric Bernhardt, Amy Bishop, Michelle Covey, Maria Aguirre, Ilona Brusil, Shawn Fallah, Ro Leaphart and M.K. Wong. They went beyond the call of duty and this book would not be what it is without them. I would also like to thank, in no particular order, the following for their contributions. Rick Dawe of the Fritz Institute of Global Logistics, Oleg Rashupkin, Dimitry Sintsov and Max Kolchin of Fritz's Moscow office, Bob Delaney and Judy Earle at Cass Logistics, Karin Jones, Dana Siverling and Mary Kane at Boeing, Cecilia Mather and Deirdre Fitzpatrick at the International Transport Workers Union, Al Graham of AirServe, Andrea Manning and Susan Paulson of The Council of Logistics Management, Maria Rosales of San Francisco International Airport, Jane Gaboury of IIE Solutions Magazine, Airports Council International, Michelle Lewis of Lonely Planet, Paul Svindland of FastShip, Elizabeth McCall, Joe Santarelli of Mersant, Evelyn Benson of the International Air Cargo Association, Silke Roesser and Martin Bahr of CargoLifter, Tim Barker, Kiyoko Gardner of The Economist, Allison Acosta of Apex, Oscar Flores and Hadis Fearn of American President Companies, David Hoppin of MergeGlobal, Stanley Shen/Debbie DiLauri/Jeff Beason/Ron Wolf of OOCL, Craig Weicker, Wanda Bailey/Scott Mungo/David Littlejohn of FedEx, Shawn Fallah of U.S. Customs, Debbi Hall, Paula Copeland of Centennial College (Toronto), Joseph Englert of Export Assist, Inc, William Cruze/Carin Saunders of Manalytics, Claudia Wedell of Kuehne and Nagel, Tony Phillips of Transport Management Systems, Jeff Engels of US AID

(Armenia), Marcus Weiss of Lufthansa, Marcia Borja, Dale Coker of Educational Development International, Professor William Wagstaffe of Golden Gate University, Claudia Wedell, Edward Emmett of NIT League, Alejandra Rapsis of 3Com, Matheen Sait of 3M United Arab Emirates, Sharon Eckroth of GSA Transportation Management, Richard Burkhard of San Francisco State University, Hamza M. Osman, General Manager of the Sea Ports Corporation, Port Sudan, Captain Omer A. Siam, Nasser A. Sidki, Halim Mohammed, Richard Hallal of Logistics Development Corporation, John Gerber, Paul Piai and Nikki Anderwson of Lonely Planet, Steve Averett of IIE.

"The Logistics of Famine Relief" reprinted with the permission of the Institute of Industrial Engineers, 3577 Parkway Lane, Suite 200, Norcross, GA 30092, 770-449-0461. Copywright© 1977.

Lonely Plant maps reproduced with permission from Middle East, Edition 3© Jan 2000 Lonely Plant Publications and Central America, Edition 3©Sep 1997 Lonley Plant Publications.

Container illustrations reproduced by permission of World Trade Press, (c) Copyright 2003 by World Trade Press. All Rights Reserved.

OOCL photos reproduced by permission of OOCL(USA).

Graphics were created by Sharon Till and Neil Ishikawa of Sharon Till and Associates of San Francisco.

The Instructors Manual and Test Questions were written and prepared with the assistance of John Gerber.

I am eternally grateful for all the assistance and support received on this project. This book would not have been possible without their input. However, all mistakes, omissions and shortcomings are my own.

Comments and questions are gratefully accepted, and have been the cause of many improvements to this second edition. I may be contacted at www.douglaslong.com, or email at dclong@aol.com.

Douglas Long's research and work has taken him to over 120 countries. He teaches at San Francisco State University and has guest lectured in several universities worldwide. He worked for Orient Overseas Container Line's pricing department for several years. Prior to that he managed logistics in the U.S. Army throughout Asia-Pacific. In addition to his work in logistics, he has managed elections for the United Nations and other international organizations in Bosnia, Kosovo, and East Timor.

dclong@aol.com
www.douglaslong.com

Section I
Global Operations Management

1. Introduction to International Logistics

2. Global Procurement and Trade

3. Global Supply Chain Management

4. Strategic Planning

Chapter 1
Introduction to International Logistics

This chapter introduces the field of logistics, including its many parts. Much of this task involves separating the numerous terms being used in the industry. Logistics is described in the widest sense, although this book concentrates on one aspect, cargo. The section on regional logistics assessment provides a convenient way of understanding the how difficult or easy logistics will be in an area based on geography, infrastructure and institutions. Transportation is the central issue of logistics, which is why it is introduced here and explained in much greater detail in Chapter Five. The brief history of logistics is provided because it helps to understand how this industry came about, as well as to see where it is heading.

WHAT IS LOGISTICS?

Imagine you have just been hired by General Motors, one of the largest corporations on earth, to manage logistics. What does it mean to manage logistics? To begin with, General Motors acquires parts from suppliers all over the world. They assemble cars in factories all over the world, and sell them all over the world. Imagine one single model of car, and the thousands of parts that go into that one car. Now imagine all the places those parts come from. That is one model, mind you, and General Motors makes over a hundred models of cars, trucks, vans, and other vehicles. Someone needs to manage the shipment of all the parts and vehicles so that they arrive where they are needed when they are needed.

It is not one of the most glamorous jobs, but as we can see from the example of General Motors, it is far more important than most of us realize. It can even be a matter of life and death. The food we eat comes from far away places. If there were not the ability to distribute the food, the result would clearly put us all in great danger. Almost everything we have in our homes comes from different places, and logistics is what makes it possible to buy these things at a reasonable price. Management guru Peter Drucker referred to it as "the economy's dark continent" and said logistics is the most neglected and most promising business area.[1] This was written in 1962, but businesses are still learning how to use logistics effectively.

Logistics is the job of getting things to where they need to be.

The definition of logistics is not without a little debate. Logistics has been defined by the Council of Logistics Management (CLM) as "that part of the supply chain process that plans, implements, and controls the efficient, effective flow and storage of goods, services, and related information from point of origin to point of consumption in order to meet customer requirements".[2] Note that their definition does not include the movement of people. What the CLM considers logistics could more accurately be referred to as **business logistics**. When one considers the overall field of logistics, it would be necessary to include the movement of people.

Business Logistics

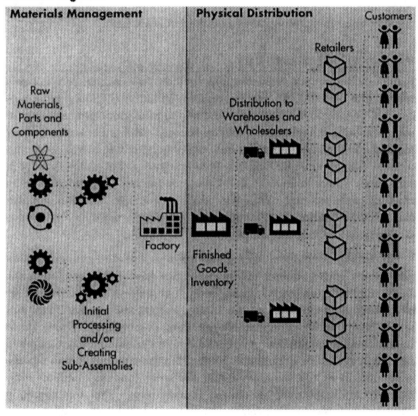

The logistics of **moving people is fundamentally different from moving cargo,** which is why it is important to distinguish between business logistics and general logistics. When moving cargo, there are issues about ownership. When people travel, they operate on their own initiative. Cargo does not move itself, whereas people make their own decisions and take their own initiative. People usually make a round trip, whereas cargo rarely does. If anything, cargo keeps moving from one location to another, often changing form in the process.

There are also the differences in travel patterns. The kind of places people travel to include vacation spots or metropolitan areas, whereas cargo tends to move from the source of raw materials to industrial areas and only at the end of the line to metropolitan areas to be consumed by people. The modes of transport are also different. The road, rail and ocean shipping industries are usually for either cargo or people. It is relatively rare that these carriers include both people and cargo. Air cargo is unique in that most cargo is moving on passenger airlines.

Council of Logistics Management's Definition of Logistics
That part of the supply chain process that plans, implements, and controls the efficient, effective flow and storage of goods, services, and related information from point of origin to point of consumption in order to meet customer requirements.

Supply chain management is logistics taken to a higher level of sophistication. A supply chain is the movement of goods not just from one place to another, but from the ultimate origin (presumably when minerals are taken out of the earth) to the ultimate destination. Imagine, for example, the book you are reading. It was produced in a print shop. That print shop bought paper from a paper company. Moving the paper from the paper company to the printers is one link in the chain. Now take a step backwards from the paper company, and notice that the paper came from a paper pulp company, which in turn got the pulp from a logging company. Put all these links together, and we have a supply chain. This important concept is the subject of Chapter Three.

Logistics Evolution to Supply Chain Management

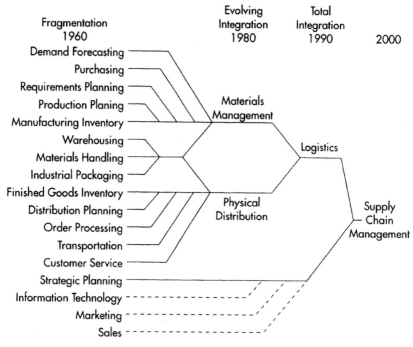

Reproduced by permission John J Coyle.

Council of Logistics Management's Definition of Supply Chain Management

Supply Chain Management is the systemic, strategic coordination of the traditional business functions and the tactics across these business functions within a particular company and across businesses within the supply chain for the purposes of improving the long-term performance of the individual companies and the supply chain as a whole.

WHAT IS THE GOAL OF LOGISTICS?

Logistics may be about getting things to where they need to be, but it is not the same as transportation. While transportation is important, logistics is much broader. According to Donald Bowersox and David Closs, noted authorities on logistics, it requires coordination of many activities that surround and control transportation, including network design, information, transportation, inventory and warehousing.[3]

Network design. The way companies organize themselves and other companies for the best logistics network.

Information. The information needed to coordinate the logistics operations.

Transportation. The physical movement of goods.

Inventory. The storage of goods.

Warehousing, material handling and packaging. The handling and physical management of goods.

What is **the goal of logistics?** Bowersox and Closs described six operational objectives of a logistics system:[4]

- **Rapid response.** A company needs to be able to react quickly to changes or new developments. Often the ability to provide a customer what they need is the key to getting their business.
- **Minimum variance.** The output, such as delivery times, should be consistent.
- **Minimum inventory.** Inventory is expensive and needs to be kept at a minimum.
- **Movement consolidation.** Transportation costs can be reduced by consolidating many small shipments into bigger, less frequent shipments. This is not always so easy, as we will learn.
- **Quality.** Not only do products need to be of the highest quality, the logistics service also needs to conform to quality standards.
- **Life cycle support.** This refers to the need not just to deliver a product, but handle returned product as well. This may be the return of defective products, or recycling of the packaging and product.

Logistics is not simply about getting things to where they need to be, but to do so in a competitive market environment where other companies are eager to take your customers when possible. According to Bowersox and Closs, "the overall goal of logistics is to achieve a targeted level of customer service at the lowest possible cost".[5] One common definition of logistics, to be discussed shortly, included 'effective and efficient'. What does that mean? There are a few standards by which we judge effectiveness and efficiency, including reliability, speed, information flow, cost and control. While this book's definition left out terms such as 'effective and efficient' for the sake of brevity, companies that are sloppy with logistics may soon find that they do not have any customers.

Value-Added Analysis of Global Logistics

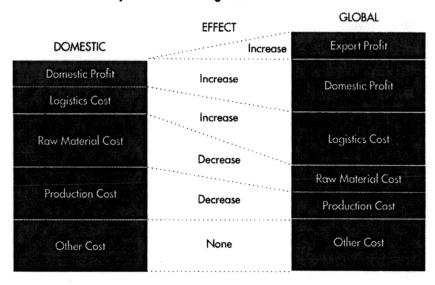

There is a limit to how good a product can be, but logistics can add further value. One can produce the highest quality product possible, but there is clearly a limit to what that product can do. For example, consider pizza delivery. While the history of pizza delivery is a little vague, there must have been a point at which pizza restaurants realized they can only go so far in making the best tasting pizza possible. Yet one thing that pizza restaurants offer today that they did not in the past is home delivery. Companies have found that once they are producing their product to the highest standards of quality possible, they need something else to gain an edge over the competition. This edge is often logistics; delivery service, on-time delivery, inventory that does not run out, and so forth.

A couple of other terms need to be clarified. **Materials management** is what a company does with inputs. In other words, it is the way they acquire materials such as raw materials or parts, how they handle them once they arrive at the company, and how they are shipped out. **Physical distribution** refers to the way a company delivers its product to the market, which could be the customers or retailers.

Business logistics emphasizes the role of logistics in the overall firm, and refers to the most advanced level of integrating the operations of a company, such as finance and marketing, into the logistics job. For example, transportation decisions will have financial and marketing effects, and it helps to recognize this and involve those areas from the beginning.

Integrated logistics is similar but emphasizes the need to coordinate with suppliers and customers. Why would it be difficult getting things to where they need to be? That sounds like an easy job. However, logistics involves other companies. This is the major challenge. Making the different parts within a company work together is difficult enough. Effective logistics requires the coordination of other companies, organizations and individuals. Making these outsiders do what is needed for the benefit of the company is why this is such a challenging and rewarding career.

Integrated Logistics

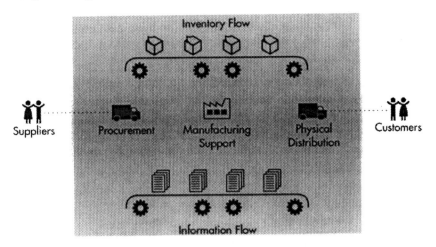

One reason integrated logistics does not always happen is that it requires a lot of work and expense. Managerial talent and time needs to be invested to bring together the different companies or the different departments within a company. Procedures and systems need to be rearranged. Such an investment needs to be justified by the benefits that will be gained.

One term that can get confusing is **operations**, the basic job of a company. If you are Siemens, the operations would be the manufacturing of electrical goods (they are a ▬▬ge and ▬erse company, so I am simplifying the example here). But if you are Siemens, functions such as logistics, human resources, and finance are not your operations. While they are important for success, Siemens does not operate for the sake of managing logistics. This confusion arises because many books and universities discuss logistics as part of operations. Actually, logistics is a **support function** to operations, just like human resources, finance and other functions. The only time logistics is the company's operations is in logistics and transportation companies.

Looking at the logistics from the point of view of a firm, there are three distinct areas. **Inbound logistics** includes sourcing and materials management. **Operations logistics**, closely related to material management, emphasizes the way logistics affects operations. **Outbound logistics**, also known as physical distribution, refers to the way the product is physically delivered to customers. Why is this distinction important? In the case of inbound, you are the customer. With outbound logistics, you are the seller. As we will discuss at length in later chapters, your ability to get things done will be drastically affected by your powers as either a buyer or seller. Also, the inbound/outbound logistics involves other actors, whereas operations are internal.

One also hears the terms **inter-firm logistics**, which refers to that which is between companies, and **intra-firm logistics**, which is within the firm. Whether it is referred to as inter-firm or inbound/outbound, the point is that the cargo is within a firm. However, that does not means the cargo is in one location. The firm may be a massive multi-national corporation with facilities all over the world.

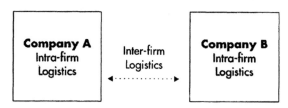

Intra-firm logistics is also known as Materials Management

There are a couple of different ways to study logistics. This book takes the view of a businessperson. What we mean by the **business view** is to look at those aspects of logistics that affect the business issues. This includes, for example, the financial costs associated with a given decision, or the role of marketing in designing a logistics system. In contrast to the business view is the **engineering/operations view**. This view puts the emphasis not on business but on the physical and quantitative aspects. It attempts to find an optimal solution for the system design, whereas the business view puts more emphasis on customer needs. They are both equally important and valid views of logistics.

International logistics is special for a variety of reasons. The business environment varies around the world the world. Cultures are different, and this affects how business and logistics are handled.[6] The role of culture is an area that has not been given the attention that it deserves. As a field of study and a profession, different countries take very different views of logistics. The French view is more like operations, in which there is little difference between the managing of the operations and the managing of logistics. Russia has not developed much of a field of logistics, although obviously Russian companies are doing logistics work. One will have a difficult time finding any courses in the universities there using the term. In Japan, logistics has been regarded as an important factor in their economic development.

How much is spent on logistics? The question seems simple, but analysts have come up with various answers depending on how they define the field. According to one study, $2.89 trillion was spent on logistics (in 1996).[7] Fortune Magazine cited 10% of the price of all goods in the US as attributed to the cost of logistics.[8] Cass Logistics estimated that in the U.S, 10.7% of GDP goes to cover the cost of logistics, and total spending in the US was $600 billion, twice what was spent on national defense.[9] William Copacino estimates international logistics costs range from 25-35% of a product's sale value, whereas domestic shipments it's only 8-10%.[10]

Spending on logistics is important because, given a global economy, controlling this cost is imperative for the success or failure of international ventures. Logistics spending seems to be going down. According to the Michigan State study, there was a 3.6% reduction in logistics spending over the years 1992-96. Ironically, this does not mean that logistics is less important, but that companies are becoming more efficient. More recent research indicates that this downward trend reversed, and there has been an increase in logistics spending. The reason is not that companies are less efficient, but that they have made a conscious choice for faster delivery using air freight over cheaper ocean shipping. In other words, logisticians are choosing better customer service (by faster delivery) over cheaper delivery costs.

The Business Logistics System
Accounted for 10.1% of Current GDP in 2000

Carrying Costs—$1.485 Trillion All Business Inventory	$ Billions
Interest	95
Taxes, Obsolescence, Depreciation, Insurance	204
Warehousing	78
Subtotal	377

Transportation Cost

Most Carriers:	Truck - Intercity	323
	Truck - Local	158
	Subtotal	481
Other Carriers:	Railroads	36
	Water (International 18, Domestic 8)	26
	Oil Pipelines	9
	Air (International 8, Domestic 19)	27
	Forwarders	6
	Subtotal	104

Shipper Related Costs	5
Logistics Administration	39
Total Logistics Cost	**1006**

Source Cass Information Systems, Inc / ProLogis

Spending on logistics depends primarily on **three factors**. First is the **level of economic activity** and trade. The more things people buy clearly has a direct effect on the need to move those things. The second factor is **efficiency**. The more efficient the logistics system, the less it is going to cost. Finally, there has been a shift in the developed countries away **from goods and toward services**. People want fewer things and more services, such as travel, entertainment, and so on. This has two contradictory effects on the demand for logistics. On the one hand, people are buying fewer goods, so there is less cargo moving. On the other hand, they want faster and better delivery of their purchases, more specialized products, and more variety, all of which means a greater demand for logistics.

Index of Physical Distribution, Transportation and Inventory Carrying Costs as a Percentage of GNP

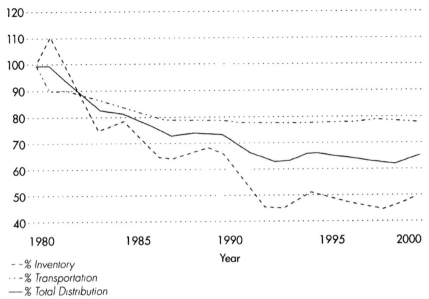

- - % Inventory
· · · % Transportation
—— % Total Distribution

Source Cass Information Systems, Inc / Prologis

REGIONAL LOGISTICS ASSESSMENT

So far we have been discussing businesses, but now we shift to a higher level. There are times when you want to know about the abilities of your business partners, and there are times when you want to know about the region you are working in. Assessing businesses for their logistics capabilities will be taken up shortly, but now we will discuss how to assess a region. Countries and regions can be assessed for their logistics assets and abilities. A **regional logistics assessment** is important because some areas of the world have excellent logistics assets that contribute to their prosperity. Other areas of the world are very deprived and suffer economically and in other ways. A region assessment looks at logistics assets including geography, physical infrastructure, and institutional infrastructure.

Geography. Some geographic features are natural aids while others are obstacles. Natural ports, such as a bay, are one of the most valuable features that any country can have. Many of the greatest cities in the world are centered on natural ports: Istanbul, Rotterdam, Bombay, and San Francisco, to name a few. Overland access is the ability to travel across the land. Flat land with firm ground is the best, while on the other extreme mountains and marshes make overland travel difficult or impossible. Central Europe and the US benefit from excellent geography; this is why industry is centered in the Ruhr Valley in Germany, and the Ohio Valley of the US. On the other hand, Chad is the poorest country in the world. They have no direct access to the ocean, and only 200 miles of all-weather roads.

Physical infrastructure. Having good geographic features does not help if there is not infrastructure, such as roads, ocean ports, or airports. Public infrastructure makes an enormous difference to a company's ability to operate. No business can avoid the consequences, good or bad, of where they are located. There are two types of infrastructure, the macro infrastructure, which includes things like bridges and roads. There is also the micro infrastructure, such as road signs.

Infrastructure needs years, and even decades, to build. However, it also lasts for a long time and is difficult to eliminate or destroy. War and civil conflict can only do limited damage to ports and roads. Road systems are often the paths begun by animals hundreds or thousands of years ago, that then become the natural route followed by humans. Infrastructure will still be there after years of neglect by a government in financial crisis.

> The greater the gap between the state-of-the-art (transport and trade facilitation) system and the system available in a trading bloc, the greater the penalty that specific regional trading bloc will pay in terms of foregone trade and economic growth.
>
> T.R. Lakshmanan[11]

Legal/business infrastructure. Once all the physical parts of the logistics system are in place, there is the institutional part. Modern businesses require a legal system with rules of trade and commerce, customs officials, and legal enforcement of business contracts. There is also a need for banks to provide financing. Trade and logistics requires a wide variety of services provided by other business and the government. Without these services it would not be profitable to do business, regardless of how good the infrastructure or ports may be. Once again, those areas that are most economically developed have proper legal systems[12], and a full supply of logistics services. Developing economies seek to build from the inside or attract from the outside banks, freight forwarders, investors and so on, to provide these needed services.

Two Countries Compared: Switzerland and Chad

When it comes to assessing the logistics environment of different countries, one cannot find more of a contrast that Switzerland and Chad. These differences serve to illustrate the role of geography, infrastructure and institutions.

Chad is landlocked in the middle of Africa surrounded by countries that also suffer from poor transportation, unstable governments and little money. With 500,000 square miles of land, it only has 200 miles of all-season road. Their per-capita GDP is $600, one of the poorest in Africa, and thus the world. Ironically, it is rich in minerals such as oil, uranium, gold and diamonds and has 50 million acres of arable land. Yet civil strife has prevented economic development since the 1980s.[13] Development for Chad includes money from the World Bank, the European Union, Germany the OPEC Fund and France. There are plans for a 700-mile oil pipeline to tap the estimated 900 million barrels of oil and bring it to a port in Cameroon.

Switzerland, on the other hand, has almost all of the resources needed for world-class logistics, though they do face some special challenges. Geographically, they are landlocked, like Chad. Instead of being surrounded by endless desert, Switzerland is crossed with high mountains. On the positive side, their infrastructure is probably the best in the world. They have an extensive and well-maintained network of roads, railroads, airports, and other infrastructure. In fact, they have overcome the barrier of the Alps with some of the most dramatic tunnels in the world, and the world's steepest railroad.[14] Their institutions include all the services necessary for trade and commerce, including the famous Swiss banks.

This does not mean that Switzerland is the easiest place in the world to manage logistics. They also have a heavy tax burden and laws that can make one wonder how anything gets done. Most work is forbidden on Sundays, which means trucks traveling through Europe often end up camped at the border waiting for midnight, Sunday, before they can continue their journey.

This is just an example of what to look for in comparing or assessing regions. Many firms, for example, need to decide where to locate a warehouse or enter a market. This assessment is useful at the strategic planning phase.

Some countries are much more competitive than others for a variety of reasons that we have just discussed. The overall situation can be viewed by the amount of money spent on logistics. According to one study, the most efficient nations in terms of logistics spending are the United States (10.5 percent), the United Kingdom (10.6 percent), France (11.1 percent), Italy and the Netherlands (both about 11.3 percent). The countries that have the greatest opportunity to improve their logistics spending were found to be Germany (13 percent), Spain (11.5 percent), Mexico (14.9 percent) and Japan (11.4 percent).[15]

The astute reader may notice that these numbers seem high. These statistics are for the **entire nation's spending**, which includes all forms of transportation, including passenger transportation. **Company spending on logistics** tends to be very different. For example, the average American manufacturer spends about 4% of their total costs on transportation, but even this varies dramatically depending on the type of industry.

The Three Logistics Environments[16]
Military
Business: Product, Service
Civil: Public, Private

The most basic unit of logistical analysis in a business is the **performance cycle**, also referred to as **fulfillment**. This is defined as "those activities involved from the point of recognizing demand from a customer through delivering the goods or services to the customer, complete with an invoice or similar instrument to facilitate the collections of payment".[17] In a simple example of a performance cycle, imagine you are working at your job and your chair breaks. You call the office supply company and order a chair. Working at a desk is pretty much impossible without a chair, so you pay extra for an overnight delivery. The next day there is an express-package delivery person at your door with the new chair. The performance cycle began when you called for the new chair, and ended when the chair arrived.

The performance cycle includes a number of different actions going on simultaneously. There are actually **three different channels** in action during any performance cycle:

- **Transaction**. This refers to the purchase and sale of something.
- **Distribution**. The physical movement of the product.
- **Documentation/communication**. The documents that go along with the transaction and the communication of related information.

We do not live in an ideal world where everything happens flawlessly. Mistakes happen. That is why another goal of logistics is to deal with problems and fix mistakes. The ultimate goal in a performance cycle is what is known as the **perfect order**, in which every order is complete all the time. Imagine if every time you went to a grocery store or a department store, they had everything you wanted, in unlimited quantities and endless colors and sizes. That, obviously, is only an ideal. Another goal, and one that is more realistic, is **immaculate recovery**, in which mistakes are fixed before the customer is aware of or inconvenienced by the mistake. Recall the last time you asked at a store if they have a certain product, and they said no, but their other store across town does have it in stock. Would you like it delivered or can you come back tomorrow? This is an example of how companies seek to offer immaculate recovery.

THE ROLE OF TRANSPORTATION

Transportation plays a central role in logistics, and especially international logistics. In order to 'get something to where it needs to be', the most important of all the logistics tasks is the physical transportation. When we shift from domestic logistics to international logistics, the role of transportation becomes even more important.

Transportation deserves to be studied for a variety of other reasons. It is a big industry in itself, with billions of dollars being spent on airlines, railroads, trucks and ships. Transportation ties the world together. There is a mistaken belief that telecommunications has made it unnecessary to travel. If people are to work with each other, it is still important to meet each other personally. The globalization of the economy is based on the ability to transport goods around the world. When we put these reasons together, it is apparent that the neglected field of transportation deserves credit for creating the 'global village'. In fact, the person who created the field of international law, Hugo Grotius (1583-1645), began writing about transportation in *Law of Prize and Booty* (1604-5) and the issues relating to freedom of navigation in *Freedom of the Seas* (1609).

There are **three perspectives** to transportation, that of the **shipper**, the **carrier** and the **consignee**. These are terms that can be confusing and need to be understood carefully. In the field of logistics, a shipper refers to the company that owns or controls the freight, and is arranging for its transportation. This could be the owner, but not necessarily. One legal body refers to the shipper as "any person, organization or Government which prepares a consignment (any package or packages or load) for transport".[18]

This term is a little confusing because most people think of a shipper as one that runs a ship. The carrier is the transportation company that actually moves the freight. Another way of thinking about this is that the carrier carries the freight and the shipper is their customer. The shipper is also known as a consignor, one who consigns a shipment to the carrier. However, the term 'shipper' is generally used.

The consignee is the entity receiving the shipment. The shipper may be the same entity, such as a company that ships something from one of its facilities to another, but they may be completely different entities. Even if the shipper and the consignee are the same entity, it is important to note that they are **performing different functions**. The things you do as a shipper, sending out the cargo, are very different from what you would do in receiving cargo. For example, the legal implications of being an importer (a consignee) versus an exporter (a shipper) are very different.

Author's Note: In all cases were I refer to an "entity", this could be an individual, a company, a government, or some other entity. In commercial shipments, in fact, the entity is usually a company rather than a person.

Now that the terms are defined, discussion of transportation may be different depending on one's point of view as a shipper or a carrier. Shippers often do not care about the details of how the carrier operated, but just wants the cargo to arrive on time. However, transportation can be seen as an important activity in its own right, which is the carrier's view. In this book, we will mostly take the point of view of the carrier, explaining in some depth how transportation companies operate.

There are some special **characteristics of international transportation**. The most important difference is that the shipment will be international, which means that it will cross a national boundary. This is a political distinction. There has been a lot of discussion about globalization and the reduced powers of nation state. However, shipping anything across national borders is still very different and creates some special difficulties. There are the political issues. Regulations will be applied from all countries that the freight moves through. The documentation required will be much greater than a domestic shipment. In fact, many domestic shipments require no documentation at all.

The average distance of an international shipment will be longer. The modes of transportation will vary, a direct result of the longer distances covered. Instead of trucks and to trains, there will be more use of ships and planes. There is also a much greater chance that the transportation will use multiple modes of transportation, called intermodalism.

A BRIEF HISTORY OF LOGISTICS AND TRANSPORTATION

Logistics was originally a **military term**, because it is a key factor in the success or failure of many wars. In fact, many military experts would say that logistics is the key to winning a war. Historically, logistics has been a leading factor in the success or failure of many military conflicts. Transportation gave countries ability to project power domestically, regionally and globally. It strengthened the economy by promoting trade that further improved the military powers. Chinese Emperors sent their fleets throughout Asia to intimidate neighboring countries and demand tribute.

Admiral Alfred T. Mahan wrote a famous book, *The Influence of Sea Power Upon History, 1660-1783* in 1890, that influenced President Theodore Roosevelt to make the US a world naval power. During the US War of Independence, the British could move along US coast faster than Continental Army could, and thus were able to attack whatever targets were unprotected. More recently, the Gulf War was won by logistics and maneuverability, not brute force.[19]

If logistics is a critical part of military operations, it is just as important to many companies. In recent decades logistics has become an important part of business success, and that is the emphasis in this book. It is important to recognize, though, that logistics is done by public and private sector organizations, and the ideas are still basically the same. Some of the techniques are different depending on the needs of the organization, but the differences are quite small. Military logistics emphasize quick response and the flexibility to adapt to changing situations. Many businesses work in an environment where quick response and flexibility are important for commercial success.

Logistics was not always a recognized field of study or a career. In a classic book from 1915, the only two functions of marketing were demand creation and physical supply.[20] Prior to 1950s, no theoretical basis for logistics existed.[21] According to Bowersox and Closs, there were three reasons integrated logistics did not come about at first:[22]

- There was no role for computers seen in integrating functional areas.
- Volatile economic conditions led management to concentrate on cost containment.

- There was difficulty in quantifying the returns that could be gained.

Since then, a variety of things have changed to the point where logistics is now a well-recognized profession, and a valuable part of the company. Why did things change? A few reasons can be identified:

Computerization. Having the ability to manage vast amounts of information is vital to many logistics functions. Previously, the problems were simply too complex and there was a limit to the level of sophistication.

Quality management. Companies have been looking at ways to improve their operations, and recognized that logistics was one area that had been historically overlooked.

Partnerships and alliances There has been a trend for companies to work closer with their suppliers, customers and other partners to improve efficiency.

Deregulation. Certain key parts of the logistics profession were heavily regulated, which prevented them from engaging in close cooperation with their customers. This included railroads, ocean shippers and airlines. There were also anti-trust rules that prevented companies from working together lest they prove to be 'anti-competitive'.

Changes in transportation have played a mixed role in the development of the logistics field. It is important not to mistake the changes in the transportation industry for the demand of the trading public that is driving those changes. Ships, trucks and trains have seen little change in how they operate over the past several decades. Aviation has seen a lot of change because it is relatively new technology. However, the major changes in transportation came not from the transport itself but its role in the larger logistics and business practices. The different modes of transport now work together (intermodalism).

As we proceed into the 21st century, logistics is continuing to affect and influence the global economy. This book seeks to explain how this industry operates, and contributes to the changes we are seeing in all aspects of our lives. It is important to understand that the industry is constantly changing, and as soon as we come to understand some of these trends, they are already moving ahead with further developments. One thing can be said with some degree of certainty. Logistics will continue to play a central role in the global economy well into the future.

[1] *"The Economy's Dark Continent"*, Peter F. Drucker, Fortune, April 1962, p. 14.

[2] *Council of Logistics Management.*

[3] *"Logistical Management: The Integrated Supply Chain Process"*, Donald J. Bowersox and David J. Closs, McGraw-Hill, New York, 1996, p. 25.

[4] *Ibid, p. 41.*

[5] *Ibid, p. 6.*

[6] *"The Shadow Organization in Logistics: The Real World of Culture Change and Supply Chain Efficiency"*, Jo Ellen Gablel and Saul Pilnick, Council of Logistics Management, Oak Brook, IL, 2001.

[7] *"Global logistics market trebles"*, American Shipper, October 1998, p. 40. This article cited GeoLogistics, but referred to a study by Michigan State University. There has been some dispute over the role of GeoLogistics and their ability to take credit for its findings.

[8] *"They've Got Mail!"*, Brian O'Reilly, Fortune, February 7, 2000, p. 110.

[9] *"Fourth Annual State of Logistics Report"*, Robert V. Delaney, Cass Logistics, St Louis, MO, 1993, figure 10.

[10] *"Logistics: International Issues"*, Chris Steven, ed., Cleveland: Leaseway Transportation, 1985, p. 101.

[11] *"Transport and Trade Facilitation: An Overview"*, T. R. Lakshmanan, in "Integration of Transport and Trade Facilitation: Selected Regional Case Studies", T. R. Laksmanan and others, World Bank, Washington DC, 2001, p. 7.

[12] *The phrase 'proper legal system' is not a value judgment of one legal system over another. It simply means that the legal system is 'proper' if it does the job it is intended to do, such as protect citizens and their property, settle disputes and so on.*

[13] *JOC, October 26, 1998.*

[14] *The world's steepest railroad is actually for tourists, a 48 degree cogged railroad going to the top of Mt Pilatus, above Luscerne.*

[15] *"Global logistics market trebles"*, American Shipper, October 1998, p. 40.

[16] *"The Impact of Information Technology on Materials Logistics in the 1990's"*, Richard L. Dawe, Penton, Cleveland, OH, 1993, p. 8.

[17] *"Keeping Score: Measuring the Business Value of Logistics in the Supply Chain"*, James S. Keebler et al, Council of Logistics Management, Oak Brook, Il, 1999, p. 108.

[18] *IMDG Code.*

[19] *"Moving Mountains"*, Lt General William Pagonis, with Jeffrey Cruikshank, Harvard Business School Press, Boston, 1992.

[20] *"Some Problems in Market Distribution"*, Arch W. Shaw, Harvard University Press, Cambridge, MA, 1915.

[21] *Bowersox and Closs, Ibid, p. 13.*

[22] *Bowersox and Closs, Ibid, pp. 13-14.*

CHAPTER 2
GLOBAL SOURCING AND TRADE

In order to understand the logistics and transportation industry, we must first identify the demand for these services. Why are companies transporting cargo in the first place? This chapter identifies and discusses the sources of demand both at the firm level and at the market level. At the firm level, we identify why companies go internally for sourcing, and how their global operations create demand for logistics. At the market level, we identify the patterns of trade.

The key part of the term "international trade" is the crossing of an international boundary, which is a political distinction. Associated with trade are a variety of issues, such as distance and differences of culture and language. Even legal systems change within a nation. However, what we are mostly concerned with in this chapter is the differences associated with moving from one nation to another.

GLOBAL SOURCING

The first step in understanding the demand for logistics and transportation is to look at an individual firm and ask, why go international? Why import or export goods? While every company has its own reasons, there are a few general trends that push companies into the global marketplace:

- **International customers** - If a company's customers are international, there is a strong incentive to follow them into foreign markets. The customer would want their supplier to be in every market that they are in. If that customer goes into a market and their original supplier is not there, this gives competitors a major opportunity.
- **International competition** - If a company's competitors are international, it may be necessary to match them by going international.
- **Regulations** - The laws in a country may make it difficult to produce a certain product there, and then it would need to be imported. Environmental regulations are a common example. When environmental protection rules are strong in a country, there is an incentive to import the product from countries with laxer regulations.
- **New, expanded markets** - The company may be looking for new markets. This can happen when the domestic market is saturated, but that does not need to happen before a company looks to foreign markets.

- **Economies of scale** - A company can gain greater economies of scale by producing more while supplying foreign markets.

When a company makes the decision to enter a new market, they are faced with the choice of how to go about this. There is a spectrum of choices, in which on the one end is exporting. In this case the company has no significant investment in the foreign market, and simply exports the product when an order is received. On the other end of the spectrum is a wholly owned subsidiary, in which the company owns a subsidiary in the foreign market. In between these two extremes are a wide range of options that vary in the degree of commitment and risk involved.

The decision on how to enter the foreign market has a major influence on the demand for logistics. Note that if the firm chooses to export, there is a lot of cargo moving internationally. If the firm chooses to acquire a local subsidiary that produces for that market, there is little if any cargo moving internationally. Not only does the entry strategy determine the demand for logistics, but the cost and quality of logistics services affects the entry decision. If transportation is expensive or lacking, exporting to that market may not be feasible.

Even if the firm chooses a local subsidiary, there is often a considerable amount of trade involved. This is because that subsidiary still needs to acquire parts and supplies. As we will discuss shortly, a large percentage of world trade are parts and supplies being shipped between subsidiaries of the same firm.

Export. The least committing method, simply making an individual sale and export the product from the home country.

Contract. This could include licensing or franchising or a long-term sales contract.

Joint venture. The firm entering a foreign market may have a joint venture with a firm in the host country.

Wholly owned subsidiary. The foreign firm buys or builds a subsidiary in the foreign market.

Global companies organize themselves in one of two ways, **national** or **stateless**.[1] A national organization scheme is one in which the company treats each national subsidiary as relatively independent. In the stateless organization, which has come to represent the new and global economy, the enterprise operates across borders with minimal regard for national boundaries. Wherever the best opportunities lay is where that company will operate. This sort of company needs managers with the authority and training to work in a variety of countries, and function well with different cultures and legal environments.

Production sharing, one of the most significant aspects of globalization, is when a company distributes different stages of production to subsidiaries or other companies, often spread far across the globe. Why would a firm do this? World-class companies concentrate their efforts on their core competency. In order to do that, many firms outsource those aspects that others can do better. Production sharing is only possible, if there is to be a gain in efficiency, with the use of world-class logistics. It would not help contracting for another firm to do a phase of the processing if the shipments were getting delayed or costing too much. The logistics associated with production sharing tend to be the most challenging, because the firms have high standards of on-time performance and coordination.

Globalization of production also affects how the logistics department controls its operations within the company. They can be centralized in one office, or they can be decentralized into the various field offices. When the various offices are spread around the world, the question of how much to centralize can be important. A study by The Ohio State University found that the percentage of centralized logistics organizations and the line authority roughly doubled in the 1980's, yet, there was no more control over internal relations of the different departments. In other words, more companies are centralizing the logistics functions. Yet the logistics function is not coordinating with the rest of the company. In fact, the increased centralization often resulted in internal conflict and competition instead of cooperation.[2]

Sourcing, also known as purchasing or procurement, is the "series of activities that results in decisions regarding from whom/where goods, materials, and services should be obtained".[3] Parts and components sourced internationally accounted for $800 billion in the early 1990s, which is about one third of total trade.[4] This is one area that has a strong and close relationship with logistics. The sourcing office makes its decision based on a wide variety of factors, such as the need for given products, the characteristics of the suppliers, and logistical feasibility. As we have already discussed, there is a limit to how good a product can be. Suppliers are now trying to distinguish themselves by value-added services, most notably logistics. Sourcing for a small or local firm is based mostly on the characteristics of the product being purchased. Global sourcing, however, places much more emphasis on logistical issues.

Dick Locke notes that a company's profits can be dramatically affected by how well it finds and works with suppliers. The corporation has the option of centralized sourcing, or delegating this role to the local subsidiaries. For companies that do a lot of overseas sourcing, there is the option of an **international procurement office** (IPO), an office located in a local subsidiary whose job is to source from that country or region. The cost to the firm to run such an office may range from 1% for large-volume offices to 5% for smaller volumes. The IPO should recognize that they are in competition with independent agents and others that could fulfill this role. Locke warns companies not make a stand-alone IPO office because it means employees have a very limited career situation.[5]

Barter is an alternative to a normal sale, but this has become increasingly rare. During the Cold War, the Soviet Bloc would use barter since they were short of hard currency. Since then, barter is a sign of an economy in severe distress. In the late 1990's, one company that loaned money to Russian farmers ended up accepting the produce as payment. Interest rates to these small and medium size farms ranged from 24% to as much as 70%, given the extreme risk in that market, which meant they could not afford normal bank financing.[6]

Three **strategies of procurement** are identified by Bowersox, Closs and Cooper. The first is volume consolidation, in which a buyer seeks to reduce the number of suppliers. This increases the buyer's ability to control the process and reduce the number of transactions. Second is the supplier operational integration, in which the buyer seeks to integrate their systems with the supplier. As we see throughout this book, it is the goal of companies to work more efficiently with their suppliers. Third is value management, which goes beyond the traditional buyer-seller relationship. In this strategy, the two parties seek to promote the overall value of their relationship, such as early discussion on new product development and its effect of their joint operations.[7]

Strategies of Procurement
Volume consolidation
Supplier operational integration
Value management

Tariff engineering is an example of how logistics can affect the sourcing decision. This will be elaborated in Chapter Ten, but this term means that a product is made with the intention of lowering tariff rates. Some products can be almost identical except for slight differences, but one may have a much lower tariff rate. Tariff engineering may be done either by changing how the product is designed and/or manufactured, or by picking one supplier over another.

The Challenges Of Exporting

A survey of 497 firms in Indiana asked to list the chief problems of exporting. In reading the following results, it helps to see what problems arise and their relative importance. Also note that there were exports in Indiana. If one were to ask the same question in different cities or different countries, the results may be very different:[8]

Export documentation	23%
Transportation costs	20%
High import duties	17%
Unable to find foreign reps with know-how to market products	16%
Delay in transfer of funds	13%
Currency fluctuations	12%
Language barriers	10%
Difficulty to service products	10%

TERMS OF SALE AND INCOTERMS

We have just identified why companies go into foreign markets, and why there is a demand for logistics and transportation services. We now focus on how these shipments are arranged, who makes the arrangements and who is responsible for the cargo.

When you buy something, you take possession at a certain place and time. When you go to a store, the product is normally there for you to pick up. If not, it is either delivered to the store for you to pick up later, or delivered to your house. In the business world, things are much more complex. When something is bought on the other side of the world, and it takes many days and thousands of dollars to deliver it, who pays and at what point does ownership transfer from the seller to the buyer? Also, the origin and destination of the cargo may be nowhere near the buyer or seller.

Selling terms in this case refers to the point at which ownership is passed from the seller to the buyer, and the arrangements for carriage and related activities. A few things are going on here. First, there is the question of transfer of ownership. Second, who arranges for and pays for shipment (carriage)? Third, who assumes the risk of loss/damage to the cargo at any point along the way?

Shipping terms: "Those provisions that define the seller's and buyer's responsibilities for making the shipping arrangements, paying transportation charges, procuring insurance on the goods, paying port charges, and bearing the risk that the goods may be lost or damaged in transit."[9]

The buyer and seller agree at the time of contract on the terms of sale (selling terms). A contract which mentions a sale price without selling terms is meaningless and not legally binding. **Destination contracts** are when the contract calls for the cargo to be delivered to a given destination. In a **shipment contract**, transfer of ownership occurs when the cargo is delivered to the first carrier (a shipment can involve multiple carriers, such as truck, a ship, and another truck on the other side). If the contract calls for the seller to ship the goods but does not state a specific place at which it is to be delivered, then ownership, according to international law, takes place when the goods are handed to the first carrier.

On one extreme, the buyer can take possession of the piece wherever it is at the time of sale. In other words, it may still be sitting in the seller's warehouse, but the buyer now owns it. On the other extreme, the buyer may only take ownership when it arrives at her warehouse. There is a wide variety of choices in between. The determining factors on the selling terms include:
- **Risk**. How much risk is the buyer/selling willing to accept.
- **Location**. The buyer/seller will be more or less able to take ownership at different locations.
- **Buyer/seller relations**. Buyers and sellers may be related parties.
- **Ability to arrange transport**. One of the parties may be in a better position to arrange transportation.
- **Cargo**. The nature of the cargo and any special transportation needs.

Based on the above general criteria, the buyer and seller agree to a specific geographic location along the shipment where ownership transfers from the seller to the buyer. There are a few logical points along the way that most people use to transfer ownership. For example, the most common points along a shipment may be at the origin, the port of export, the port of entry, or the final destination. In Figure 2.1 we see how the value of a product increases, so that the further along in a shipment one takes ownership, the more valuable is the product. In other words, if you buy something at the manufacturer's location, it is cheaper, but then you have to spend money transporting it. The further it is delivered before you take ownership, the more you are going to pay for the product, since it also includes the service of delivery.

Costs and Times in International Move

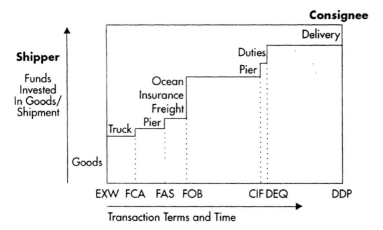

Transaction Terms and Time

Incoterms

We have been talking about ownership, but there is also the issue of carriage. Generally speaking, ownership and control of carriage should be the same, but this is not always the case. For contract of carriage, there is a well organized set of rules, known as **Incoterms**. These terms were developed by the Paris-based **International Chamber of Commerce** (ICC). These terms are legally defined and include obligations and duties on the part of the buyer and seller. This way, traders can simply refer to one term instead of having to describe in exhaustive detail what that means. When an arrangement has been chosen, the sale documents will state the selling term. This then obligates one of the parties to arrange for transportation up to the point where ownership changes hands.

Incoterms are for contracts of carriage, NOT contracts of sale

The ICC is an organization that deserves some explanation. They are a private organization, not part of any government or inter-governmental organization like the UN. Yet they have a long history and are considered extremely credible in such matters as establishing legal and business standards. When they develop the Incoterms, they bring together the top minds in the industry, establish agreement on what the terms should include, and publish them as their own suggestion. Governments and industry are in no way obligated to follow the Incoterms, but since the business world needs to agree on one set of terms, these are the global standard. Governments generally do recognize Incoterms as a legal standard.

The ICC issued a new set of Incoterms effective January 1, 2000, the sixth time Incoterms have been updated since their original publication in 1936. Technology and business practices change, which is why traders' needs change. This means that the definitions also need to change. Creating these terms was not always easy. A group of 40 experts from around the world was assembled and discussed at great length such questions as:

- Who is a 'shipper'? The seller of the goods or the buyer of the transportation?
- What is meant by 'deliver'? Does it mean the container that is supposed to hold the goods has arrived, or does it mean that the buyer has accepted a container which is supposed to have the proper cargo?
- At what point is a cargo considered 'delivered'? It could be the port of destination, an inland port, the buyer's location, or somewhere else.

The exact meaning of these terms needs to be precise because of the wide variety of business practices. Another challenge for those who developed the terms is the fact that the book is translated into as many as 20 languages and applied in almost every country in the world and very different cultures.

One of the biggest changes in Incoterms 2000 resulted from the dramatic increase in intermodalism.[10] For example, under the old Incoterms, FOB, CFR and CIF mentioned the 'ship's rail' as the point where the seller transfers responsibility to the buyer. Yet in the age of intermodalism, sellers are not taking their cargo to the port, but giving it to a carrier such as a rail head or trucker. This means they would still be liable for the cargo until it reached the ship. Now the terms refer to that point where the cargo is handed over to the carrier. What matters most is not the mode of transport, but who performs the transportation and where.

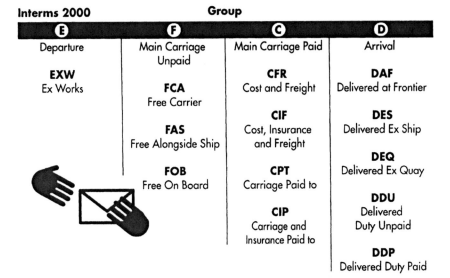

Interms 2000		Group	
E	**F**	**C**	**D**
Departure	Main Carriage Unpaid	Main Carriage Paid	Arrival
EXW Ex Works		**CFR** Cost and Freight	**DAF** Delivered at Frontier
	FCA Free Carrier		
		CIF Cost, Insurance and Freight	**DES** Delivered Ex Ship
	FAS Free Alongside Ship		
			DEQ Delivered Ex Quay
	FOB Free On Board	**CPT** Carriage Paid to	
			DDU Delivered Duty Unpaid
		CIP Carriage and Insurance Paid to	
			DDP Delivered Duty Paid

Previous terms also assumed that sellers were more sophisticated about customs clearance matters, and so gave them more responsibility. Now it is assumed that the importer is just as capable of clearing customs, including performing the formalities and paying the duties.

There are thirteen terms, divided into four groups, A, E, C, and D. These groups refer to the four most likely places for transferring control. The A groups refers to the seller's location, where they buyer takes all responsibility. The E group makes the seller responsible only for carriage within the country of origin. The C and D groups are often misunderstood, because it seems in both cases that the seller is responsible for shipment to the destination country or final destination. There is a major difference between C and D. With D terms, the seller is responsible for providing the cargo to the named destination. If the cargo is lost or damaged, the seller must replace it in order to fulfill the contract. With the C terms, if the cargo is lost or stolen, the seller is not responsible for replacing the cargo.

Why are there thirteen terms? It may seem that only a few would be needed. There are so many terms because of the different modes of transport, ownership arrangements and carriage arrangements. Sometimes cargo changes ownership enroute and it may be exported without knowing who the importer is going to be. Some terms only meant for carriage by sea (FAS, FOB, CFR, CIF, DES, DEQ). For manufactured goods, which are usually containerized, one would want to use FCA, CPT or CIP (and not FAS, FOB, CFR, or CIF).

Incoterms have some limitations. They are not intended to apply to all aspects of a shipment. Most importantly, they only apply to contract of carriage and export/import clearance. In other words, they are **not terms of sale**, and therefore they have nothing to say about transfer of ownership. The terms do not deal with relief of obligations and exemptions of liability for any reasons. They have nothing to say about enforcement or consequences of breeches of contract. Contracts should state that terms are based on Incoterms 2000 because, while the terms are used all over the world, they are not universal. There has been some confusion with the last edition, Incoterms 1990.

INTERNATIONAL POLITICAL ECONOMY

We have been discussing how firms do business across the globe. Yet in order for them to do business, there must be a legal and political system that allows this. Without a proper legal system, the risks would be so great that commerce would soon come to a stop. Recall from the first chapter that the logistics of a region depends, in addition to geography and physical infrastructure, an institutional infrastructure that includes laws and the means of enforcing them.

Trade is influenced first and foremost by economic forces. Political influences are important, and often play a role in how business is conducted globally. However, politics does not determine the overall pattern of trade. A river provides a good analogy. The economic forces that create trade are like the gravity that makes a river flow. Political forces are the bumps and channels that may influence the course of a river, but do not make it flow. Why is this important for logistics? This industry is influenced primarily by economic forces. Political issues often seem to control or determine what firms do or how trade patterns occur, yet this is misleading. If you want to understand what drives trade, look first at the market forces, and then note how these forces are changed (distorted) by political forces.

International political economy is where business/economics meets politics. It refers to the arrangements made by countries to regulate, encourage or restrict international commerce, and by the way firms seek to influence governments. Note that political economy concerns both domestic and international issues, but we are mostly concerned here with the latter.

International law is often poorly understood. Law by itself, without enforcement, is almost useless. In order for trade to occur, there must be a certain degree of agreement between the governments on how this trade is to be regulated. Since there is a lot of trade, and yet there is no global police force, how does this work? Each nation is enforcing their own legal standards. In order to participate in the global economy, governments soon learn that they need to set standards roughly the same as the others. Yet it would be misleading to say that there is any true international law as we understand our local law.

International law is a group of treaties and other agreements that are used as a common standard. Yet to become a truly effective law, a government must establish it as a law within their own country. Some countries have a **monist** system, in which as soon as a treaty is signed, it becomes the local law. In a **dualist** system, the signing of a treaty does not make it law in that country. Instead, the government must pass a law implementing the treaty.

There is the idea that something becomes international law when most of the world agrees on it to the extent that everyone becomes subject to it. In other words, if the vast majority of the world agrees that seizing ships is wrong, then those few places that allow it become, in the eyes of the rest of the world, subject to this new 'law'. In reality, a treaty or other agreement is needed to make it clear when there is agreement. For a treaty to become effective, it is first signed by the countries at the meeting where it is written. Each representative then returns to their respective nation where, depending on their legal system, it may become effective automatically or they pass a law putting it into effect. A treaty normally states that it comes into effect when a certain number of nations have ratified it. Thus a treaty may be signed at the meeting, but may take years to become effective.

For trade and commerce, one must distinguish between **public international law** and **private international law**. Of the treaties and agreements, some deal with public issues such as war, while others deal with private matters such as commerce. Virtually all trade is done by private parties, so what we are mostly concerned with is private international law.

In the US, the **Uniform Commercial Code** (UCC) has been since 1951 the basic law governing sales and contracts of sale. However, the UCC is being replaced by the **United Nations Convention on Contracts for the International Sale of Goods** (CISG), which became effective in 1988. The CISG will be referred to often in the chapters on Customs and Documentation.

International logistics faces some unique challenges because of the nature of international law and its enforcement. Countries cannot enforce their laws on trade, but only on what happens on their side of the transaction. The case of gray market trading is a good example of the problems associated with different standards.

The gray market refers to the practice of importing a product when you are not the official dealer of that product for that market. For example, Levi Strauss has an official importer in Europe who is the only one authorized by Levi Strauss to import and sell their clothes. Yet other companies found that they can buy their jeans and import them into Europe cheaper than the official agent. Levi Strauss won a case in the European market when the European Court of Justice ruled that Levis could in fact control who retails their jeans. Tesco, a British supermarket chain, was selling Levis jeans at about half the price of what they are offered in specialist stores authorized by Levis. Their argument was that by selling in low-scale stores their brand name was being degraded. Consumer groups criticized the ruling stating that the interests of consumers were being hurt.[11]

Levels of Economic Integration

No Trade Agreement
Country A   Country B

Free Trade Agreement
Country A   Country B

Barriers are reduced but not eliminated. Only Countries A and B are affected.

Customs Union

Barriers between A and B are eliminated, and all other countries can trade with A and B as if they were one country.

Common Market

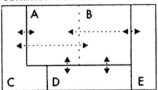

Nothing changes from Customs Union except now workers can move freely between A and B.

Is gray market trading legal? Governments cannot agree on this. Some governments, such as the European Court, believe that international contracts can and should be enforced. Others, notably the US, believe that such only domestic contracts in the US may be enforced. If a foreign firm has a contract with Levi Strauss, for example, then that is of no concern to the U.S. courts.

International trade faces special treatment in other ways. Trade within a country may be regulated depending on the ability of that administrative ability of that country to enforce rules. When cargo is shipped across borders, it is relatively easy to track and tax. For this reason the rules related to taxing international trade tend to be very different than domestic commerce.

Efforts to **liberalize trade** are of fundamental interest to the logistics industry. Free trade efforts dramatically increase the amount of cargo being moved, but it also determines where it is moving. It also affects corporate strategy. Countries can be integrated, politically and economically, at different levels or not at all. Even if there is no 'free trade agreement' intended to promote trade between the countries, they would still have some sort of arrangement on how to handle trade. There would still be the need for agreement on matters of documentation, transportation, business regulation and so on. There have been cases where nations with no formal ties, such as Taiwan and China, were forced into some sort of accommodation simply to make the movement of cargo possible.

There are many different special arrangements, but there are a few main **stages of regional economic integration**.[12] They are usually done as a formal treaty between the two (or more) governments:
- **Free trade agreement** - Makes arrangements to reduce trade barriers between countries. These arrangements can include reduced tariffs, simpler documentation, tax breaks and so on.
- **Customs union** - The countries agree that cargo can be shipped between the two countries as if there were no borders. This is important because something can be imported into one country and then enter the other country. The parties to a customs union must be confident that the other country will enforce similar standards on imports.
- **Common market** - In addition to the customs union, a common market may allow for labor to move freely among the countries. The movement of

labor is fundamentally different from cargo, as already discussed, which is why the shift from a customs union to a common market is a major leap.

- **Economic union** - The difference between this and the common market is that under an economic union, economic policies are coordinated, and administered by a common body. In other words, the individual nation loses significant independence to decide its own economic policies.

Tariffs And Taxes
A 'tariff' is simply another name for a tax, even though in some industries you hear one or the other being used. There is no significant difference in the two terms.

WHY NATIONS TRADE

International trade occurs for a variety of reasons, some of which is still not well understood. Trade can be explained at an economic level, where the economic forces affect the global trade, or at the national level, where the country's laws or economy influence their imports/exports. Trade can also be explained at the business level, why a specific company might decide to import or export.

Up until the mid-1700's, countries followed a **mercantilist** policy of trade. This was a belief that trade was a zero-sum game; one country's gain was another's loss. It was believed that the way to become wealthy was to save precious metals. Exports were encouraged but imports were discouraged with trade restrictions. These restrictions naturally caused other nations to react by putting up their own import restrictions.

The first explanation for international trade, known as the 'classical trade theory', was the idea of **absolute advantage**. Adam Smith described his idea in the famous 1776 book *The Wealth of Nations* attacking mercantilist philosophy. Absolute advantage means that countries will export whatever they do better than any other country. The theory of absolute advantage did not explain why countries still produce things even if they are not the best at this. **Comparative advantage** helped explain this phenomenon. Based on 18th and 19th century economists, notably David Ricardo in his 1817 book *Principles of Political Economy*[13], comparative advantage states that a country will specialize in those products in which it has the greatest comparative advantage and import those for which it has a comparative disadvantage. In other words, a country does not just trade in whatever they do better than anyone else. They trade in whatever they do best.

Comparative advantage, according to Ricardo, assumes **constant returns to specialization**. This means that the amount of resources needed to produce one unit of product will stay the same regardless of how many units are produced. In reality, there is more likely to be **diminishing returns to specialization**. This means that the more units of a product a country produces, the more resources will be needed, and thus the efficiency will be reduced. This is because not all resources are of the same quality. As a country tries to increase output, it needs to draw upon more marginal resources. Another reason is that different goods use resources in different proportions. This means that a country can gain from specializing in a certain product, but only up to a point.

The **New Trade Theory** emerged in the 1970's in response to criticism of the diminishing returns to specialization. In some cases we are seeing increasing returns to specialization. This is because increased output results in lower fixed costs, such as in building jet aircraft. The increased costs of resources did not seem to be having the effect that they had in other industries. The classical example is the airplane manufacturing industry. The new Boeing 777 cost an estimated $7 billion in development, which means they would need to sell 200 of these planes just to recover their costs. As more planes are produced, the average fixed cost is reduced.

30-50% of US trade by dollar value is by airfreight,
90% by tonnage is by ships
68% of US shipping imports by tonnage are petroleum and petroleum products.

Some industries require such large economies of scale that only a few companies can survive. To use the airplane industry as an example, the costs of starting up a new company to build planes are so great, and the size of the market is so limited, that those companies that already exist will hold a great advantage over any others. This is known as the **first mover advantage**.[14] This helps explain why some companies are located in places that seem to contradict other trade theories. This theory has been used to justify government support for certain industries.[15] It is too early to see how well the New Trade Theory holds up.

Michael Porter published an influential book in 1990 called *The Competitive Advantage of Nations*. He had studied 100 industries in 10 countries to see why some nations succeed and others failed at international competition. Porter suggests that four broad national attributes shape the environment in which local firms competes, which may help or hinder their creation of a competitive advantage:

- **Factor Endowments.** Advanced factors such as technology are more important than basic factors.
- **Demand Conditions.** Companies are most sensitive to the needs of their closest customers, so firms do best when their domestic demand is sophisticated and demanding.
- **Related and Supporting Industries.** Industries tend to be clustered and support each other. For example, having a group of computer companies in a country is much better because they benefit each other.
- **Firm Strategy, Structure, and Rivalry.** Different nations have different 'management ideologies'. For example, major manufacturers in Europe have many engineers in high positions, whereas in US one sees more finance experts. This affects the company's behavior and strategy. Having strong rivalry domestically among the different companies in an industry forces them to compete and innovate.

INTERNATIONAL TRADE PATTERNS

Some very distinct patterns can be seen where trade moves around the world. Rich countries do most of their trade with other rich countries. This would seem to contradict trade theory that states that rich countries would produce capital-intensive goods and trade with poor countries that produce labor-intensive goods. Instead, rich countries produce different goods, and the major markets are those countries that can afford them. Neighbors tend to trade with each other. This is because transportation is relatively less, and there is more contact among nearby businesses. There also tends to be more trade between ex-colonial powers and their ex-colonies, and among common cultures.

Trade Trends
- Rich countries with rich countries
- Colonial networks
- Neighboring countries
- Common cultures

Cross-trade is the practice in which a country acts as a hub for trade between other countries. This may be the result of geography, in which some port or region is located in the center of a trading area. The cargo generally does not leave the ports. This means that in a country that does considerable cross-trade, the ports may be developed for a very high level of trade, but not the interior. Singapore, for example, acts as a hub for cargo moving through Asia. The percentage of cargo that is actually going into or out of the country is relatively low. Another example is Panama, thanks to the Panama Canal. Those countries or ports that become established as cross-traders usually worked very hard to become that way. This is a highly lucrative position to be in.

Europe's Vital Axis

Europe's Vital Axis

Major Axes of the Future

Coherence of the English Channel/North Sea Area

Areas of Development with Net Migration Gains and New Activities

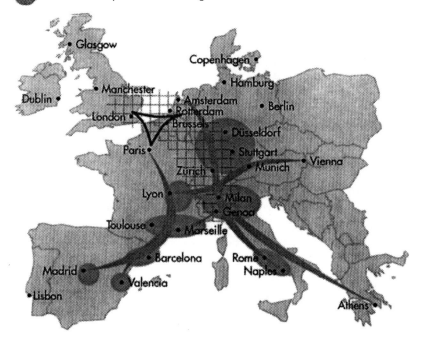

Cross-trading from the perspective of the shipper is known as transshipment. This is when cargo is shipped through a port. One reason may be logistical efficiency. Another may be to avoid trade restrictions. If a nation restricts imports from another nation, one way to get around this is to ship the cargo to a third country, and make the documents look like that was the origin of the cargo. This practice is highly illegal. Chinese textiles into the US are heavily controlled, so Macau has been a common port for transshipment.[16]

Regional Trading Blocs Around the World

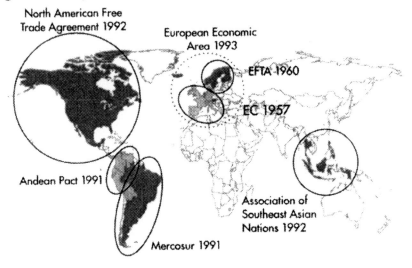

North American Free
Trade Agreement 1992

European Economic
Area 1993

EFTA 1960

EC 1957

Andean Pact 1991

Association of
Southeast Asian
Nations 1992

Mercosur 1991

Trade Lanes

A **trade lane** is a route for trade, connecting the major cities or industrial areas. On land, we see some roads used as trade lanes. On the seas and in the air, trade lanes are invisible but they are still very real. There are certain paths across the seas and the sky used by carriers to move cargo and people between the major markets.

This point is important because there are certain major trade lanes in the world as a result of the patterns of trade. For example, between the US west coast ports and Japan are major trade lanes. Between the major Asian ports, through the Suez Canal to Europe is another major trade lane. Transporting something between two points along a trade lane is much easier and cheaper than moving something to a place that is away from the major trade lanes.

There are **primary trade lanes**, which have heavy volume and are usually longer (across the ocean), and there are **secondary trade lanes**. An example of a secondary trade lane is the path between a small local port and the nearest major port. The difference is all relative, though, and what may be a secondary trade lane for global trade may be a primary trade lane for that area.

There is a strong path-dependency to trade lanes, in that they are often the result of history, and once established are hard to change. Why would anyone want to change a trade lane? If there were a port or country, for example, that wanted to increase their traffic, they would have a difficult time attracting cargo away from an established trade lane.

Trade Liberalization in Transportation

International transportation is fundamentally different from any other industry in that the essential nature of the industry is to cross international borders. The WTO has had little success in getting any major agreement on liberalizing ocean shipping and aviation. This is a complex field and will be discussed in later chapters.

There have been ongoing attempts at liberalizing trade in transportation and other industries. The Millennium Round of the General Agreement on Trade and Services (GATS) dealt with many issues not covered in the World Trade Organization's mandate. These issues included agriculture, environmental services, trade facilitation and electronic commerce.

One of the more interesting trade disputes has involved the carriers themselves. Specifically, Boeing and Airbus, the two companies that overwhelmingly dominate the jet airplane market, has long been the beneficiary of government support. They have also been the target of dispute from the other side of the Atlantic. Ironically, each group spends about $5 billion a year on parts and services from each other's home market, supporting about 100,000 jobs. Even more ironic is that they were at one time talking about a deal to produce a new plane together, a project that fell apart only because they could not agree on the size of the market for that plane.[17]

---

[1] *Bowersox and Closs, Ibid, pp. 140-143.*

[2] *"The 1990 Ohio State University Survey of Career Patterns in Logistics", Bernard LaLonde and James M. Masters, The Ohio State University, Columbus, OH, 1991, p. 13.*

[3] *Keebler et al, Ibid, p. 108.*

[4] *"Entering the Twenty-First Century: World Development Report 1999-2000", World Bank, Oxford University Press, New York, 2000.*

[5] *"Global Supply Management: A Guide to International Purchasing", Dick Locke, McGraw-Hill, Boston, MA, 1996.*

[6] *"US lenders farm out Russian produce", JOC, February 16, 1999, p. 7A.*

[7] *"Supply Chain Logistics Management", Donald J. Bowersox, David J. Closs, M. Bixby Cooper, McGraw-Hill, New York, 2001, p. 139.*

[8] *Steven Ibid, p. 101.*

[9] *"International Business Law and Its Environment"*, Richard Schaffer, Beverley Earle and Filiberto Agusti, West Publishing, St. Paul, MN, 1996, p. 210.

[10] *JOC*, September 9, 1999.

[11] *"Trouser suit"*, The Economist, November 24, 2001, p. 58.

[12] *"International Business"*, Charles Hill, McGraw Hill, Boston, MA, 2000, pp. 233-235.

[13] *"The Principles of Political Economy and Taxation"*, David Ricardo, Irwin, Homewood, IL, 1967, first published in 1817.

[14] *"First-Mover Advantages"*, M.B. Lieberman and D.B. Montgomery, Strategic Management Journal, vol. 9, Summer 1988j, pp. 41-58.

[15] *"Does the New Trade Theory Require a New Trade Policy?"*, Paul Krugman, World Economy, vol. 15, no. 4, 1992, pp. 423-41.

[16] *"Customs clamps down on Macau imports in battle against textile transshipments"*, Paula Green, The Journal of Commerce, September 16, 1998.

[17] *"Super-jumbo trade war ahead"*, The Economist, May 6, 2000, p. 63.

CHAPTER 3
GLOBAL SUPPLY CHAIN MANAGEMENT

Of all the issues in logistics, none is more important yet less understood than a supply chain. The supply chain is a theoretical concept, the result of multiple independent entities that affect each other. In this chapter we describe supply chains. However, the most important part of this section is separating fact from fiction. Logistics is not supply chain management. Finally, the section on marketing channels explains the various ways that products get from producers to customers and why it is important to manage the various channels.

THE SUPPLY CHAIN CONCEPT

Henry Ford created what is one of the best examples of a supply chain. He started with a car assembly factory. Then he needed car parts, so he made a car-parts factory. The car parts were made out of smaller parts, so he has more factories to make all those little bits and pieces. Then those pieces needed to be made out of steel, so Ford Motor Company included a steel foundry. Ford was so concerned about self-sufficiency that he bought 2.5 million acres in Brazil to develop a rubber plantation, and grew soybeans used to manufacture paint.[1] Each of these companies in his empire supplied the other in one long chain that went from the mining of iron ore to the final assembly of cars. Even that was not the end of it, because he also controlled the retailers who sold the cars.

A **supply chain** is a system of entities that supply the next one. These entities may be independent companies, or the chain may be entirely within one firm. In the case of Ford, he tried to bring his supply chain entirely within his company. The point of this story is not that companies should buy up their suppliers. What Ford created was a very inefficient empire that would eventually be disassembled. Companies have found it much more practical to specialize in what they do best, their core competency, and buy from others what they do best.

Supply Chain Management Defined
The integration of key business processes from end user through original supplier that provides products, services, and information that add value for customers and other stakeholders.[2]

What Ford did was bring attention to the way materials flow through a system. A supply chain is made up of many distinct entities, linked together by the process of buying and selling pieces and services that will eventually create a final product. It is not always clear what is the 'final product'. What may be the 'final product' in one chain can become a raw material in another chain. For example, a hammer is the final result of a supply chain that goes back to wood handles, cut wood, lumber supplies, and forestry. Yet that hammer may then be put into a tool kit, so the kit maker views it as a component part.

A product gains value as it works its way through the supply chain from raw materials into finished inventory. This is called the 'value added' process. If a link in the supply does not add any value, then market forces should eliminate that link. For example, if a company bought a product, did nothing to it, and resold it at a higher price, its customers would eventually find the original seller and buy it at the lower price.

The Supply Chain

| Ore | Steel | Engine Blocks | Car Engine | Car Maker | Car Retailer |

Supply chain management (SCM) is the way the links are integrated to promote efficiency. Another term for this same idea is **supply chain integration**. The reason this is important is because well-integrated supply chains can increase the value of the whole process for all the companies involved, and creates superior customer value. **What is the difference between SCM and logistics?** Logistics is getting things to where they need to be, but SCM takes this process further by organizing the overall business operations and the way it interfaces with other companies and organized ongoing logistical operations. In other words, when you want to get something to a customer once, you are doing logistics. When you are organizing the company for ongoing logistical functions, you are more in the realm of SCM. These terms are overlapping and vague, and you should not be too concerned about this fine point. The problem arises because the professional world uses both terms, but they are usually talking about the same issues.

A distinction must be made between a supply chain consisting of independent firms and a supply chain within a firm. When the chain is within a firm, the job of managing the process is entirely within the control of the firm. Intra-firm supply chains are very different, in that each firm only controls its own behavior. Yet the chain is a system of relationships between the firms.

Recall from the first chapter the difference between inter-firm logistics and intra-firm logistics. Supply chain management can be challenging because each of the links in the chain are normally independent companies. Getting independent companies to work together is rarely easy since they each operate in their own best interest. What may be in the best interest of the overall supply chain is not necessarily in the best interest of a particular member. This is known in the social sciences as the **collective action problem**. Also, companies may find it easy to work with their customers or suppliers, but when they reach out to their suppliers' suppliers, and then back further along the chain, things get very complicated.

Supply Chain Management: Webster Motor Company

Imagine Webster Motor Company is assembling a new model of car. The company buys engines from another company that just makes car engines. The Car Engine Factory buys the parts needed to make car engines, such as engine blocs. The Engine Bloc Company, in turn, buys steel from The Steel Company, who in turns gets steel ore from The Mining Company. This is an example of a supply chain of independent companies.

Now consider an example of a supply chain within a company. Webster's assembly plant needs the engines, so they send their production schedule to the Sourcing office. The Sourcing office contacts The Engine Company.' When the engines are shipped out, they are trucked to a train station, where they ride the railroad to the station nearest the factory. From there they are put on a truck and taken to the factory. Once at the factory, they are held in the warehouse until needed. Once needed, they are placed next to the assembly line, and put in a car as it works itself through assembly.

In the next section we will see how the Webster Motor Company is faced with the challenge of waste in the supply chain.

EFFICIENCY IN THE SUPPLY CHAIN

The more difficult the integration process may be, the more there is to be gained. If a company can create a highly efficient supply chain where their competitors failed, it has a major competitive advantage. How does supply chain integration benefit its members? Cooperation between the chain members reduces risk and improves the efficiency of the overall logistics process. Secondly, waste and redundancy is eliminated from the supply chain.

What kind of **waste** and **redundancy** are we talking about? Imagine a car manufacturer that needs car parts. If this company cannot afford to run out of a certain part, then it will keep large inventories, which is expensive. That car part company needs parts from its suppliers, so to prevent any problems it too keeps inventories. Through the supply chain every company is keeping inventory because they are not sure that their supplier will be able to fulfill their needs. This example only looks at inventory. There are many other aspects to risk, waste and redundancy.

Time is another major aspect of a resource that may be wasted. A study showed that the average dry grocery product took 104 days from supplier's line to supermarket checkout counter.[3] This same study showed that it took 66 weeks for a product in the apparel industry for raw material to the retailer.[4] In many industries, such long timelines are totally unacceptable. Sometimes the difference of a few hours makes all the difference in the world, such as shipping fresh flowers. Computers are introduced into the market and have a product life of, on average, six months. Such a product needs to move through the various companies of the supply chain very quickly.

While the supply chain is a basic concept of logistics, it is also somewhat of a **myth**. It applies to movement within a firm, but rarely does it operate between more than two firms. We rarely see well coordinated logistics between three or more unrelated companies. Why is this? Imagine you are representing your company, and you ask your supplier for their supplier's information. You are a forward thinking logistics professional and you want a thoroughly integrated supply chain. What do you suppose would be the supplier's response? If she provides you information regarding her supplier, you can eliminate her company and deal directly with the other company, or you can push for lower prices.

A company usually protects information about its suppliers for a good reason. Therefore, the idea of integrating a chain of independent companies for logistical efficiency is unpractical. Researchers for the book *Keeping Score* interviewed 22 companies and could not find a single case in which two or more companies were working together to measure logistics functions.[5] The authors refer to a **Supply Chain Orientation**, in which companies in the supply chain are aware of how they affect each other even if they cannot control the situation. This they defined as "the recognition by an organization of the systemic, strategic implications of the tactical activities involved in managing the various flows in a supply chain."[6]

Social scientists noticed long ago that organizations value **autonomy**. Companies and governments want the freedom to make their own choices and do what they want without needing to get permission from others. This seems to be one of the most fundamental rules of organizational behavior. In SCM, the challenge is that each company would like to conduct their operations without concern for others, but in the business world, this is clearly not possible. A company, such as Webster Motor Company, for example, must coordinate with customers and suppliers. Ideally, the customers and suppliers would take their orders from Webster. But as we can see, this is not realistic. Still, companies will resist taking any actions that will reduce their autonomy. When we see strategic partnerships, they are examples that have overcome the urge for autonomy to take advantage of a profitable opportunity. Yet many companies see such opportunities but prefer to operate independently.

A supply chain is not the same thing as a **value chain**. A value chain is a management concept in which each link in the chain must create some value to the product. If a link does not provide any value, it must be eliminated. While a supply chain is a type of value chain, there are many value chains that do not involve the movement of any goods. The service industry offers such an example of a supply chain without anything that we could consider logistics.

Supply Chain Efficiency: Webster Motor Company

Webster Motor Company is scheduled to produce 1000 of their new model car, the Florence. In order to make 1000 cars, they need to order 1000 engines. However, running out of inventory in any one part would be a major problem because the entire assembly line would shut down. Maybe some of the engines will be defective, so they order 1050 engines.

The Engine Bloc Company gets an order for 1050 engines, so they need 1050 engine blocs, plus a few extra just in case there get defective blocs. They order 1100 engine blocs. The Engine Bloc Company needs to make 1100 blocs, so they order 1150 blocs worth of steel from The Steel Company. Finally, The Steel Company orders 1200 "blocs" worth of steel ore.

In order to make 1000 units at one end of the supply chain, it is necessary to produce 1200 units at the beginning of the chain, a 20% difference. This does not even consider the time factor, or the transaction costs.

Waste in a Supply Chain

Ore	Steel	Engine Blocks	Car Engine	Car Maker
1,200	1,150	1,100	1,050	1,000
200 Extra	150 Extra	100 Extra	50 Extra	

CHANNEL RELATIONSHIPS

Supply chain management is partly to do with a company's internal operations, and partly to do with its relationships with outside companies. The company's internal operations need to be made compatible with partner companies. But the critical part is the relationship with the outsiders. There has been a major change in the way companies manage their relationships with other companies over the past few decades. Instead of competition and confrontation, we are seeing a shift toward cooperation. Note the difference between competition within and between industries. Competition is to be expected between different companies producing similar products and competing for customers. But what we are referring to here is the competition between buyer and seller for the best price and terms.

What is it that channel members are competing or cooperating over? They are competing over price and quality of service, as well as liability for problems, especially inventory. We have just seen an example of waste in a supply chain. When business slows down, companies tend to get stuck with large inventory, which is costly. The details are to be discussed in Chapter Fourteen, but for now it is important to understand that avoiding large inventories is very important. The result is often a lot of arguing among channel members over who is responsible for this inventory.

Channel members are also competing for the best price and terms from their suppliers and customers. They do this by getting information that will help them. That is why we said SCM is something of a myth. Recent experience has indicated that even the most cooperative arrangements tend to break down during recessions. Harriet Green of Arrow Electronics notes, "I am alarmed about the lack of ownership of the massive buildup of inventories. Everyone says it's yours".[7] This is why the supply pipeline (another term for supply chain) has been referred to as the "glass pipeline". One business executive notes, "When I talk to customers, they flat out don't want...the retailer tapping into your inventory levels to see what's there. If they see you have lots of inventory, they can drive down the price. Or, if you're low on inventory, they might panic and go to another supplier to fulfill an order."[8]

Delaney and Wilson, in their "State of Logistics"® report of 2001 cited three reasons for breakdowns in cooperation during a recent recession. Demand forecasts are often inflated. Manufacturers have given contractors more control over procuring parts, but it is unclear who owns excess inventory. Finally, software programs, despite their overall benefits, still tend to be difficult and costly.[9]

Primary channel members refer to those companies that are willing to devote resources and take on risk to manage the supply chain. For example, the car maker holds inventory, which is expensive, and will have a strong incentive to develop its supply channels to reduce this cost.

Assortment
One of the functions of a supply chain is to change the composition of shipments, known as **assortment:**[10]
Concentration- multiple small shipments are combined into larger shipments.
Customization- a shipment of different pieces is assembled.
Dispersion- large shipments are broken down into smaller shipments. [11]

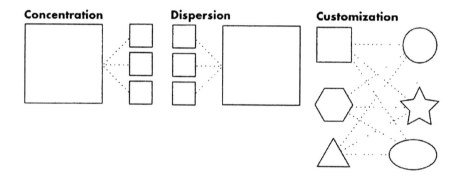

Not all channel members have an equal stake in the success of the arrangement. In other words, in a supply chain, not all the links are the same. A channel member is **specialized** if a large percentage of its work is within one supply channel. In the Webster Motor Company example, The Motor Company would be highly specialized if they only made motors for Webster.

Risk is another basic issue with supply chain management and a reason why it can be such a challenge. Any time an enterprise attempts a collaborative effort with its partners, there is the chance that it will end up losing. The time, money and effort put into developing collaboration may be lost, or the other partners could gain inside information and use it against the enterprise.

The more specialized an enterprise's role in the supply chain, the greater the risk. If a company produces something for just one supply chain, or relies on one supplier, they are extremely vulnerable. If the company sells its product to many companies and has multiple suppliers for the same product, there is less risk of disruption.

Power is another fundamental issue of supply chain management. Each link in the chain has different degrees of relative power to influence the other partners. Recall that what may be good for one channel member may not be good for others. This is where power becomes important, the power to make the channel work to one's own best interest.

Some general trends can be seen in the business world in the distribution of power between types of channel members. Over the past decade, US retailers have increased in power over the manufacturers. This is because consumers have shifted loyalty from products to providers. Instead of looking for a certain brand, for example, customers tend to look for a certain store. This trend is the result of a few changes in the marketplace:[12]

- Ownership of retail outlets has consolidated into a few major firms.
- Retailers have vital knowledge of what is happening in the marketplace. They are on the front line and can be the first to see changes in consumer demand.
- Manufacturers finding it more costly and difficult to create new brand franchises.
- Whole process of logistical replenishment has shifted from push to pull.

However, there is also a countervailing force, the spread of new channels. Manufacturers are selling their products through more and more channels, which mean it is more possible that any one retailer can be replaced. Marketing channels will be discussed later in this chapter.

Push And Pull Systems

A supply chain 'pushes' products when an enterprise makes something and then attempts to sell it. A 'pull' system is when an enterprise looks at what customers want and then seeks to satisfy their demand. Pull systems are inherently more efficient because there is no doubt about the demand for the product. However, that is not always possible. Push systems are used when customer demand cannot be known. For example, some products are new on the market, so it takes some time and research before their demand patterns are learned.

Leadership is necessary for supply chain management. There is no clear guideline who in the supply chain should take leadership. It **should be the company that has the most to gain in the whole supply chain.** Often the company that has overwhelming power and influence does not want to take leadership. But someone needs to take the initiative to bring the channel members together and develop ways to manage the system if there is to be any true supply chain management. Obviously companies work together in the supply chain just by pursuing their own best interest, but this is not 'management' and it is unlikely the channel will operate at its most efficient.

Channel relationships can be viewed as a spectrum, with little or no dependence on one end and complete dependence on the other. A **single transaction** between two parties involves almost no dependence. A **conventional channel arrangement** is one in which there are multiple transactions going on, but there is no attempt to make them organized or managed between the parties. For example, you go to the store every week to buy groceries. You and the grocery store know this is almost certainly going to happen but you do not discuss with the grocer your future shopping plans. In the business community, companies often do ongoing business without bothering to organize the process.

The next stage is an **administered system**, in which there is some organization to the process, but it is informal and unwritten. This could be as simple as communicating to the supplier or customer of your plans. Following an administered system is a **contractual system** in which the parties to the supply chain engage in a formal agreement, such as a written legal agreement. This would be, for example, a contract in which the buyer is obligated to purchase a specified amount during the contract period. Finally, in a **partnership/alliance**, the parties have more than just transactions in mind. They are working as partners and have a vested interest in each other's success.

Channel Relationships With Increasing Dependence
- Single Transaction
- Conventional Channel Arrangement
- Administered Systems
- Contractual System
- Partnership/Alliance

MANAGING THE SUPPLY CHAIN

Establishing the supply chain is only the first step. Managing the system is an ongoing and never ending process. International supply chains pose some special issues. The performance cycle is longer as a result of the distance between the different companies, and the added complexity of the arrangement. Many operational aspects are different; channel members are working in different languages, different currency and different legal/political systems, to name a few. There is also added complexity in any international transaction. This results from the regulatory issues of crossing an international border.

One goal in managing the supply chain is to reduce the number of transactions, called the **Principle of Minimum Total Transactions**. Since each transaction has a cost, rearranging the system to reduce the number of transactions should result in savings. Another basic goal is **velocity**. This refers to the speed at which materials move through the supply chain. It may be possible to increase the velocity, but there is then the question of cost. The manager must determine what is an optimal velocity given the costs.

Inter-company collaborations need to comply with the law, especially in matters of anti-trust. The Cooperative Research and Development Act of 1994 signaled the change in Federal regulations to encourage cooperation while increasing overall competitiveness. The laws of developed countries generally try to prevent collaborations that would restrict competition, but it can be difficult to know exactly what is or is not permitted. That is why the US law was changed so that efforts at collaboration would not be confused with efforts to restrict competition.

The job of managing a supply chain can be contracted out in some situations. The companies that do this, known as intermediaries, will be discussed at length in Chapter Fourteen. At this point it should be noted that contracting out aspects of a supply chain can solve some problems, but can create other problems. The Principal of Minimum Total Transactions told us that the more actors involved in the supply chain, the less efficient it will be. Using intermediaries should only be done after a careful analysis of the benefits and costs.

One of the major goals of a supply chain is to minimize what is known as the **bullwhip effect**. This is when changes in demand cause increasingly greater changes as one goes up the supply chain. Imagine the car maker who orders parts. They cannot be sure how many they are going to need, so they order some extra. The parts supplier needs parts from their supplier, so they too order a little extra. And so on. By the time one goes up the chain a few steps, these little extras are getting increasingly larger. Now imagine the car maker's demand is less than expected, so they reduce their order size. Now the parts company already had extra parts, so their revised order is going to be even smaller. At the beginning of this chain, that company's orders are drastically different from the original customer, the car maker.[13] There is also what is known as **rationing** and **shortage gaming**, in which companies anticipate shortages or excess inventory, and try to counteract it. This gaming usually only aggravates the situation.

How does a well-managed supply chain control this bullwhip effect? While eliminating it entirely may be impossible, the following are the best ways to control it:
- **Demand forecasts** - Accurate information from the source of demand, usually the retailer, is the most effective method.
- **Information sharing** - The more information channel members have, the better they can make decisions.
- **Shorter lead times** - Provide more flexibility by allowing production to be based on most recent information on inventory levels.
- **Order batching** - Reduce batch size and more frequent orders. This makes it easier and faster to let changes in demand run through the supply chain.
- **Price fluctuations** - Price changes trigger changes in demand, so by keeping prices steady you keep a smooth supply chain.

Just-in-time (JIT) management systems have come a long way from the original plan to reduce inventory. JIT is an outgrowth of the Kanban system, also known as the Toyota Production System. It was originally an inventory and operations management system, but is also relevant to the supply chain.[14] The essence of JIT is to have supplies delivered only when needed. This eliminates the need to keep inventory. It has many other effects, most importantly that it forces a company to make accurate forecasts, and to emphasize the quality of their operations. JIT can be viewed in a number of different ways:

A production strategy. Eliminates waste and makes more effective use of company resources.

A philosophy. Getting the right materials to the right place at the right time.

A program. Eliminates non-value-added activities, improve quality (zero defects), improved productivity, lower inventory and long-term relationships with channel members.

JIT is most appropriate for repetitive manufacturing. It is not a solution for every company. While it solves some problems, it creates others. Production scheduling becomes much more complex. Supplier production schedules are an important issue yet not under the control of the JIT company. The supplier may be far away, making close coordination difficult. JIT shifts inventory to the supplier. It requires supplier cooperation. In terms of logistics, integration is even more important, and transportation becomes more important. The warehousing function shifts from storage to consolidation, since warehouses are less of a place to leave inventory.

It is not necessary for a company to be entirely JIT. It can be phased in or partially implements. For example, a study of European logistics managers showed that at the time of the survey (1988), 18% of their shipments would arrive under JIT conditions, and they were planning to increase that to an average of 45%.[15] In other words, the overall operation was not necessarily geared for JIT, but as much as 45% of their shipments could take advantage of this technique.

MARKETING CHANNELS

Earlier, we looked at international trade as the source of demand for logistics from a global perspective. Now let us consider demand from the perspective of marketing. The first thing students of marketing learn is the **4 Ps: product, place, promotion and price**. While all of these work together, we are most concerned with Place. Marketing a product requires that customers have access to the product.

This is not jut a matter of the point at which the product is handed over to the customer, but how it gets to that point. Marketing channels, as defined by Bowersox and Closs, are "...systems of relationships among businesses that participate in the process of buying and selling products and services.[16] Note that the terms *channel, pipeline,* and *chain* are in many ways the same concept in this case.

Putting together producers with customers is a massive industry. The US has over 1.5 million retailers and over 460,000 wholesalers.[17] When one looks at the business to business market (commonly known as B2B market. There is also the business to consumer, B2C), there are many more players, all of whom need decide how they will get their products to their markets.

B2B: Business to business
B2C: Business to consumer

An example of a simple product and its channels is Coca-cola®. The product is essentially the same, yet you can buy it from any one of a wide variety of sources. It is sold in grocery stores, convenience stores, hospital cafeterias, vending machines, a sports stadium and many other places. Yet consider the difference that a channel makes. In a hyper-market, where you go to buy things in large quantities for the lowest possible price, a can of soda may cost 50 cents. Go to the soccer game in a stadium and how much do you pay? Maybe $4. Why the difference? They provide different utility.

Utility is the benefit provided by the product. The four economic utilities in a product: form, possession, time and place. **Form** is when a company processes something to change its form. This is primarily what manufacturers do, but another example is a restaurant that changes the form of raw food into a cooked meal. **Possession** utility is the ability to transfer ownership to the customer. **Time** utility means the company is presenting you with their product at a certain time. For examples, some stores provide utility by being open 24 hours a day. **Place** utility is similar, in that the product is offered at a convenient place. Local markets offer place utility because they are located where it is easy for people in that community to reach. Retailers provide possession, time and place utility, but rarely would they be offering form utility. Factories would generally offer form utility, but little of the other three.

There is another utility, which is ownership, that has relatively little to do with logistics, but it should be noted since we are on this topic. Ownership utility is the ability to transfer ownership from the buyer to seller. Some companies may provide any of the above three utilities, but not be involved with the actual sale of the product. Some companies only handle the sales and offer little of the other three utilities.

In the example of Coca-Cola®, you want that can of Coke®, but just as importantly, you want it at a certain place at a certain time. When you buy the cheap can in the hypermarket, you are paying in time, effort and inventory costs to go to the market, stockpile large quantities. In a restaurant, you are paying for the chance to enjoy your Coke® served to you with the meal.

Utility
• Time   • Place   • Form   • Ownership

The logistics of marketing channels will vary dramatically depending on the product. Coca-Cola® has a wide variety of channels to reach customer at as many times and places as possible. Each channel will require different logistical support. Delivering large cartons of soda is easy and cheap. Delivering individual cans, served cold, is very expensive (note, however, that Coca-Cola® generally does not do the local distribution, but instead licenses other intermediaries). On the other end of the spectrum, recall the example of the Webster Motor Company. They will probably have a simple marketing channel, since almost all cars are sold through dealerships.

Rohm and Haas, a chemical company, offers another example in a very different industry. This company has three channels. There is the mass market, products that they always keep in stock, and their production strategy is for make-to-stock. Next is the Special Order, which is the Make-to-Order strategy. The company would commit to an order date for special customers. Finally, they have what they called the Turbo Channel, in which they focused on particular customer/product combinations, in which there are high-volume customers and products. These products are produced according to the customer's plan, more of a partnership.[18]

A major issue in channels is known as the **"last mile problem"**, referring to the challenge in accomplishing the final leg of delivery to the customer. Parcel carriers, for example, have extensive home delivery infrastructures, but for other industries the last leg of the supply chain is the biggest problem. These include "big ticket" and bulky products like furniture, appliances, white goods, and perishables.

Not only do different products require different logistical support, the marketing strategy will change over time. The **product life cycle** describes how a product's position in the market changes. A new product on the market offers something that nothing else can offer. Over time, the product moves into a growth state, where its sales grow. At some point, competing products appear on the market and the sales level off, the maturing phase. Eventually, the product will become obsolete, known as the decline stage.

The logistical support required will vary as the product moves through this life cycle. In the beginning, a new product means that the channels are still being developed. It is not always obvious how a product will get to market. It may be best to sell it through department stores, or the company may find that it does better to only sell it through specialty stores. On a larger scale, a product may do better in one region instead of another.

As the product matures, the channels will also develop and become more efficient. This is important because as competitors appear on the market, there will be a need to reduce costs. This is why established products generally hold an advantage over new entrants to the market. Once a product enters the decline stage, it is essentially a commodity that is best offered as the cheapest of its kind.

Now that we have looked at the development of a single product, let us consider the view from a retailer, in which there are many products being offered by suppliers, and customers demand a variety of choices. Retailers must make decisions about what to offer and what to decline. An individual product, in a given size and color (and whatever other distinguishing characteristics it may have), is known as a **stock-keeping unit**, or **SKU**. There is much more variation in SKU's, they are changing much more often, and they are spending less time in stock. This makes the retailing sector of the economy a much more challenging place to be.

Computers are an excellent example of the complexities of retailing logistics. Personal computers (PC's) have a shelf life of about nine months, and the selling price erodes every two months. PC's come with lots of different configurations. For each PC model, there can be two chassis, three processors, two operating systems, eight memory configurations and five hard-disk drives. By one estimate, there are 70,000 different configurations on the market.[19]

Another implication of the increasing demands of customers for a variety of choices is an increased need to service many different markets through different channels. A company can no longer offer its product through only one channel. The different markets use different channels. For example, people used to buy books from bookstores, and that was pretty much the only way it was done. Now books are sold on-line, in schools, and specialty stores. The publisher must then develop all these different channels. It would be disastrous to expect the customers to only come into a bookstore if they want that particular publisher.

Earlier in this chapter we introduced the concept of power and leadership of the supply chain. To be more specific, channels members have one or a combination of the following **four types of power. Legitimate** power refers to that which is considered by all involved to be the 'proper' leader. For example, when the law states that a given company is in charge, that company has legitimate power. **Economic** power is when a company has more money to spend on the channel. **Expert** power goes to whoever knows the most about the channel or its products. Finally, **Reward** power is the ability to reward other channel members.

A study by Bowersox and Calantone found that loyalty has shifted from manufacturers to retailers.[20] This is because retailers possess the most recent information on buying habits of their customers. When you buy something and the store scans in your purchases at the check-out counter, they are obtaining a vast amount of information on you. For example, not only do they know everything you bought and how you paid for it, but they capture information such as the time of day and what combination of things you bought. Stores can thus react to changes in customer preference much faster than manufacturers can.

E-commerce has created a new channel, which is still in rapid change. The role of the Net has resulted in much confusion, which is discussed in Chapter Sixteen. As a channel, e-commerce has some special characteristics. Whereas normal retailers have stores close to their customers (place utility), **etailers** (retailers that sell on the Internet) do not. This gives them the advantage of going after a much wider market, but the problem is the "last mile problem". Recall from the discussion of the last mile problem that some industries have a problem delivering to their customers. The solution has been to provide customer access so they can do their own pickup and delivery. Etailers generally lack destination presence and without the corresponding density lack the ability to deliver in a cost effective fashion.[21]

---

[1] *"Henry Ford: The Wayward Capitalist"*, Carol Gelderman, St Martin's Press, New York, 1981, pp. 226-270.

[2] *"Supply Chain Management: Implementation Issues and Research Opportunities"*, Douglas M. Lambert, Martha C. Cooper, and Janus D. Pagh, International Journal of Logistics Management, vol. 9:2, 1998, p. 1.

[3] *"Efficient Consumer Response: Enhancing Value in the Grocery Industry"*, Kurt Salmon Associates, Inc, January 1993, p. 26.

[4] *Presentation by William C. Copacino, Anderson Consulting, before the Health and Personal Care Distribution Conference, Longboat Key, FL, October 21, 1992.*

[5] *Keebler et al, Ibid, p. 60.*

[6] *Keebler et al, Ibid, p. 79.*

[7] *Quoted in "Managing Logistics in a Perfect Storm, 12th Annual State of Logistics Report", Robert V. Delaney and Rosalyn Wilson, Cass Information Systems and Prologis, Washington D.C, June 4, 2001.*

[8] *Robin Roberts, Stephens, Inc, quoted in JoC Week, April 16-22, 2001.*

[9] *"Managing Logistics in a Perfect Storm, 12th Annual State of Logistics Report", Robert V. Delaney and Rosalyn Wilson, Cass Information Systems and Prologis, Washington D.C, June 4, 2001.*

[10] *Bowersox and Closs, Ibid, p 94.*

[11]

[12] *Bowersox and Closs, Ibid, pp. 104-5.*

[13] *"Global Operations and Logistics", Philippe-Pierre Dornier et al, John Wiley, New York, 1998, p. 216-25.*

[14] *"Fundamental of Logistics Management", Douglas M. Lambert et al, McGraw-Hill, New York, 1998, pp. 197-203.*

[15] *"Logistics: Perspectives for the 1990s", Bernard J. LaLonde and James M. Masters, International Journal of Logistics Management, volume 1(1), 1990.*

[16] *Bowersox and Closs, Ibid, p. 92.*

[17] *"County Business Patterns-1989", United States Department of Commerce, Bureau of the Census, August 1991.*

[18] *"Supply Chain Innovation in a Global Multi-Business Enterprise: Techniques Applied, Culture Change Challenges and Benefits Realized", James J. Curry and Robert P. Petrich, Annual Conference Proceedings, Council of Logistics Management, 1998, p. 51.*

[19] *"Dead in nine months", Robert Mottley, American Shipper, December 1998, p. 30.*

[20] *"Global logistics market trebles", American Shipper, October 1998, p. 40.*

[21] *Thanks to Richard Hallal of Logistics Development for explaining this.*

CHAPTER 4
STRATEGIC PLANNING

This chapter looks at how companies plan for the future, with an emphasis on how those plans affect their logistics. Once the company has made their strategic plans, then there is the logistical planning process. Logistics planning is done at the strategic level, as we explain in this chapter, and at lower levels, as we explain in the next chapter.

This chapter includes planning as done by a company, but it also looks at some of the major issues that affect planning. ISO9000 and finance are two of the most important considerations. Location and network issues are important because companies must make their plans based on past and future plans of others. Material resources planning is the standard method for manufacturers to plan their operations. The sections on forecasting and benchmarking offer some practical tools of planning. Finally, we consider reverse logistics, a topic that is becoming increasingly important for legal, political and environmental reasons.

THE PLANNING PROCESS

Strategic planning is the way an enterprise seeks to answer three basic questions, who are we, where do we want to be, and how are we going to get there?[1]

Who are we?
Where do we want to be?
How are we going to get there?

All enterprises should make strategic planning a basic part of their work. While many fall short in this regard, high performing organizations frequently credit their success to a good plan. A CLM study found that three-fourths of the companies they surveyed reported doing strategic planning, and two-thirds had formal written plans.[2] Strategic planning may be defined as:[3]

The process of identifying the long-term goals of the entity (where we want to be) and the broad steps necessary to achieve these goals over a long-term horizon (how to get there), incorporating the concerns and future expectations of the major stakeholders.

Taking this definition one step further, logistics strategic planning may be defined as:[4]

A unified, comprehensive, and integrated planning process to achieve competitive advantage through increased value and customer service, which results in superior customer satisfaction (where we want to be), by anticipating future demands for logistics services and managing the resources for the entire supply chain (how to get there). This planning is done within the context of the overall corporate goals and plan.

Planning is an ongoing activity that is done at many levels, and may be divided into strategy, tactics and operations. Each of these levels is defined by their 1. time horizon, 2. the people doing the planning, and 3. the activities being planned.

Strategic planning is done at the highest levels (though it involves input from all levels of the organization). The strategy is long term, generally three to five years. However, what is meant by 'long-term' differs depending on the enterprise. The dot com industry may consider a one-year time horizon to be long term, whereas a slow-moving industry may plan many years out.

The time horizon chosen is partly to do with the financing of capital assets. If a company has factories or other major fixed assets, it may develop strategic plans depending on the time required to turn a profit on those assets. One factor not always mentioned is that the company managers doing the planning need to see results within the time they are in charge. It is rare that we see planning done by people knowing that they will not be around to see the effects of the work. It does happen, but leaders prefer to enact plans that they can see through to the end. Finally, the stock market and market watchers feel more comfortable with plans that are within the range of their abilities to forecast. Since forecasting beyond five years is almost guesswork, managers do not have an incentive to do things that will not be rewarded in the company's stock value.

Tactical planning is when the strategy is translated into more specific plans that the different parts of the organization can use. **Tactics** are the specific activities that must take place to enact the overall strategy. Tactical planning is done at the middle and lower levels of the organization. Tactics are short-term or medium-term, in relation to the strategy. For example, a 5-year strategic plan may have yearly tactical plans. The important thing about tactical planning is to give the operating units of the organization instructions that are specific enough that they can put them into operation.

At the lowest levels, **operational planning** is when the operating units of an organization take the strategic and tactical plans and translate them into instructions that are specific enough for individual workers and units to follow on a daily basis. The CLM's book, *Keeping Score*, has one of the best explanations of how the small tasks build into bigger tasks.[5] The authors' point was that we often focus on the smallest tasks when we should be looking at the overall result.

Keeping Score: Definition of Terms

The following terms are defined in *Keeping Score: Measuring the Business Value of Logistics in the Supply Chain*

Task: a coherent piece of work that can be assigned to an individual or small team and completed in a reasonably short amount of time.

Activity: a collection of tasks that have a common purpose, produce a common output, or address a common theme.

Function: a grouping of related activities contributing to a combined result, where trade-offs between the tasks and activities can be made under unified management.

Process: a series of linked, continuous, and managed tasks and activities that contribute to an overall desired outcome or result. Processes have a specific starting point and ending point and often, but not always, cross functional boundaries. Customers of the process are always at the end point of the process, and they are also often at its starting point.

Integration (as applied to Process): the uniting, combining, or incorporating of two or more functions within a company, or two or more processes between two or more companies into a compatible or unified process. This presupposes that joint definitions and agreements concerning the separate functions and processes have been defined and articulated between all parties.

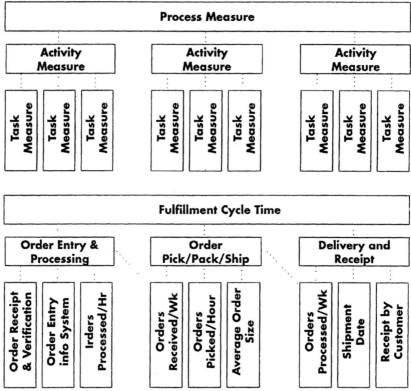

Source: Keebler "Keeping Score", Council of Logistics Management, Oak Brook, Il, 1999, p142

In this example, we could have the Process Measure of "Fulfillment Cycle Time",
which is a very common concern in companies. The Activity Measure would be
"Order Pick/Pack/Ship", and the Task Measure would be any of the many tasks
necessary to pick, pack and ship an order.

Transportation planning is a special area, which is of particular importance to international logistics and transportation. It is normally done at the tactical and operational levels. It will be discussed in Chapter Five.

Now that we see how a strategic plan ends up on a factory floor, let us step back to the strategic level, which is the subject of this chapter. Logistics strategic planning does not stand by itself, but is part of the overall business plan. Since logistics is partly about coordinating the activities of different departments of an organization, it is critical that the logistics strategy be based on the business plan. Some organizations put logistics within this business plan, while others leave it to the logistics department to create their own plan. In manufacturing, logistics is a support activity that addresses the *what*, *where* and *when*, but not the *how*. Logistics plans are *not* telling the manufacturing manager how to make whatever it is they make.[6]

It is widely recognized that while a strategic plan can be valuable, it is the planning process itself that is most important. This book does not go into the strategic planning process because it is already discussed in just about every management book ever written. What we will look at is what special aspects of logistics will affect the planning process.

One of the first differences we see between logistics planning and other types is the **number of dimensions involved**. Planning normally involves **time** and **functions**. For example, a company is planning on what its different functional areas are going to do across the given period of time. Logistics planning is more complex because it also involves **space** (as in geography) and **coordination with other parties** (suppliers and vendors). This is because logistics, being the management of getting things to where they need to be, is fundamentally involved with geography. SCM is fundamentally about coordinating with other parties.

Four dimensions of Logistics Planning
• Time   • Functional areas   • Space   • Other parties

Given that strategic planning can be important to giving an organization direction, **why is it that many do not do any strategic planning**? One reason is that the planning process **takes time**, especially the time of executives who already have busy schedules. Planning also requires the organization to make some important and **difficult decisions** about where they want to go in the future, decisions that they may not want to make for internal political reasons. Also, the organization may be in a situation where their **future is uncertain** that a strategic plan would not be worth the effort. For example, a company that is about to be bought out would probably wants to wait to see what its new owners want first.

In the first chapter we noted that logistics is not highly regarded by the business world or top management. This becomes evident when we look at how rarely companies and institutions consider logistics in their strategic planning. One of the most widely used and respected university textbooks on strategic management is over a thousand pages long, and the word 'logistics' cannot even be found in the index.[7]

What has been published on strategic planning for logistics assumes that the company does logistics in the first place. Some companies have minimal logistics needs, so even if they are doing strategic planning, that would not entail logistics planning. Professional corporations, such as an accounting firm or an investment bank, does not move much around and thus would have minimal need for logistics.

A strategic plan does not necessarily mean change. It may be that the organization is already heading where it wants to go. Radical changes at a strategic level is rare and usually means that something is very wrong or that conditions have changed. According to one logistics analyst, "Companies which are really making substantive changes tend to be totally innovative or totally desperate. You see very few in the middle doing anything but tire-kicking".[8]

While every enterprise will want to develop its own strategy, there are some general paths that can be seen. These are a general class of strategies. Richard Dawe identifies five strategies for logistics:[9]

Cost minimization attempts to reduce the overall cost of logistics while providing a minimal standard of customer service, and would be most appropriate in an industry where customers put more emphasis on cost and less on service. Raw materials, minerals and bulk cargo are some examples where it is most important to have low transport costs.

Value-Added emphasizes the coordination of the different logistics activities and channel members. It would be more appropriate where profit margins are higher and customers are willing to pay for additional services. The luxury goods market has increased over the past decade, and this is a prime area for value-added logistics.

Channel Integration is commonly used in industries where close coordination with other channel members can result in significant gains. For example, a car assembly plant using a just-in-time inventory system needs to have its component parts delivered promptly, which means it is important to work with their suppliers.

Quick Response emphasizes rapid response, and is best used when the product is very time sensitive, such as fresh cut flowers.

Total Enterprise attempts to optimize the organization's overall performance even if it means suboptimizing individual units. This strategy would be most appropriate in an organization with centralized control and decentralized operations.

A company's involvement with strategic planning, and especially logistics planning can be seen as an evolutionary process. Historically, logistics was not always seen as being important. Many companies simply did not recognize logistics as an area that required much attention, and not much planning. As organizations recognized the gains that could be made by effective logistics, they then noticed the benefits of planning. One can see four evolutionary stages in an organization's involvement in logistics:[10]

Stage 1: We focus on today because of a crisis.

Stage 2: We focus on this month because of budget pressure.

Stage 3: We focus on this year because of an improvement program.

Stage 4: We focus on the long term because we want to gain strategic advantage.

In Figure 4.2 we see these four stages as they are referred to by Richard Dawe. He shows how companies may be at any one of the four levels, defined according to how effective they are in fulfilling their service goals (in other words, do they have the inventory or can they fill the orders that they receive) and their efficiency (how much does it cost per unit of output. There is increased sophistication in logistics as one moves up. Controlling operations becomes increasing difficult as the speed of change increases. The goal, which is obviously an ideal, is perfect efficiency and effectiveness, which requires strategic planning and advanced logistics.

Materials Logistics Performance Evaluation
Impact of Information Technology on Logistics Performance

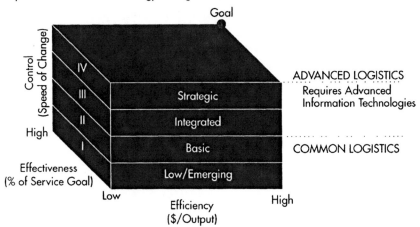

One aspect of logistics that make it unique in the planning process is the **task of coordinating with different channel members**. A strategic plan is by definition done at the organization's highest levels. Integrated logistics emphasizes coordinated operations among the departments and with outside channel members. Although this may sound quite obvious, it is important to note that a strategic plan is developed by an organization for its own purposes. Other organizations are doing their own planning. How does one plan for the future knowing that its partners and channel members are doing their own strategic plans?

There is no easy answer, but we can make some observations. First, before an organization enters into any kind of partnership it should carefully consider the other **partner's strategic interests, goals and plans**. Nobody can be sure about the future but at least one can make sure that the partner has roughly similar goals. Second, a plan is always unsure and should be able to **anticipate changes**. There are different aspects of logistics, some which entail outside parties more than others. Manufacturing support logistics, for example, is within a firm, whereas the other elements such as transportation are subject to outside parties.

Planning Authority

Authority is the right and power to make certain decisions. The distribution of authority is an important question that should be addressed. This is not the same as responsibility, which is the obligation to make decisions and be accountable for the results. It is often said that one can delegate authority but not responsibility, referring to the management principle that no matter how much leeway a manager gives her subordinates to make decisions, she is still responsible for the final results.

The basic decision in the distribution of authority is where to give what powers. This generally means **centralization** versus **decentralization**. Some firms are highly centralized, in which important decisions may only be made at headquarters, versus decentralized systems in which the operating units make more decisions. There has not been any definitive study on whether logistics firms differ significantly from other industries. It may be said that for the international transportation industry, because it is inherently distributed across a wide geographic area, the question of how authority is distributed is of great importance. Professor Dicken shows in the illustration some of the ways that multinational enterprises may arrange their operations.

The issue of centralization also depends on the type of industry. Some industries might require centralized authority, such as a politically sensitive industry. Others may decentralize authority to keep in touch with local markets. Gus Pagonis, who was the general in charge of logistics for the US in the Gulf War, believes in centralized control, decentralized execution. While his experience is from the military, he adamantly believes that the logistics are not significantly different from civilian organizations.[11]

A. Globally
   Concentrated
   Production

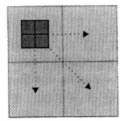

All production occurs at a single location. Products are exported to world markets.

B. Host-Market
   Production

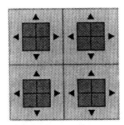

Each production unit produces a range of products and serves the national market in which it is located. No sales across national boundaries. Individual plant size limited by the size of the national market.

C. Product Specialization
   for a Global or
   Regional Market

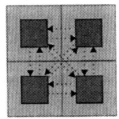

Each production unit produces only on product for sale throughout a regional market of several countries. Individual plant size very large because of scale economies offered by the large regional market.

D. Transnational Vertical Integration

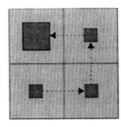 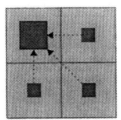

ISO9000

ISO9000 is a certification program to improve quality in organizations (the vast majority being businesses, but some are public enterprises). ISO is the International Standards Organization, based in Geneva, Switzerland with local organizations in each country. The ISO establishes standards, mostly voluntary but many are passed into law, on everything from the size of the holes in binder paper to the shape of a light bulb socket. ISO9000 is a certification that organizations can win that shows they adhere to a recognized standard.

The standards are based on business practices that are associated with quality. These include things like having procedures manuals, written business practices and quality control systems. To become ISO9000 certified, an organization pays an ISO-certified agent to come in and evaluate them. This is a long process that can cost hundreds of thousands of dollars and weeks or months of inspections. Many organizations have said that the process of gaining ISO forced them to improve their internal operations, so the cost was well spent.

Many companies now require that their suppliers also be ISO certified. If a company is following certain internal procedures, it is much easier for them to work with other companies that are also following similar procedures. This means that being ISO9000 is not only valuable for a company's internal efficiency, but it can also give it a competitive edge in its industry.

Three common misunderstandings of ISO9000 are that A. it is a guarantee of quality, B. that it is a certification that the company has achieved a given result and C that it is legally required. First, certification means that the company has identified a goal and has procedures to achieve it. It is assumed that the goal is a high-quality product or service. But those providing the certification have no way of knowing if the company has picked the right goal. In other words, certification means the company is doing things right, but not that it is doing the right thing.

Closely related to this point is that to gain certification the company shows that it follows a given set of procedures. These procedures are the well recognized steps that high-performing organizations usually follow. However, as they say in advertising, results may vary. The certification does not mean that the organization achieved any particular goals.

Finally, no government laws require ISO9000 certification. Companies get accredited to gain credibility and be more competitive. Some governments require their suppliers be certified. Many contracts require that the parties to the contract be certified. But in no case that the author is aware of has any government passed a law that requires a company to be certified.

Accuracy versus Specificity

Anytime we are discussing management planning, it is important to recognize the difference between accuracy and specificity. This is especially important in planning, where there are fundamental differences between strategy, tactics and operations.

Accuracy means that a description is close to reality, or free of error. Specificity means that the description is at a high level of detail. Why is this difference important? In planning, one must anticipate the future. This is an inexact science. One often hears forecasts like "sales for the next year will be 325,448". Why is that incorrect? It is very specific, but it is not accurate. If one were talking about last year's sales, it may be appropriate to give a number down to the single digit. Yet by doing so in a forecast, it gives the misleading impression that the forecast is accurate down to the single digit.

Therefore, statistics should be rounded off to the level of specificity that is accurate for the purpose. Statistics that are too specific, or too general, are not as accurate as they could be.

FINANCIAL ISSUES

We will now look at some financial issues in the planning process that are of particular importance to logistics. Keep in mind that finance is a language, a way of understanding a given phenomena, and with any language, there is always the difficulty of translation. Logistics is about moving things. Finance seeks to explain this in terms of money. Understanding and managing logistics, particularly in the business world, is best understood in terms of finance, but it is not the only way. Some logistics issues, such as customer service or humanitarian relief, do not lend themselves to financial analysis.

To repeat, logistics is about coordinating operations in different organizations. Whereas the finance department of a firm describes things in terms of money, other departments may track units of product, time or other measurements. There are special challenges presented in trying to tie together the different departments as well as the concepts we are about to describe.

The first step in understanding finance is to understand prices. This is the amount a product may be sold for. This requires a market, in which there are many buyers and sellers who collectively, yet operating for their individual best interest, determine the price. When something is bought and sold, there is a price associated with it. However, as products move through a supply chain, they are not always being sold. The cargo may be transferred from one facility to another, or some other **non-commercial transaction**. That is why internal pricing and transfer pricing is an important concept to logistics.

This happens constantly as large organizations transfer things among their operating units. For example, imagine Ford's Japan subsidiary sending parts to Ford's Indonesia subsidiary. The company may choose to have their subsidiaries **operate at arms length**, which means anything transferred must be bought or sold as if the two units were not at all related. The parts would be sold at market rates.

Transfer pricing is the way a company assigns a price to such things as if they were sold at arms length. Most countries require a transfer price be designated for accounting purposes. The challenge is that since this price is hypothetical, so companies can adjust them to benefit themselves. For example, a product coming from a high tax country would be priced very high, as if that company had to pay a lot. This reduces their tax obligations. Transfer prices are determined based on a combination of competitive market prices, costs and legal restriction.[12]

Transfer pricing affects logistics in a number of ways. In strategic planning, a company often makes decisions such as whether it is worthwhile to import a part from a foreign subsidiary or acquire it locally. This is partly a logistics question, since the price of importing the parts will partly determine the answer. The transfer price will also affect the decision because the company is comparing a real price to be paid for acquiring the parts locally versus paying a hypothetical transfer price to their foreign subsidiary.

Another way that transfer pricing affects logistics is in documentation. Import documentation asks for the cargo value, used to assess customs duties. This is why customs authorities look very closely at transfer pricing which may be used to avoid paying duties.

Pricing leads into the next area of concern for logistics, which is **costs**. Understanding the true costs of logistics is a critically important and difficult task. Without a clear understanding of costs, it is almost impossible to see the benefits associated with logistics. It has long been recognized that while the costs associated with logistics are easy to see, the benefits associated with those costs are much harder to see. For example, providing better customer service by increasing the number of deliveries clearly increases costs, yet seeing a return on that improved customer service is not so easy to measure.

The most accurate measure of materials logistics costs is the concept of **value-added**, according to a 1984 study by A.T. Kearney for the Council of Logistics Management. After the "uncontrollable" costs of material, supplies and energy are excluded from a product's cost, what is left is the value-added cost. Materials logistics are a quarter of the cost of most products. Petroleum and chemical products have the highest logistics costs, while tobacco, furniture and apparel have the lowest.[13] Value added costs are harder in reality to determine. Each component has its own value added aspect. Martin Christopher identified two principles of costing:[14]

1. **Accounting systems should mirror the material flow to identify the cost of servicing customers.** Accounting systems are often organized around the company, which produces different products for different customers, all with very different costs/revenues. This can give misleading costs.
2. **Accounting systems should be able to analyze costs and revenues by customer type and market segment or distribution channel.**
   Imagine a company that makes two products, a high school math textbook and an easy-to-read math reference book for the general public. The textbook is shipped out maybe twice a year, once for each semester. The customers are schools, a limited number of customers each of which buys large quantities. The reference book is shipped to bookstores, and they are shipped out whenever the store sells the copies in stock. Deliveries are made many times through the year, and the order quantities are very low. In this simple example, an accounting system that followed the different logistics associated with these products would have very different costs associated with the two different customer markets.

Christopher goes on to explain how there are some problems with conventional accounting when applied to logistics operations. There is a general ignorance of the true costs of serving different customer types, channels and market segments. Costs are captured at too high a level of aggregation. Full cost allocation still dominates. Conventional accounting systems focus on functions rather than outputs. In other words, an output (that is, the final product) may create costs from the factory, the accounting department, the human resources department, and so on. Finally, companies understand product costs but not customer costs. For example, it is rather easy to identify how much it costs to produce a product, but difficult to identify the cost of losing a loyal customer.[15]

Problems with Costing and Logistics
Ignorance of the true costs
Aggregation is too high
Full cost allocation
Focus on functions rather than outputs
Product costs versus consumer costs

Part of the reason that accounting systems are not designed for logistics is that logistics is a relatively new field for executive-level attention. There has not in the past been as much interest in identifying logistics costs. This is changing as more companies see the competitive advantages to be gained through logistics. Another reason is that logistics is a complex activity for accounting, crossing traditional departmental and company boundaries.

Types of Costs[16]

Controllable versus Non-controllable. Controllable costs are those associated with the amount of effort expended. Uncontrollable costs are those that vary regardless of what the organization does.

Direct versus Indirect. Direct costs can be attributed to a specific output, such as the raw materials used to make something. Indirect costs are things like general administrative costs.

Fixed versus Variable. Variable costs change depending on the amount of output, whereas fixed costs stay the same. For example, a manufacturer will have variable costs for raw materials depending on how much product is made, while the cost of the factory itself is fixed since it will cost the same regardless of how much is produced inside it. Many costs are semi-fixed or semi-variable. For example, the factor may be partly variable if its maintenance costs increase or decrease depending on how much it is used.

Actual versus Opportunity. Actual costs are based on activity that actually occurs. Opportunity costs are what are sacrificed by making a different choice. For example, if a company offered less truck deliveries and thus was able to use eight trucks instead of ten. The extra two trucks could then be leased out to earn extra money. The money that could be earned by using two less trucks would be the opportunity cost associated with using all ten trucks for deliveries.

Relevant versus Sunk. Relevant costs change with future actions. Sunk costs have already occurred and are not going to change. If a company wants to decide what to do with a truck, the price paid is 'sunk', while the future costs (resale price, insurance, operating costs etc) are 'relevant' costs.

Activity-Based Costing (ABC) has been a major development in the past decade to more accurately capture the costs of given activities. Traditional accounting follows the costs of each functional area. ABC looks at the costs related to a given activity, which is a combination of different types of costs. For example, preparing a shipment for delivery may entail work for the customer service department, the warehouse, the delivery trucks. Each of these departments would normally account for their own activity. ABC, on the other hand, would look at the delivery and calculate what costs were incurred from each of the departments affected.

LOCATION THEORY AND NETWORK DESIGN

Location theory seeks to explain why things are placed where they are. This is done at two levels, the firm and the place. In the first case, firms use (consciously or unconsciously) certain criteria for deciding where to locate their headquarters, their office space, factories, warehouses and so on. In the second case, one looks at the overall pattern of industrial development and seeks to explain why certain places have a lot of factories, for example, while other regions are used for office space.

A classic work in location theory was done by German estate owner **Heinrich von Thunen**, who's book *The Isolated State* was published in 1826. He described a hypothetical town, surrounded by farmland that was all of the same quality. The only source of transportation was a horse and cart. The primary difference between different plots of land was the cost of transportation to and from the town, which was the market for the products grown in the farmland. He suggested that bulky or heavy products that were expensive to transport would be grown only in the land closest to the town. Perishable products would also be grown here. As one moved farther from the town, other products, those cheaper to transport, would be grown. The outermost lands would be used for livestock. Another implication of this model is that land closest to the market will be the most expensive, and land prices fall in relation to their distance from market.

Von Thumen's "Isolated State"

Cows

Farming

Industry

Town Market

Higher Land Value

Lower Land Value

Alfred Weber first described **industrial location** in 1929. He suggested that manufacturers can be either **material-oriented**, which means that they are located near their source of raw material, or **market-oriented**, which means they are located near their customers. *The best location for a manufacturing plant is the place where transportation costs are minimized.* This depends on the nature of the raw material and finished product. Industries that reduce the bulk of their materials during processing tend to be located near the source of the materials. For example, mineral processing is done near the mines because the ore is heavy but the output is less so. He classified different types of raw material depending on their transport characteristics:

- **Ubiquitous raw materials**, such as air, are everywhere, and there is no transport cost.
- **Localized raw materials**, such as coal, are found only in certain places, and their transport cost depends on how far they must be moved.
- **Pure raw materials** lost no weight in processing, such as an automobile part.
- **Gross raw materials** lose weight during processing, such as fuel, and thus their transport cost is a combination of how fast they are consumed in the production process.

Weber's work was only a start, and has become vastly more complex given modern conditions. Transportation costs have been declining, and transportation costs of finished goods are now more expensive than the cost to transport raw materials.[17] His model only looks at the cost of inputs and not production costs (i.e. labor). Land costs vary widely. Historically, factories were located near towns so workers could get to their jobs, and other suppliers could get their products to the plant. However, land costs in the city are higher than in the suburbs. Now with improved transportation, workers can commute to work, which means firms can be located further away from the town, thus using cheaper land. This is why there has been such an increase in suburban industrial parks in recent decades.[18]

The cost of labor also influences firm location. Wages vary widely from one place to another for a variety of conditions, one of them being the cost of living. Yet another reason that wages differ is inertia. Even if people can move to areas with higher wages, that does not mean they are going to. We all have our family and friends where we live and are not always eager to leave just for a better paid job. Skilled labor is particularly prone to this inertia. Older industrial areas in the US tend to be highly unionized, while areas of newer development have less unionized labor and thus offer cheaper labor.

Capital comes in two forms, **liquid capital** such as money, and **fixed capital** such as buildings. The cost of capital also influences where a firm will locate. While liquid capital can be easily moved almost anywhere in the world it is needed, buildings and other capital investments are expensive, last a long time, and cannot be easily moved. Therefore a firm must choose between the capital facilities that are already in existence, or spend the money to build new ones in a new location.

Managerial and technical skills are needed to operate a firm, and people with these skills are not evenly distributed around the world. Professionals prefer to live in 'nice' places that are safe and attractive, while firms can find it very difficult attracting the skilled individuals to other places. Oil production in Central Asia has been booming, yet oil companies have an extremely hard time convincing their engineers and other professionals to live and work there. This usually means paying them a large bonus. In another example, consider the computer industry. Certain places in the world have a high concentration of computer-related industries: Silicon Valley, Taiwan, and even Ireland. It then becomes easier to locate a computer-related firm there knowing that hiring specialists in that area will not be a problem. On the other hand, opening a computer chip plant somewhere all by itself means the technicians must be brought in from elsewhere.

So far we have been looking at a firm's location decision as if it operates by itself. Companies rarely do this now, but instead operate with a variety of locations such as the headquarters, office space, manufacturing locations, storage locations and distribution centers. The **decision of where to locate a facility is based on the location of suppliers, other resources inputs, competitors and the market**. For example, a factory would want to be near its suppliers, but they would also want to be near the market (perhaps a tradeoff needs to be made here). But they also want to be near their suppliers, and near other resources. What do I mean by other resources? These can be the companies own supplies. A company with many facilities spread across a country (or around the world) may have its own resources, but they may be in another facility far away.

   International borders have a major influence on location decisions. Different countries vary widely in all the factors listed above, and then there is the role of the border itself. Crossing borders can be costly, and there is always the possibility of disruption for political reasons. Even if there are little direct costs associated with border crossing, there is that extra complexity which carries a cost.

Traditional Cross-Border Shipping Method

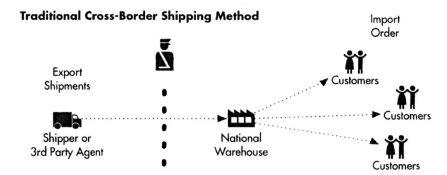

Direct Cross-Border Shipment

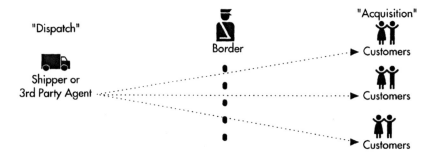

Organizational decisions have changed the nature of firm location. JIT influence where the facilities are depending on how they interact with each other. If the shipments from one plant to another need to be extremely reliable, then they should either be very close or have very reliable transportation between the two. Internationally, consider the effect of a border crossing on a JIT operation. This creates the possibility of delays that could be costly. For this reason, suppliers often locate in the same country as a major customer.

Network design is a highly abstract tool for things like designing distribution routes or a corporate chain of command. Networks are sometimes the result of historical accident, such as a country's road systems. The emphasis on networks is not so much the location of any one facility, but how they are all connected with each other. As we see in the illustration, there are a wide variety of networks.

Connectivity in a network refers to the ratio of routes to hubs, and gives us an idea how easy it is to move from one place to another. An area with more connections offers easier and more efficient movement. In one interesting case, ex-colonies have a transport network that looks like a drainage system, all leading to the trade port. This was because the colonial administrations designed the system to bring export commodities from the interior to the port for export, with fewer connections within the country to serve domestic needs.

Network Design

The following are some of the most common methods.

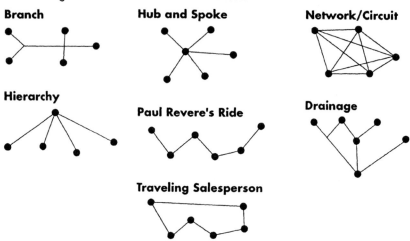

Branch

Hub and Spoke

Network/Circuit

Hierarchy

Paul Revere's Ride

Drainage

Traveling Salesperson

Route Design

In many shipments, there are a couple of stages. The cargo is sent to a hub, and from there to another hub or the final destination. The primary move is called the **line-haul**. The delivery to that hub is called a **feeder**. This concept is closely related to the discussion in Chapter Two on trade lanes. The line-haul is usually on a primary trade lane, and the feed routes are usually on secondary trade lanes. The difference between line-haul/feed routes and primary/second trade lanes is that the former refer to an individual shipment, and the latter refer to overall trade patterns.

Linehaul = primary trade lane
Feeder route = secondary trade lane

This is confusing and deserves an example. If one were to ship a car from Seville, Spain to Moscow, Russia, it is very unlikely that any transport company offers direct service between these two cities. So the shipper looks for the nearest city that does offer direct service. That would probably be Barcelona, the hub. From Seville to Barcelona is the feeder route, because cargo is feeding from all the little towns around Barcelona. The shipment from Barcelona to Moscow is the main part of the shipment, the longest distance and the largest shipment, probably a train-load of cars. That is the line-haul.

Trade Lanes

Barcelona

Seville

Ocean

St. Petersburg

Primary Trade Lane
(Line-Haul)

Secondary Trade Lane
(Feeder Routes)

Moscow

Note that the distinction between a line-haul and feeder is relative. In this case, St. Petersburg to Moscow is a feeder route. If you were going from a farm to St. Petersburg, where the cargo was put on a train to Moscow, then the St. Petersburg/Moscow leg of the trip is now the line-haul.

The distinction between primary and secondary trade lanes is important because, when making a decision on where to cut costs, one should look at the largest costs first. If a shipment involved $100 in trucking to the port and $2,000 for the ocean carrier, one would not want to spend too much time trying to get that trucking cost down. It would be better to work on the ocean shipping costs.

MATERIAL REQUIREMENTS PLANNING

A major tool of logistics planning is **material requirements planning** (MRP)[19], defined as "a system of forecasting or projecting component part and material requirements from a company's master production schedule (MPS) and the bill of material (BOM). The time-phased requirements for components and materials are then calculated, taking into account stock in hand as well as scheduled receipts".[20] Another way of thinking about MRP is described by Vollman, Berry and Whybark's classic book on the subject, in which they refer to it as "the right part at the right time to meet the schedules for completed products".[21]

MRP systems put together computerization and a manufacturing information system that ties together all the inputs of production. Remember the last time you bought something that was in many parts that required assembly. Included in the instructions was a picture of the product "blown apart". This is essentially what an MRP program does. It looks at the final product, then steps back to show how each piece fits in. The process includes all of the pieces needed for assembly, the sequence of assembly or processing, and the timing. Also included may be the place of each step and what action is being performed.

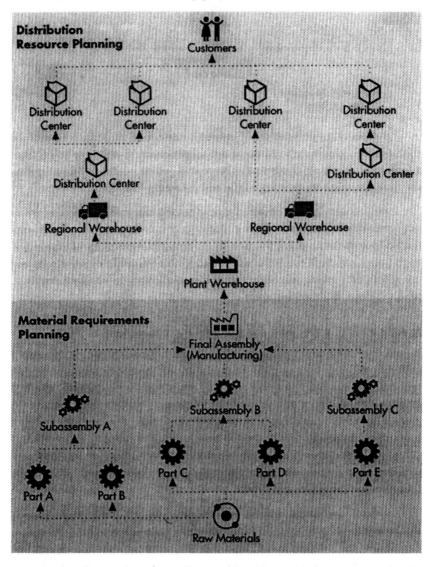

Source: Based on Vollmann, et al "Manufacturing Planning and Control Systems, 3rd edition, Irwin, New York 1992, Copyright the Mc Graw-Hill Companies

Manufacturing Planning and Control System

Front End

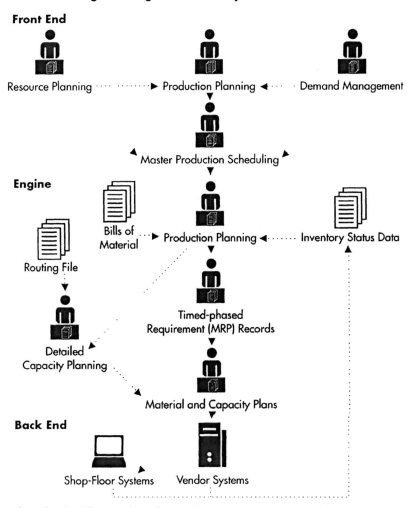

Resource Planning ·········▶ Production Planning ◀···· Demand Management

Master Production Scheduling

Engine

Bills of Material ···▶ Production Planning ◀······ Inventory Status Data

Routing File

Detailed Capacity Planning

Timed-phased Requirement (MRP) Records

Material and Capacity Plans

Back End

Shop-Floor Systems Vendor Systems

Source Based on Vollmann, et al "Manufacturing Planning and Control Systems, 3rd edition, Irwin, New York 1992, Copyright the Mc Graw-Hill Companies

The manufacturing company will have a **master production schedule,** based primarily on marketing plans. This schedule states what is to be produced and when. MRP is based on the schedule, since this is what tells the MRP planners what to do. Based on the MPS, there will be a bill of materials, a list of all the inputs needed. In other words, the MPS says what is going to be produced, and the BOM is the list of all the parts and other materials needed to produce them. One can think of the BOM as a shopping list needed for production. The MRP then plans out when each and every piece is needed.

Below we see an example of an MRP system. Imagine we are making two types of cars, sedans and compacts. In order to make these cars, you need engines. The factory has a six week plan in this example. A **planned order release** is when the factory sends an order to its vendor for that part. For sedans, they plan to make 30 in Week 1, 25 in Week 3, and 25 in Week 4. You see how the orders of mini-vans are scheduled. Following the arrows, we see that in Week 1, there are **gross requirements** of 45 engines. A gross requirement is how many engines are needed, but we do not know if we already have them in inventory. **Scheduled receipts** are the amount of engines expected to arrive from the vendor, in this case 95 in Week 1. We had 18 already in inventory, leaving us with 68 in inventory. We refer to the inventory level as **projected available balance**. Given this demand, they plan to order 100 engines in week two, which creates demand for blocs and pistons.

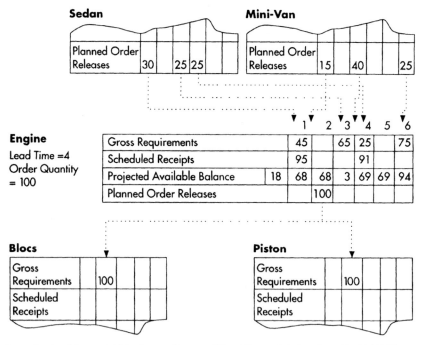

This system has been taken one step further, to include financial, marketing and logistical elements. In other words, manufacturing is not a process by itself, but affects the entire enterprise. MRP II, known as manufacturing resource planning, takes the basic system and ties it into the other departments.[22] For example, an MRP II system can calculate the financial impact of any changes in the manufacturing process.

Distribution Requirements Planning (DRP) is related to MRP but it looks at the distribution requirements. The main difference is that MRP is based on internal production requirements, whereas DRP is based on external customer demand. DRP systems serve the following purposes:[23]

- Coordinate inventory replenishment
- Select transport mode, carrier and shipment size
- Schedule shipping and labor
- Develop the master production schedule

FORECASTING

An ability to see into the future would be priceless, but the best we can hope for is forecasts, a list of all possible outcomes and the probability of any given outcome. Often we do not even know all the possible outcomes or their probabilities. In logistics, fulfilling customer needs that remain constant would be extremely easy. The problem is that customer demand is frequently changing. That leaves us with excess inventory, unfulfilled orders, or expensive last minute deliveries. All of these are potentially expensive alternatives.

We are concerned with two types of forecasting, the shippers that are forecasting their sales and thus creating the demand for logistics, and the carriers that are forecasting the shipper's demand and thus their sales demand. The previous section on MRP/DRP was essentially a forecasting technique, but a very special type of forecasting that deserved a section of its own. Now we are discussing general forecasting issues that apply to all companies.

Forecasting demand for transportation is not as easy as looking at the increase in trade volume and then increase the volume of transport demand. A growth in the overall volume of trade does not mean equal growth in the different modes of transport. The increased trade may all be in one sector of the economy, using only one mode of transport, or it may be geographically concentrated. Growth will occur where there is room for growth and, secondarily, in whichever mode offers the cheapest alternative.

Note the experience of the European Union, in which increased trade occurred on road and sea transport and had less effect on rail or air.[24]

For shippers, an increase in sales does not solve the question of how that increase will be accommodated. For example, a small shipper that normally uses containerized cargo may, when sales increase to a sufficient volume, switch to bulk carriers (containerization and bulk cargo will be explained in Chapter Seven). Economists warn that forecasting is not as simple as extending the line on the chart from past experience.

There are many types of forecasts done by organizations, not all of which affect logistics. What we are most interested in is future demand. This tells us how much will need to be delivered, where and when. **Demand is either dependent or independent**. Dependent demand is when the demand is based on demand for something else. For example, Levi Strauss and Company's demand for fabric is dependent because it is based on demand for their clothes. Independent demand is not based on anything else in particular. The public's demand for bread is relatively independent since it is not based on anything else in particular. However, no demand is entirely independent. The quantity of forecasted demand is composed of a few different parts:

- **Basic demand** - Also known as historical demand, this is the average of what has been demanded in the past.
- **Cyclical adjustments** - Also known as seasonal or economic demand, this is the variance that happens from regular cycles. Every summer we can anticipate an increase in demand for soft drinks, for example.
- **Long-term change trend** - Some long-term shifts can be seen, such as an overall increase in demand for luxury goods as overall wealth increases, or an increased demand for telecommunications products.
- **Promotional factor** - Some changes in demand can be directly attributed to promotional efforts.
- **Buffer range** - This is the margin of error that every forecast includes since it is an inexact science.

There are two approaches to forecasting, **top-down** or **bottom-up**. A top-down approach is when the executives give their forecast and the rest of the organization uses that. A bottom-up approach is when those at the lowest levels of the organization are asked to forecast their anticipated demand, or the sales people are asked to give their expected sales, and this is then aggregated at the higher levels to come up with a forecast. Both methods have their uses depending on the enterprise and the use of the forecast.

The bullwhip effect was discussed in Chapter Three, a problem that is based on forecasting. Research indicates that channel members' errors increase, resulting in over and under adjustments.[25] The longer the supply channel, and the greater the error, the more of a bullwhip. To repeat, the best solution is for more effective forecasts, and also to have more frequent and shorter performance cycles.

Forecasting accuracy has been the subject of extensive study because it is such an important area. One concern is the **cost of forecasts**. A very reliable method that costs too much is not much help. Another concern is whether the forecast can be applied to operations. It would not help much if one knew the future business conditions, but were not able to do anything about it.

There are a wide variety of forecasting methods and tools available. For example, there are many economic forecasting techniques that could be used to determine the price of your home. One could look at new home prices, "house starts", or the cost of home mortgages. Logistics professionals also have a large number of **forecast indicators**. If you work for an ocean shipping company, you might look at the number of new ships being built to determine capacity, which in turn will affect prices. You could also look at the number of firms signing long-term service contracts, which may indicate that they anticipate a change in conditions. In a classic study, Makridakis and Wheelwright offered the following criteria for evaluating the applicability of a forecasting technique: [26]

- Accuracy
- Forecast time horizon
- Value of forecasting
- Availability of data
- Type of data pattern
- Experience of the forecaster

One **warning** is in order for anyone trying to make forecasts. It is not always possible, and it is always a probability estimate. There are many events that cannot possibly be forecasted, and in such circumstances, one should not try. If you provide a forecast, you are telling people that you have some good reason to believe this is going to happen. If you are simply guessing, say so. For example, logisticians often make plans based on forecasts made by people on demand trends. When things do not go as planned and questions are raised about how the demand forecast was made, it turns out they were simply guesses. For more on forecasts, see Chapter Twelve on insurance and risk.

BENCHMARKING

Benchmarking is the process of comparing one's operations to a 'model', or benchmark, to judge performance. In Japanese *dantotsu* (striving to be the best of the best) and, like many business practices that go with Japanese names, benchmarking has been a popular way of improving logistics. Another way to think of this is a role model or a standard by which an organization can judge its own performance. In a competitive environment, it does not help if a company is doing well, if its competitors are doing even better.

Benchmarking was pioneered by Xerox in 1979 which found that competitors were selling their products at Xerox's cost of producing them. Xerox had four layers of distribution while others had two, three or three and a half (a half layer would be emergency storage). By comparing one's own practices to others, it becomes possible to see opportunities for improvement. The key to effective benchmarking is to find the right model. It does little good, and maybe even some harm, to use a role model that is performing worse than the current standard. There is also the chance that the benchmark is not operating in the same environment and is thus not an appropriate model.

Benchmarking is sometimes done as a singular event, but more often it is an ongoing process. There are different types of benchmarking:

- **Internal** Compares one part of a business with another doing the same or similar functions. This is important because internal consistency and documented work practices are needed before conducting external benchmarking.
- **Competitive** The competitor is the benchmark.
- **Functional** Comparing a functional area with that of another company that may not be in the same industry.
- **Generic Process** Benchmarking the basic business processes that all companies do.

A major challenge in benchmarking is to find out what the competitors are doing. Some aspects are easy to see, but others are internal and secret. Of the different types of benchmarking just described, competitive benchmarking can be the most difficult for this reason. Do not use macroeconomic statistics as your own benchmark. For example, it was noted in one book that "Using the fact that the U.S. logistics costs...were equal to 9.9% of nominal gross domestic product (GDP), firms should first get themselves to that level as a percentage of revenues and then attack further reductions."[27] The national or international economy is not a good guideline by which to judge the performance of an individual company.

Another question is how to decide what to benchmark. It would be very rare for an organization that does benchmarking to apply this to every aspect of their operations. That would be expensive. Instead, they may only pick those areas that are most important.

Once a benchmarking program is started, a couple points are in order. This is not a program that is dictated by the executives onto the workers. Have those doing the process do the benchmarking. They understand best how the benchmark would or would not apply. Document everything. When many changes are being made, it helps to see where they started.

REVERSE LOGISTICS

Reverse logistics includes several activities, all of which entail materials moving in the reverse direction of the normal supply chain. According to the CLM, "Reverse logistics is a broad term referring to logistics management skills and activities involved in reducing, managing, and disposing of hazardous or non-hazardous waste from packaging and products. It includes reverse distribution...which causes goods and information to flow in the opposite direction of normal logistics activities".[28]

The global economy has created the possibility of returning cargo, such as defective products, to their original shipper. This is most common in the B2B market in which factories shipping in large volumes also need to plan for the return of some percentage of their output. There is also a growing trade in shipping materials. For example, the automobile industry uses packaging materials designed specifically for their parts. This material may be very expensive and thus needs to be reused. In fact, there are prices in the tariff books of carriers just for these packing materials.

One of the largest areas of reverse logistics is **recycling** and **reuse**. This new emphasis is the result of environmental regulations in many countries. Also, many companies have realized that, regardless of regulations, reuse and recycling can reduce costs and improve efficiency. Waste management, the job of disposing of waste, has become a huge and expensive industry. Landfill has traditionally been the way to dispose of garbage, but land fills are rapidly filling up. Well-run landfills are far more sophisticated than simply a place to dump garbage. For example, they must be designed so that run-off does not contaminate local water supplies, the dust does not create air pollution, and the smell does not foul the air for local communities.

Even before recycling and reuse, **source reduction** is used to reduce the amount and toxicity of materials used so that eventually there will be less waste created. For example, packaging should be minimized. Source reduction then has other logistics implications. Recycling may mean that materials are returned to the manufacturer and turned into other products, but more likely the waste is turned over to recycling companies. When materials are reused, however, it is more likely that they will go back to the original manufacturer. For example, Kodak takes used disposable cameras, takes them apart, and reuses many of the parts for new disposable cameras.

Hazardous materials are a particularly important aspect of reverse logistics. Safety regulations are so strict in the handling of some materials that manufacturers often find it cheaper and safer to take an active role in its disposal. In many cases the manufacturer is required to take on the job of disposal because they are most knowledgeable in some special, highly dangerous materials.

The European Union has been very progressive in their approach to reducing waste. Germany has taken the policy one step further and requires that all businesses must take back their packaging. A private firm, Duales System Deutschland, offers services to act as an intermediary to take care of the packaging requirement. They accept the sales packaging waste and recycle it. The Green Dot (Der Grune Punkt) is a symbol of their work. Non-German companies are exempt from the requirement, but conscientious shoppers look for the green dot on their purchases.[29] The European Commission has been trying to find agreement on a plan to require that companies take back their packaging. EU countries would need to recover half of the packaging and recycle a quarter of it, with at least 15% of each material. Burning the material as a source of energy is considered the same as recycling it. Not all countries are prepared for such rigorous programs, which is why Ireland, Greece and Portugal have been given lower standards.

[1] "Strategic Planning for Logistics", Martha C. Cooper, Daniel E. Innis, Peter R. Dickson, Council of Logistics Management, Oak Brook, IL, 1992, p. 2.

[2] Ibid, p. 9.

[3] Ibid, p. 3.

[4] Ibid, pp. 4-5.

[5] Keebler et al, Ibid, Oak Brook, Il, 1999.

[6] Bowersox and Closs, Ibid, p. 51.

[7] "Strategic Management: Concepts and Cases", Arthur A. Thompson, Jr, A.J. Strickland III, 9th edition, Irwin McGraw-Hill, Boston, 1996.

[8] "Logistics in the new FDX", Robert Mottley, American Shipper, April 1998, p. 28. "Tire-kicking" is when you inspect something but in a very superficial manner, such as when you look a car and all you do is kick the tires.

[9] Richard L. Dawe, "The Impact of Information Technology on Materials Logistics in the 1990's", Penton, Cleveland, OH, 1993, pp. 24-25.

[10] Distribution, October 1987, p. 14.

[11] "Moving Mountains", Lt General William Pagonis, with Jeffrey Cruikshank, Harvard Business School Press, Boston, 1992.

[12] "International Marketing and Export Management", 3rd ed, Gerald Albaum, Jesper Strandskov, and Edwin Duerr, Addison Wesley, Reading, MA, 1998, p. 386.

[13] A.T. Kearney, "Productivity in Physical Distribution", p. 22.

[14] Martin Christopher, "Integrating Logistics Strategy in the Corporate Financial Plan", in James F. Robeson and William C. Copacino, eds, "The Logistics Handbook", Free Press, New York, 1994, p. 243.

[15] "Integrating Logistics Strategy in the Corporate Financial Plan", Martin Christopher, in James F. Robeson and William C. Copacino, eds, "The Logistics Handbook", Free Press, New York, 1994, p. 257.

[16] "Logistics Cost, Productivity, and Performance Analysis", Douglas Lambert, in James F. Robeson and William C. Copacino, eds, "The Logistics Handbook", Free Press, New York, 1994, pp. 263-265.

[17] "The World Economy", Frederick P. Stutz and Anthony R. de Souza, 3rd edition, Prentice Hall, Upper Saddle River, NJ, 1994, pp. 344-353.

[18] Stutz and de Souza, Ibid, pp. 353-4.

[19] Joe Orlicky is regarding as the originator of modern MRP.

[20] "Effective Logistics Management", Martin Christopher, Gower Publishing, Aldershot, 1986.

[21] "Manufacturing Planning and Control Systems", 3rd edition, Thomas E. Vollman, William L. Berry, and D. Clay Whybark, Irwin, Burr Ridge, Il, 1992, p. 14.

[22] The term was coined by Oliver Wight.

[23] "Fundamental of Logistics Management", Douglas M. Lambert et al, McGraw-Hill, New York, 1998, pp. 207.

[24] "Transport Integration in the European Union", William P. Anderson, in T.R. Lakshmanan et al, "Integration of Transport and Trade Facilitiation", The World Bank, Washington DC, 2001, p. 41.

[25] Jay W. Forrester, "Industrial Dynamics", The MIT Press, Cambridge, MA, 1961.

[26] "Forecasting: Issues and Challenges for Marketing Management", Spyros Makridakis and Steven C. Wheelwright, Journal of Marketing, vol. 55, October 1977, pp. 24-37.

[27] "eBusiness", Bauer et al, p. 83.

[28] "Reuse and Recycling-Reverse Logistics Opportunities", Council of Logistics Management, Oak Brook, IL, 1993, p. 3.

[29] "International Marketing and Export Management", 3rd ed, Gerald Albaum, Jesper Strandskov, and Edwin Duerr, Addison Wesley, Reading, MA, 1998, pp. 344-5.

SECTION II
GLOBAL FREIGHT TRANSPORTATION MANAGEMENT

CHAPTER 5
TRANPORTATION PLANNING

Transportation is the primary and most important part of logistics. In international logistics, the transportation function is even more important because the distances are greater and the difficulties associated with transport are greater. In this section we look at a variety of planning considerations, some of which are strategic; others are at the tactical and operational level. The previous chapter on strategic planning provides some background, but the transportation function is important enough to merit closer attention.

TRAFFIC MANAGEMENT

Cisco, the maker of high tech products and at one time the largest company in the world, offers a good example of the challenges of transportation planning. Not only do they need fast and reliable delivery of their products, but the deliveries need to be carefully coordinated. When they would ship 'routers' (a device that controls the flow of information on computer networks), their customer's order might be fulfilled from factories in the US, from Mexico or Asia.

Sometimes the order would be split, in which case they arrive in different packages. Customers hate getting split shipments. They get an order and think it has been botched because it is not all that they ordered. What if the customers have the technicians on hand for a project but only part of the shipment arrives? It was estimated that it would cost carriers $100 million complying with Cisco's needs, but then again, it would also eliminate the need for entire warehouses.[1]

Transportation provides two things, **physical movement** and **storage**. The physical movement aspect is obvious. Storage is provided since the cargo is being held during transit, which could be days, weeks or even months. It is also common practice to slow down transport to keep the cargo in storage longer. This is commonly done with bulk ocean ships, where early delivery would simply mean that the cargo would need to sit in storage on land.

Transportation has some profound effects on our society, not just in terms of logistics. Transportation replaced absolute location (**site**) with relative location (**situation**). This means that economic development has less to do with one's relationship to nature and more to one's relationship across space.[2]

Transportation has allowed for some important changes to our civilization, things that are not at first obvious. It allows for **geographic specialization**. If it were not for transport, a community would need to produce everything locally. Instead, a community can produce what it does best, ship out the excess and ship in those things that are better produced elsewhere. You may notice that this is the same logic as free trade among nations. Specialization can continue as long as the production-cost savings is greater than transport costs. **Large-scale production** is also made possible because the product does not need to be produced close to the market. This also improves overall efficiency. Finally, **increased land values** are a direct result of the fact that land can be used for its most efficient purpose.[1]

Transportation Allows for:
- Geographic Specialization   • Large Scale Production
- Increased Land Values

Transportation can be looked at as the means to an end, the way logistics is done, but it is also a major industry in itself. Transportation, including passenger traffic, has traditionally accounted for between eight and nine percent of GNP, but this has been reduced in recent years due to more efficient use of transport.[4]

There is a job position in many companies that specifically handles transportation, known as the **transportation manager**. A description of this position gives one a good idea of what transportation planning is all about.

CLM's Job Description of a Transportation Manager[5]
Directs the effectiveness of private, third party and contract carriage systems. Manages staff and operations to assure timely and cost efficient transportation of all incoming and outgoing shipments. Plans and assures adequate equipment for storage, loading, and delivery of goods. Responsible for scheduling, routing, budget administration, freight bill presentation, and contract negotiations. Works with international carriers and freight forwarders to streamline the flow of goods across international borders and through customs.
Key Duties
- Ensures that operations are conducted safely and within the law.
- Manages fleet and drivers
- Solicits, evaluates and analyzes contractual bids.
- Negotiates and administers dedicated contract agreements.
- Budgets and controls expenses.
- Determines economical traffic patterns and specific routes.

Transportation planning requires an understanding of three key principles, speed, consistency and control. **Speed** is the ability to go from origin to destination as fast as possible. **Consistency** is the ability for shipments to arrive at the same time, every time. One important consequence of consistency is inventory requirements. The more consistent the transport, the less inventory needed. **Control** is the ability to make changes before and during transport. Telecommunications has revolutionized transportation with the ability to communicate with the driver and possibly change the routing.

Principles of Transportation
- Speed   • Consistency   • Control

There are **tradeoffs in cost versus service** that must be considered. The total cost of transportation is more than just the freight bill (the direct transportation costs). One must consider how much is gained from transporting something. Sometimes it is cost effective to ship something around the world. Other times something must be acquired locally or not at all.

In a famous study, Emery Air Freight commissioned Stanford Research Institute to study how carriers may identify potential users of air freight. In this report, they established the concept of **"lowest total cost"** and **"total company benefits"**. This means that the choice of mode depends not simply on the transportation costs. That would mean that the cheapest mode of transport would always win, and air cargo would never be used. Instead, a shipper must look at the total costs and benefits of a given shipment. Air cargo is much more expensive than any other mode, yet for some commodities, the overall benefits are greater. This is because certain factors, such as high value cargo, insurance, the time value of money, and other considerations make a fast but expensive trip more cost effective.

After considering the economic impact of an individual shipment, one should also look at the overall transportation network to find the most efficient method of moving multiple shipments. We already discussed facility location network design in the previous chapter, which is an important concept in this chapter. Facility location/network design is a long-term decision that is not easily changed. That fixes many of the transportation decisions. In other words, they will limit the available options for transportation.

How does this affect an individual company's ability to compete in the marketplace? A company's costs are not particularly significant by itself. What is more significant is their costs *relative to their competition*. A company will have an advantage within that area where their delivery costs are less than the competition. It can certainly sell profitably elsewhere, but beyond their **market area**, the competition will have a cost advantage and their profit margins will be higher. A higher profit margin gives that competitor more flexibility in reducing price, promotions and so on. The market area of two competing firms will extend to that point where their **landed costs** (the total cost of a product at the final destination) are the same.[6]

This includes several assumptions, the most important of which is that there is no qualitative difference in the products. Note that we are only talking about transport costs here. There are many other variables that determine a company's ability to compete. For products with high transport costs, like heavy bulky products, transport costs can be the most important factor. For others, such as small, light consumer products one would find in a department store, the products characteristics and so on would be more important.

Economies of scale and **economies of distance** are two key transportation principles. Economies of scale refer to the fact that per unit costs go down as the size of the shipment increases. Small shipments are more expensive per unit than large shipments. Economies of distance means that as the distance of a shipment increases, the cost per unit distance go down. In other words, the per kilometer cost of transporting something goes down as the trip distance increases.

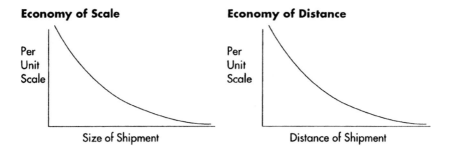

Economy of Scale

Per Unit Scale

Size of Shipment

Economy of Distance

Per Unit Scale

Distance of Shipment

Lardner's Law[7]

Lardner's Law, also known as the **Law of Squares in Transportation and Trade**, states that if the transport costs are cut in half, the market area where goods can be offered is now four times greater. Imagine a factory that can economically deliver its product anywhere within a 10-kilometer range. That's a circle with a 10-kilometer radius. If the transportation costs are cut in half, that radius increases by a factor of two, and the area of the circle increases by a factor of four. This concept can also be applied to time. If transportation speed is doubled, the area that can be serviced within the same amount of time increases by a factor of four. Why is Lardner's Law important? It means that a difference in transportation costs can have a dramatically different effect on a company's market area.

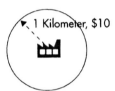

1 Kilometer, $10

A product cost $10 to deliver. It's market is a circular area extending kilometer from the factory.

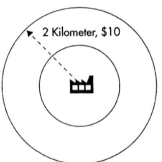

2 Kilometer, $10

The area of its market is πr2, or 2.14 square kilometers. If the transportation cost is cut in half, $10 can deliver the product anywhere in a 2 kilometer radius, πr2, which is 8.56 square kilometer

The primary reason for this is that **transportation involves fixed and variable costs**. Driver's wages, fuel etc are variable and will increase in relation to the distance and volume moved. But the fixed costs stay the same, so the fixed costs as a percentage of the overall cost go down as distance and volume increase.

For example, imagine a truck making a delivery. If there is one box being moved, the entire truck is used to move that one box. If there were an entire truck-load of cargo, the per unit (per kilogram or per piece, however you want to measure it) costs go down as a result of economies of scale. Now imagine the truck travels only five kilometers to its destination. All the costs of loading, unloading, scheduling and so on were incurred for this one short trip. Now if the truck travels across the continent, all those fixed costs are relatively much lower, economies of distance.

Economies of scale and distance have important implications for companies. **Small shippers** face some different issues from **bigger shippers.** Their transportation planning is often different from a company shipping the same product, from the same locations, only in larger quantities. For example, some of the administrative functions, such as customs clearance or freight forwarding, can be done in-house for a large company, but need to be contracted out by a company too small to afford their own person. A new class of rail car is coming into use, known as the 286 (because they can hold 286,000 pound net weight), which has 11% more capacity than older cars. Yet these new cars seem to be used mostly for the largest shippers. Smaller shippers do not have the volume and are thus forced to use the smaller, more expensive old cars.[8]

TRADE AND EQUIPMENT BALANCE

The concept of equipment and trade balance is quite simple but the implications for efficient logistics are enormous. **Equipment balance is when there is the same amount of cargo or equipment going in both directions of a trade lane**. For example, trade between Europe and Asia is balanced when there is the same amount of cargo going from Europe to Asia as there is from Asia to Europe. At the company level, imagine an ocean shipping company going between Asia and Europe. If there is the same amount of cargo going in each direction, there is a balanced trade.

Equipment Balance

1000 Containers

Balanced

Balanced

1000 Containers

Why is this so important? Balanced trade makes for the most efficient use of transportation assets. If the trade is balanced, there is the same amount of cargo going into a port as coming out, so the vehicle (ship, plane, etc) is being fully utilized (assuming the vehicle is being filled in each direction). When there is more cargo going in one direction than the reverse, either cargo is being left behind (excess demand), or the vehicle is moving at less than capacity. In either case, this inefficiency costs money.

Equipment balance is known as a "conservation system", just like accounting or physics. There are inputs and outputs that must balance. A conservation system assumes a closed system, and the units being exchanged are assumed to be durable. They do not get consumed to any significant degree in the process.[9] With a truck, the problem of under utilization is limited to the value of that truck, but ships and planes are major investments and need to be used as efficiently as possible.

Another consequence of imbalanced movement can be seen with intermodal containers. Intermodalism is discussed in the next chapter, so if you are unfamiliar with this, you may want to look ahead. Imbalanced trade in containerized cargo means that there is a surplus of containers on one side, and a deficit of containers on the other side. In order to fulfill demand on the deficit side, empty containers need to be moved from the surplus side, known as **empty repositioning**. This can be very expensive; moving an empty container can cost several hundred dollars because of the port fees loading and unloading that container.

Equipment Imbalance

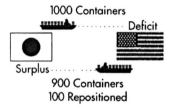

1000 Containers

Deficit

Surplus

900 Containers
100 Repositioned

Imagine containerized cargo trade between the US and Japan as an example. Every week one thousand containers are shipped from Japan to the US, and one thousand containers are shipped from the US to Japan. This is a balanced trade. The ship has a capacity of one thousand containers and is being fully utilized on both trips.

Now imagine there is a currency crisis in the US, and suddenly US products are much cheaper for Japanese to buy, but Japanese products are now too expensive for the Japanese. There are 1000 containers being shipped from the US to Japan weekly, while there is only 900 containers from Japan to the US. This is a trade imbalance. The ship going to the US has 100 empty slots, which means foregone revenue.

Now imagine there is demand for more than 1000 containers to Japan. The ship from the US to Japan has too much demand, which means customers who expect to put their cargo on the ship discover they cannot. This can be a serious problem for some companies, when critically needed cargo is left on the dock. The shipping company is faced with explaining to these customers why their cargo was not taken, and may lose customers. In addition, there are now 100 surplus containers in Japan. In order to fulfill the demand in the US, they must be repositioned at the carrier's expense.

In reality, the situation is much more complex than this. Instead of a two-port example, ports trade with many other ports. It is possible to measure whether trade is balanced between two ports, but this is not that interesting from a logistics perspective. What is more important is whether a given port or region is balanced overall. Shipping companies are constantly looking at whether a given market area is balanced from the overall trade coming into and out of the port. If there is an imbalance, it may help to see if one or a few other ports are the reason. In practice, there is frequently some imbalance. The goal is not to create balanced trade in each trade lane, but only to have it balanced overall.

A More Realistic Equipment Balance Scenario

Japan: 100 surplus
US: 100 deficit
Panama: 400 deficit
Indonesia: 400 surplus

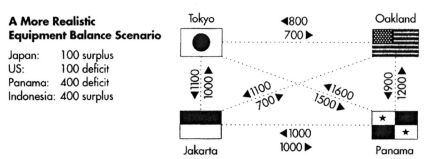

The most immediate effect on imbalance is on **prices**. When a port is surplus, that means there is more cargo coming in than going out. That means the carrier should decrease price for exports and increase the price for imports. If a port is deficit, there is too much cargo leaving and not enough coming in. The price on imports should be decreased, and the price on exports increased. Note that we are referring to the price of shipping services, not the price of the cargo itself.

Using Price to Resolve Imbalance

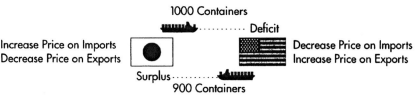

Increase Price on Imports
Decrease Price on Exports

Decrease Price on Imports
Increase Price on Exports

If the carrier could only change prices in one direction, is it better to change prices on imports or exports? There are a couple of considerations but it depends on the situation. Imagine if the ships used in this example were bigger, so there were unused slots going one way, but in the other direction it was full. If you had the choice to change prices in one direction, reduce the price on the trip that is going under-utilized to fill the ship, instead of reducing the price on the trip that is filling the ship. The most important thing when dealing with large, expensive assets like ships or planes is to have them go full.

Another consideration is the value of the customers. Increasing a long-term customer's rates may not be advisable even if it could help in the short term. The problem is that these customers are often both importers and exporters, so there is no way to increase rates without affecting favored customers.

Besides adjusting price, the other major way of handling imbalances is to adjust transport capacity. This obviously is done at the strategic level. A carrier that operates several ships or planes may take one out of that trade lane and assign it to another region. This is only done to address long term imbalances, not to fix a week-long or month-long imbalance. Air cargo rates in Europe dropped to extremely low levels at one point because of over-capacity. Ironically, this occurred because many of the freighters from the Asian market were transferred to the Atlantic as a result of the Asian economic crisis.[10]

The above example dealt with containerized cargo, but equipment balance issues can be seen in any transportation, with any mode of transport, any cargo, and with people. That is why it is such an important issue to understand. Containerized shipping companies deal with many different customers, so they have many ways to balance out trade. Oil tankers are the classic case of extreme imbalance. An oil tanker can only carry one thing, oil. It comes from those parts of the world that have oil, and ships it to metropolitan areas that consume oil, and there is no return trade. Countries that import containerized cargo also export containerized cargo (usually different types of cargo but even then not necessarily). Oil exporters virtually never import oil. That means an oil tanker goes one way full, and returns empty.

This means that in order to be profitable, the oil shipment must not only cover the costs of the trip, but must also calculate the cost of returning empty. Now we can see the big cost in a trade imbalance. Half of the traveling done by a tanker is not earning any money at all. A container ship can expect to be earning money all the time.

Trade balance also has **seasonal variations**. The summer season is generally slower, which means that carriers will move less than full. Price changes have some effect, but it is hard to attract more cargo if there is overall less demand. Reducing capacity is sometimes done but the summer slump is worldwide, so it does little good to shift equipment elsewhere.

Seasonal variations can be seen in passenger airline prices. In the summer airfare from the San Francisco to Paris is around $800. In the winter when Paris is cold and rainy, rates are around $400. By contrast, rates from San Francisco to New York do not vary nearly as much because the traffic is not seasonal. While some of it is tourist traffic, most of it is business, which is not affected by the weather or season.

How do carriers address equipment balance issues in their marketing efforts? Except for the case of oil tankers, it is rare that any individual shippers will affect the situation. That is why price changes apply to all customers. On the other hand, shippers can find business opportunities by looking at those routes that are coming from surplus ports or going to deficit ports. These will have cheaper freight rates. Commodity traders move cargo wherever in the world they can see a profit margin, and they look for these imbalances.

In one case, which will probably be copied by others, a shipping council offered round trip contract to carriers.[11] A shipper's council is a group of shippers who join together to get better rates from carriers. This will be described in detail in Chapter Fourteen. They offered carriers a contract that promised the same amount of import cargo as export cargo, which solved some important problems for the carrier. This happened at a time of economic crisis in Asia when there was a severe imbalance across the Pacific.

Another common method of managing equipment imbalance is **container pools** or **chassis pools** (for those not familiar with this equipment, they are discussed in Chapter Six). Containers and chassis tend to pile up wherever there is a surplus, and they are hard to find wherever there is a deficit. Yet it is likely that not all companies are faced with the same situation. Maybe at a given port one company is surplus and one deficit. This is because companies tend to have very different operations and regional coverage.

The participating companies agree to share their equipment. Sometimes this is a straight equipment share, or they may credit or debit accounts. At the end of a period, such as a month or a quarter, the companies look at the balance and pay for any difference. Most common for equipment pools are carriers that have service partnerships. Containers and chassis are the most common equipment pooled internationally, but it could extend to other pieces. Pallets are often pooled, but this only occurs at a local level because of the inexpensive nature of pallets.

MODE SELECTION AND CARRIER MANAGEMENT

The shippers' task of working with carriers is an important aspect of running a smooth logistics operation. This section describes the different modes of transport, their main characteristics, and how a shipper works with the carrier. These decisions are based on four variables, cargo, shippers, carriers and consignees. The shipper is the one making the transportation arrangements and obviously has a lot to say about how it happens. Yet the shipper's choice is constrained by the cargo, the carriers and the consignee. Cargo characteristics determine the best way to ship it. The shipper can only choose from the carriers that are offering their services, and often the choices are quite limited. Finally, the consignee often influences the method of delivery when the purchase was made.

Four Factors in Transportation Decisions
• Shipper   • Cargo   • Carrier   • Consignee

The choice of the three arrangements (private, contract, common carriage) depends on a variety of factors, but the most important is the level of **commitment** and **risk** the shipper is willing to accept. In a stable environment, where the cargo volume, origin, destination and timing of shipments are stable, it is more likely that private or contract carriage is more cost effective. For example, companies that ship out from the same location to the same destination without any foreseeable change in the future would be able to commit to having their own trucks. There are even companies that own their own ships, such as oil companies. In one case, even a glass manufacturer felt secure enough to buy its own ships. Private carriage is more likely to be used for inter-company transfers, between facilities that are all in the same company. This naturally would have more stability than shipments to customers, who may switch to a different supplier at any time.

The carrier characteristics are one part of the equation. The other part is the **cargo characteristics**. In the previous section, we mentioned some factors that affect transport economics. These are some similar factors used to determine cargo transport characteristics:[12]

- **Size**. The dimensions and volume.
- **Weight**. The absolute weight of the cargo.
- **Density**. This is a way of looking at both the size and the weight together.
- **Stowability**. The above three factors affect the stowability. One large, very heavy piece is more difficult to stow that a few moderate sized pieces. Many small light pieces may be easier to handle but there is a lot more handling required.
- **Handling**. Some types of cargo come with handles or other ways to move them. Others do not. Live animals are a good example of cargo that is very difficult to handle. Containerized cargo is extremely easy to handle. Refrigerated cargo, a very high-growth sector in international trade, requires special handling.
- **Liability**. This is the likelihood and the cost of damage or loss. Some cargo is prone to theft, or easily damaged. Consumer electronics or fresh fruit are good examples. Others, like wastepaper not going to present such problems.
- **Dangerous Goods**. See Chapter Thirteen for a discussion of dangerous goods. Many commodities that we do not think of as 'dangerous' can be so while under transport.
- **Special Service Requirements**. Some cargo have special needs or issues. Cow hides are soaked in a solution that is messy and mildly toxic, and the containers often leak. Live animals require feeding enroute.

By comparing the shippers' business needs, the cargo characteristics and the modal characteristics, the shipper can then pick a carrier or carriers. In the modern business world, there is an increased interest in partnerships and longer-term relationships between businesses. This reduces risk and enables both parties to learn how to work with each other more efficiently. The role of intermediaries, discussed in Chapter Fourteen, is relevant to this issue.

The various types of carriers can be represented along a spectrum, in which the cheapest and bulkiest cargo is normally carried on bulk ocean carriers or rail, and the most expensive and lightest cargo tends to use air cargo. The graph shows the major modes of transport and where they belong along this spectrum, but this is only a general rule, and there is much overlap between some of the modes. The bulk cargo may go on a ship if is crossing the ocean, or on a train if it crossing land.

Choice of Transport Mode

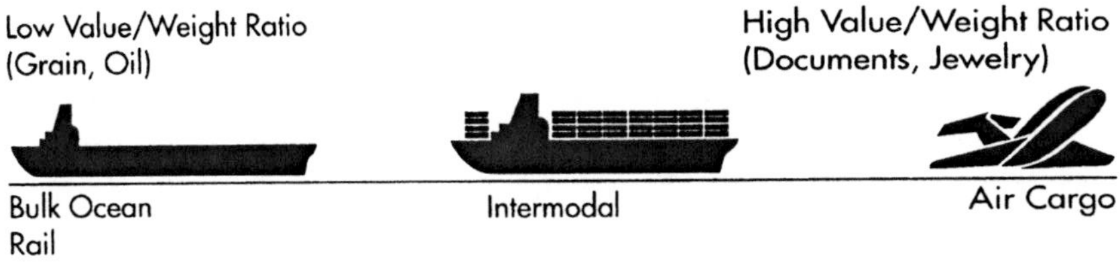

Low Value/Weight Ratio
(Grain, Oil)

High Value/Weight Ratio
(Documents, Jewelry)

Bulk Ocean
Rail

Intermodal

Air Cargo

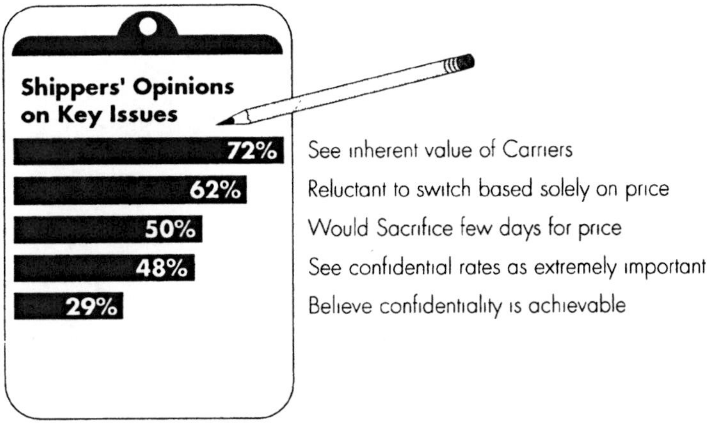

Shippers' Opinions on Key Issues

72%	See inherent value of Carriers
62%	Reluctant to switch based solely on price
50%	Would Sacrifice few days for price
48%	See confidential rates as extremely important
29%	Believe confidentiality is achievable

Each company has its own policy for picking carriers. A company should have a clear and written policy, although many do not. Much of the research in logistics looks at how companies do or should pick carriers, and how to manage the relationship.

Shippers are becoming much more demanding of carriers, for things like special services, information on their shipments and so on. The information aspects of logistics and transportation are discussed in Chapter Sixteen. Some shippers want a lot of information on their cargo. They want to know where it is at any one point of time. Others do not care. In fact, there is some dispute in the industry whether much of the information being offered by carriers is relevant to the shipper. Robert Glaviano, Executive Director of the Hi-Tech Forwarders Network, says shippers only want to know time of pickup and delivery. The rest is unneeded. He thinks shippers should let the forwarders worry about the details. Forwarders, on the other hand, need more details.[13]

In deciding between the different modes of transport available, there is **competition** not just between the different carriers, but **between the different modes of transport**. A shipper has the choice of putting cargo on a ship to cross the ocean, or on plane. Overland, there is the possibility of putting it on a ship that goes around the land, on a truck, a train or a plane. As we will see from the next few chapters, each mode of transport competes with other modes in certain sectors. Some cargo or markets can only go with one mode. Oil from the Middle East to Europe can only go via ship, economically. High value computer chips traveling across an ocean would only go via air cargo. But there are some markets that have a choice.

Transportation Ownership

Private. The shipper owns its own vehicles

Contract. The shipper does not own them but they control and operate them for a specified amount of time. This is essentially like a lease.

Common carriage. The shipper is one of many using a carrier. Common carriage also has important legal issues that are discussed below under Transportation Regulation.

An interesting case study in competition between the different modes can be seen in the US rail and trucking industry. Any increase in interstate trucking would almost certainly come at the expense of rail traffic. That is why the rail industry has been actively influencing government for rules that will protect them from the trucking industry. At the railroad industry's urging, the US Government required that trucks be no longer than 65 feet long, and no heavier than 80,000 pounds. The railroad industry notes that trucks do not pay for the full costs of the highways that they use, while the trains pay far higher fuel taxes yet they are using their privately owned tracks. In Europe, there are many heavy weight trucks that would not be allowed in the US.[14] This will be discussed further in Chapter Six.

One of the ways that carriers correct for imbalances in demand is through **rationalization**. This is where companies, usually those that would normally be competing with each other, share their assets. This is done to either **improve customer service or reduce costs**. The best example of improved customer service is in the ocean shipping industry, in which container shipping companies share space on each other's ship. For example, imagine there are seven shipping lines (lines or another term for shipping companies) that each leave the Port of Singapore once a week to go to Hong Kong. If they all shared space on each other's ships, they could offer their customers daily service instead of weekly service. Costs can be reduced when two carriers are both leaving port only half full. If they rationalized their services, one of them would carry the cargo of both companies, and thus they are using their assets more efficiently and saving costs.

Criteria For Selecting Transportation Provider[15]

In a recent survey, shippers were asked how they feel about carriers. The results show that while price is the most important factor, it is not the only factor:

Pricing	31%
On-time performance	22%
Customer Service	13%
Document quality and accuracy	13%
Shipment tracking	11%
Global coverage	5%

The shippers were also asked if they agree with the following statements:

See inherent value of carriers:	72%
Reluctant to switch based solely on price:	62%
Would sacrifice a few days for price:	50%
See confidential rates as extremely important:	48%
Believe confidentiality as achievable:	29%

TRANSPORTATION REGULATION

Government regulation is becoming less of an influence in some ways, and more of an influence in other ways. Regulation of business relations is being reduced. Controlled economies such as the communist countries are all but dead, and free markets are more the standard. On the other hand, safety, environmental protection, labor conditions and other social issues are becoming more regulated. This section will review the major types of regulations that affect transportation, the reasons behind the regulations and their effects. Almost every chapter after this discusses the specialized topics of regulations, which is why this section is quite general in nature.

Government regulation comes mostly from national governments. There is often reference to 'international law', which is highly misleading. International law has been written for many areas such as aviation safety, shipping safety or documentation standards. There are areas where it is important to have worldwide agreement on the standards. The United Nations has many agencies that have been active in developing these standards. The specific agencies will be discussed in the relevant chapters. But these regulations depend on national governments to be translated into their laws and enforced. A few countries have a monist legal system, which means that any treaty signed automatically becomes national law. The great majority of countries have a dualist system, which means that international law does not apply until a national law is passed adopting the words of that treaty.

There is also competition among countries based on their business laws. Often, a shipper has a choice of which country to ship through or do business with. If one country is highly restrictive or expensive, the shipper will look for another country. This is most important with hub ports, which will be discussed in Chapter Nine. Singapore and Hong Kong are both used as hubs for cargo moving through Asia. Whenever one government thinks of increasing port fees or changing a regulation, they need to think about what the other country is doing, and whether the overall effect is what they want. There is no easy answer here. In another example, ports in Canada are sometimes used for cargo destined to the US. When US ports such as Seattle or New York increase their fees or there is a labor strike, Canadian ports can be used as an alternative.

Government bodies below the national level also have regulations that affect transportation or regulation. Since this book emphasizes international logistics and transportation, local laws defer to national law in virtually every country. Governments generally prevent local governments from making decisions that would conflict with the national law, and this is particularly relevant in terms of international transport or trade. However, many local rules affect local transport, and as we noted earlier, all transportation is local wherever it moves.

Gerard Verhaar notes that there is **no overall legal system** for transportation and logistics. Maritime law is covered by the Hague Rules, air transport by a completely different set of laws, and the same for rail or truck traffic. All the other business practices associated with logistics, such as contracts, are covered by the local law. Yet logistics is meant to provide an integrated process across industry, transport mode and borders. In a traditional business transaction, there is only one (or a specific number) transaction, yet in logistics, problems have a chain reaction. That meant liability could be much wider than in other situations.[16]

Legal systems can vary not just in what they say, but in **what they do not say**. A government's rules may be silent on some issue. For example, among the NAFTA countries, some member nations specify gross weight of the truck, weight per axle, the way the weight is distributed to the front and back axles, and the distance between the axles. Truck length regulations may specify the overall length, the length of the tractor and trailer independently or the length of the trailer beyond the rear-most axle. In other worlds, sometimes we need a rule to clarify a contradiction but the laws are not there.

Why do we have regulations in the first place? While we often complain about regulations, they would not exist if there were not the demand for them. First is the demand from the public for certain social outcomes, such as a healthy economy, clean air, and so on. More important for our purposes is the demands of the transportation community itself. Shippers and carriers have interests and they use regulations to achieve those goals. There is a tendency for shippers to want as strong a competitive environment as possible to keep transportation costs low. Carriers want less competition, for obvious reasons. There is thus an inherent tension between these two communities.

In the US, the National Industrial Transportation League (NIT League) is an organization that looks after the interest of shippers. For example, they encourage legislation that increases the level of competition for carriers. Recently there has been a trend in the logistics world where cooperation and partnership between shippers and carriers that has made this traditional antagonism less practical. There is also less distinction between what is a shipper and what is a carrier. UPS, for example, is a major carrier yet they are also the largest customer of railroads in the US. Thus they are both a large carrier and a large shipper. For this reason the NIT League has been discussing admitting carriers as members.[17]

"It is always better to be able to collaborate and come up with a solution that benefits the entire industry"
-Edward Emmett, National Industrial Transportation League

Competition regulation is one of the most contentious areas in the regulation of international logistics and transportation. Lack of government regulation leads to one of two extremes, monopoly or destructive competition.[18] In almost every country there is ongoing debate over the ideal level of competition. Not only are countries unable to clearly define an ideal regulatory system for competition, governments and companies are constantly making claims that someone else is doing something that violates the spirit or the law of free markets and open competition. As we will see in the chapters on maritime shipping and air transport, there is no such thing as a completely free market, and governments are intimately involved with economic affairs.

Gourdin describes four **types of market structures**, pure competition, monopoly, oligopoly, and monopolistic competition.[19] Some industries, by their very nature, tend toward one of these four models. A government's ability to regulate an industry is based on the natural tendencies of that industry. This is why we see very different rules for air cargo, ocean shipping, trucking and so on. The needs of the companies, of society and government are different in each mode of transport.

One way that governments ensure a competitive market in the transportation industry is to see that there are multiple companies competing in the market. However, it is hard to say how many carriers are needed to qualify as a competitive market. There is a difference between lots of local carriers operating in their own market, and lots of carriers competing side by side. Even when we see multiple carriers operating side by side, there are other ways that a given carrier has a captive market.

A common way that governments undermine competition is to grant **cargo preference**. The US and many other countries have rules that require any cargo shipped by the government to go on carriers of that nation. In other words, US government cargo can only go on US carriers. The effect of this is quite obvious; the US pays much more to ship its cargo than any other shipper. Sometimes they pay two or three times as much as other shippers. There is a reason for this, which will be discussed in Chapter Seven.

Governments are usually concerned that their companies are able to compete internationally and export as much as possible. This is the motive behind much of their regulation in logistics and transportation. They want the logistics and transportation industry to be competitive and offer the lowest possible rates. Furthermore, they do not want to see their companies being discriminated against by other countries or the carriers of other countries.

Non-discrimination is a legal concept designed to protect competition and unfair business practices. In terms of international transportation, it **means two things**. A carrier **cannot charge different prices** for **similarly situated shippers**, and they cannot refuse service to any shipper. A similarly situated shipper means a shipper that is moving the same commodity, and the shipment is roughly similar. For example, two chemical companies are shipping the same product from Wilmington, Delaware to Tokyo, Japan. They are similarly situated shippers. This term is a little vague because if there were another chemical company in, for example, Louisiana, would they be similarly situated? The important thing is that a carrier needs good reason to charge very different prices to competing companies. The second part of non-discrimination means that a carrier **cannot refuse service**. In other words, a carrier may not accept one company's cargo and refuse to carry a competitor's cargo.

If the law did not include non-discrimination rules, this would be a highly effective method of hurting the competition. Imagine a major company shipping their product. They approach a carrier and offer them a large amount of business, in return for a promise that that carrier would not accept the competitor's cargo. That competitor would now be restricted in who they ship with. In some cases, this could increase their costs or make it almost impossible to ship their cargo. Governments would prefer that all their companies have open access to foreign markets, which is why virtually every modern government has non-discrimination rules.

Common Carriage

Non-discrimination does not apply to every carrier. There is an important legal distinction known as **common carriage**, which means that the transportation service is provided to the entire shipping community on a non-discriminatory basis. Generally speaking, a common carrier is a carrier that operates on a regular route, publishes its schedule and its rates. If it does this, it is legally considered a common carrier, whether it wants to be so or not. Common carriage means that as a common carrier, it must provide its services on a non-discriminatory basis, as just described. There are some other requirements and benefits to being a common carrier, but this will be discussed in later chapters.

Common carriage also applies to other industries, such as telecommunications. The phone companies in many countries are open to competition, but there is only one wire going into each house. In order to allow other companies to offer their services, the primary company is considered a common carrier, and must offer the use of their system (the wires going into each house) on a non-discriminatory basis.

One thing to note here is that some modes of transport, or types of carriers, are always common carriers, others may or may not be, and others are never so. For example, a containerized shipping operation is always a common carrier. There is nothing that says it must be, but by the very nature of its service, it needs to 1. Have many different customers, 2. Provide a regular scheduled route and 3. Publicize its rates. On the other hand, an oil tanker is chartered, works for only one customer at a time, and has unique rates established for each shipment. It is hard to see how such an operation could ever fit the legal description of a common carrier.

There is a gray area in the definition of who is a common carrier. Ships that only carry cars, known as car-carriers, may or may not be common carriers. There have been court cases on this. Intermediaries, discussed in Chapter Thirteen, may or may not be. Courts, governments and industry leaders have been actively discussing whether freight forwarders may act as "shippers" by consolidating cargo for lower rates from carriers.[20]

Non-discrimination by common carriers does have two limits, at least under US law. They only need to provide their service **up to the physical ability of carrier.** In other words if they cannot physically offer their services to a competing company, they can refuse service without it being considered discrimination. Second, a similarly situated shipper must be part of the same market. A company that ships its product on bulk ocean ships cannot say it was discriminated against by a competitor that ships with container ships.

Common carriage has some other interesting legal effects. For example, a few states in the US required all beer imported into the state to be transported on common carriers, believing that this would aid them in enforcing the tax collection. The issue went to GATT, which ruled this requirement illegal because other states found a less intrusive method of collecting taxes. This ruling was a major precedent in giving GATT more powers.[21]

Cabotage

While governments generally encourage open competition, one area where this is not the case is domestic transportation. Shipping cargo between two points in the same country is known as **cabotage**. It is extremely rare that any countries allow foreign carriers to do this. Often a foreign carrier will call at two ports in a foreign country; they may pick up or drop off cargo, but they may not pick up cargo at one port and drop it off at the other port. For example, a Chinese ship may have a route in which they call at Los Angeles, then up to Oakland, and then back to China. The ship may pick up or drop off the Chinese cargo at Los Angeles or Oakland, but the cargo picked up in Los Angeles may not be dropped off at Oakland.

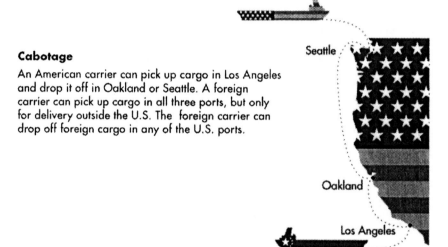

Cabotage

An American carrier can pick up cargo in Los Angeles and drop it off in Oakland or Seattle. A foreign carrier can pick up cargo in all three ports, but only for delivery outside the U.S. The foreign carrier can drop off foreign cargo in any of the U.S. ports.

There is a difference between the use of labor (the truck driver or the crew on the train/plane) and the equipment. In the US, Canadian trucks may carry domestic freight, but only with US drivers. Trains may cross the border but the crew needs to change at the border.

Restrictions on cabotage reduce competition and increase prices, but almost every country feels that it is more important to have a healthy domestic transportation industry. Part of this is to promote local jobs, but a major part is for national defense and military reasons. If there were ever military conflict, it would be dangerous to have foreign carriers working throughout one's country. Even if the foreign carrier were not from the aggressor country, that carrier would probably want to get out of the way in case of conflict. That means that a country that is about to be in a conflict would find its transportation companies leaving the country.

Cabotage restrictions are being reduced in some isolated cases. The US makes a minor exception for Canadian truck drivers. If the truck is already making a delivery, it may take on cargo along the way from within the US. Europe may be the biggest example of eliminating cabotage. This has been one of the more controversial parts of European unification, but the benefits would be large. It has been estimated that reduced cabotage restrictions will save US companies 10-15% on intra-European shipping costs.[22] Note that in the European example, there are other agreements on military matters that address the national defense concern with cabotage.

Deregulation

At the beginning of the section, it was noted that most governments usually encourage competition and free markets. The key word here is 'usually'. Much of the anti-competitive regulation is still in place, though there has have been major progress in recent decades to reduce regulations. Of all the areas of an economy that experience anti-competitive protection from governments, transportation and logistics tends to be among the last to get deregulated. International transportation has tended to be the first to get deregulated, but national transportation tends to be the last to get deregulated. Even the WTO had to leave transportation out of the initial agreement, and is still struggling to find agreement on open competition.

In the US, every major transport sector has been deregulated in the past few decades. The rail industry used to have fixed prices and many other protections. This was eliminated in the 1970's. The same applied to interstate trucking. The deregulation of the airline industry began with domestic airlines, but has since spread around the world. This will be discussed in Chapter Eight. Ocean shipping was never heavily regulated except for their pricing system, which has already been mentioned and will be elaborated on in Chapter Seven.

What exactly is meant by deregulation? In almost every case, price restrictions intended to guarantee a higher-than-market price were eliminated. The services provided were limited in order to prevent carriers from offering new routes or new value-added services. Under deregulation, the US railroads are now allowed to start service to new markets or stop providing services where they were previously required to operate. Airlines are now able to offer new routes. US common carriers historically were only allowed to offer a new route if they could prove it was 'necessary'. Practically speaking, this was very rare. Under the deregulated environment, the carriers only need to show their ability to provide the service. Specifics on different regions and different modes are discussed in later chapters.

Deregulation entails turmoil and uncertainty in the logistics industry. In a newly competitive environment companies that are not efficient can and often do go bankrupt. Prices and services fluctuate. This often has operational results. For example, these uncertainties are one reason for an increased demand for service contracts to stabilize relationship between shipper and carrier.

The Undercharge Controversy
The US deregulated the trucking industry in the early 1990's, resulting in a flurry of competition. However, even after deregulation, motor carriers (as trucking companies are known) were still required to file their rates with the government. Given the wild market, many companies offered under-tariff rates, which was illegal. The customers often did not know about the rules or were told they were being offered a published rate. Many of the carriers went bankrupt, at which time the company is placed in the hands of trustees. These trustees, who usually had no relation to the customer and not much interest in the trucking industry, went after the customers to charge the filed tariff rates. These rates were often ridiculously high because they were never used but never deleted (known as paper rates). Customers, who had paid what they thought was the fair and legal charges, were faced with massive bills, to the point where nation-wide $27 billion was involved.[23]

The government intervened with the Negotiated Rates Act of 1993, which gave some relief. There were settlements of 5-20% of what should have been owed, and complete relief in some cases, such as with charitable organizations. But motor carriers also gained with regulations and stiff fines for off-bill discounting and strict enforcement of filed tariffs.

TRANSPORTATION ECONOMICS

In the transportation industry, we often get distracted by the large, impressive ships, planes and other vehicles. We sometimes mistakenly think that they are their own reason for existing. Actually, the only reason they exist is to fulfill a business need. They owe their existence to economic needs, which is the topic of this section. We will look at the demand for logistics and transportation in economic terms, and how logistics companies fulfill these demands in terms of pricing.

Pricing is a powerful tool used to achieve a variety of business and regulatory goals. It entails both the result, a price, but also the process, how prices are developed. The terms 'rate' and 'price' are, for our purposes, the same. The transportation industry tends to use the term 'rate' most often, but the process of developing rates is called rate-making or pricing. The regulatory role of pricing has been mentioned and will be discussed further in other chapters. This section will be an introduction to the types of rates used.

The prices charged for transportation are known as **freight rates** or **freight charges**. The **freight bill** is a similar term, but it usually refers to the overall bill for transportation services, not for a specific shipment. How are prices developed for transportation services? Like most industries, carriers have a pricing department, normally within the marketing department. International transportation is unique in that rates are based on the commodity, even when it makes no difference in the costs. For example, container lines charge a different price for shipping a cargo container depending on what is inside that container. Shipping bottled beer in a container costs more than shipping beer in kegs in a container.

This practice goes back to previous centuries when shippers would show up at the dock and the ship's Captain would look at what is to be moved, and change his price depending on the shipper's ability to pay. Someone shipping hay, which is very cheap, would pay very little. If someone showed up with gold bars, they could obviously pay a lot more. Now the idea is that the shipping charges are based on a percentage of the cargo's value. This is a polite way of making shippers pay more just because they can afford to pay it. This practice also has the effect of making the high-paying customers subsidize the low-paying customers.

This practice is the basis of **yield management**, a common practice in many industries including transportation, refers to a policy in which the maximum amount of revenue is gained from each customer. A common example is airline tickets. Two people sitting next to each other on the same flight may have paid very different air fares. The carrier determines the fare based on a long list of variables. Those purchasing a ticket at the last minute, flying somewhere for a short period of time, or flying during the week pay much more. These are the symptoms of business travelers, who are not paying for their ticket and thus are not as price sensitive. The same sort of thing happens in the cargo industry.

Is yield management fair? Is it fair that one customer is paying more than others? One way to think of it is that this way, each person pays what they are willing and able to pay. Those who pay higher prices are essentially subsidizing those who are paying less.

The pricing process has become increasingly complicated because of the trend toward alliances and partnerships. The freight bill, instead of being based simply on the nature of the service, is often influenced by the overall partnership. These partnerships can include trade in services or products that offset freight rates. Service contracts are being used more often, because they are a way of increasing the level of cooperation and commitment between shippers and carriers. Alliances and partnerships is a special area that will be discussed later. What we will do here is look at the pricing process from a basic level.

The following are a few main variables. These include the cost of providing the service, the value of the service to the customer, the ability of the merchandise to support the transport expense, economic conditions in general, and supply and demand trends. These are described in more detail:[24]

Cost of providing service: The cost structures are very different and are discussed in the relevant chapter. There are some definite boundaries that restrict the range that prices can vary. One of them is regulations. Many governments have price supports, or competition rules may influence how much a price can be increased or decreased. This usually does not mean a specific rule that states a price; more likely the rule states that beyond a certain limit, the carrier may be asked to justify that price change. For example, if a company was being charged much more than a competitor, the government would probably not say, "X is an unfair price". Instead it would ask that the rate in question be justified or, if it was unjustifiable, that it be changed.

Value of service to customer: One of the most basic laws of economics is that one would not engage in a transaction if they would be worse off. In other words, one would only buy something if they were going to get something in return that is at least as valuable to them.

Ability of the merchandise to support transport expense: Some commodities are not shipped, or only shipped within a certain distance, because they cannot sustain the cost of transport and still be sold at a profitable level. The price for shipping must be less than the spread of commodity prices at the two different ports. There is much more potential for trade that does not occur due to high transport costs.

Elasticity for Transportation

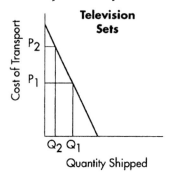

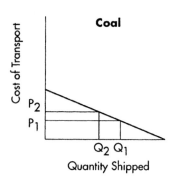

Economic conditions in general: This mostly affects the volume of trade, which is the demand side of the equation. Note that demand can and usually does shift much more rapidly than supply. In a recession, trade levels can drop quickly but ships and planes do not disappear, which leads to excess supply.

Supply and Demand: Supply: This includes the number of carriers in a given market, and the number of vehicles. When we talk about supply in the air cargo market, for example, we look at the number of planes carrying cargo on a given route, and their total capacity. Supply can be increased by having those same vehicles do more trips, but usually they are already working at capacity. Capacity can also be increased by bringing in new vehicles, or transferring them from other trade lanes.

Demand: refers to the total amount of cargo that is to be shipped. It can vary when the economy does well and more overall trade is occurring. It can also be affected by changes in competing transportation services. For example, the introduction of larger cargo planes can attract cargo that otherwise would have gone by ship. The supply and demand relationship is constantly changing, and influences each other in a variety of ways. One factor can cause changes in other factors, or vice versa. For example, changes in price may increase or decrease demand, or changes in demand can put upward or downward pressure on rates.

Rate Types

Now let us look at rates from a more operational perspective, the types of rates. The two basic types of rates are class rates and commodity rates. We already noted that some rates are based on the type of commodity. Yet this is not always the case. **Class rates** are used when it is impossible for a special rate to be established for every possible commodity, from every destination, to every possible destination. In order to reduce the number of possible rates, classes are designed. For example, carrier may group together all cargo coming from a given province, going to a given province, with different rates for every 1,000 kilogram increment, and so on. This results in a rate sheet that has a manageable number of possible rates.

Commodity rates, as already discussed, are based on the commodity. Note above that this may be because a given commodity entails special costs, or simply because of market considerations (for example, shippers of high value cargo are willing and able to pay more). **Freight All Kinds** (FAK) is the term that means 'none of the above'. Since the attacks of September 11, one of the proposed changes for security has been to eliminate FAK as a cargo description. Instead, shippers would need to identify each of the many commodities in the container.

Special Charges

Special charges may be assessed for many reasons, including the following special services:

Transit services. Allows for products to stop enroute. For example, this may be done when a half-full container stops enroute to pick up more cargo.

Diversion. When a shipment is redirected to a different destination while enroute.

Reconsignment. When a shipment is redirected to a different consignee.

Split delivery. When a shipment is split and part goes to one place, and another part goes somewhere else.

Demurrage/detention. In containerized cargo, the customer has a certain amount of time to load the container, and on the other end a certain amount of time to unload it. If they need to hold onto the container for a longer period of time, this is known as detention (when the shipper holds the container) or demurrage (when the carrier is holding the container).

Accessorial charges. These can include things like currency charges, fuel charge and so on. This is normally done when there is a special cost that may fluctuate widely and the carrier wants to make it obvious why that charge is being changed.

PACKAGING

Packaging is one of those areas that most people do not think too much about. Yet it plays an important role in business for marketing, safety, transportation and other reasons. There are **three levels of packaging**: retail, group packaging and transport.[25] While all three levels serve for both promotion and protection, the promotion role falls mostly on the retail end. Packaging at the group and transport level also serves a promotion role, but for security reasons the containers are usually left blank. A **package** refers to packing and the contents, while **packaging** refers to "a receptacle and components necessary for containment function".[26]

Three Levels of Packaging
• Retail   • Group   • Transport

Some of the **markings** on packaging are required by law, for things such as customs, security, etc. Language requirements on packaging vary, but it would be prudent to include the native languages of the origin and destination, and in all cases, English. The marketing role of packaging means that it should let everyone know what was in the box, in as appealing a manner as possible. This raises an obvious security problem. If everyone knows that a given box holds a high value product, it is now a target for theft. For this reason, packaging should be anonymous.

There is a concern with supply chains when it comes to packaging. The goal of each individual channel member is to pay as little as possible for packaging. Packaging only needs to get the product through the member's responsibility and on to the next without damage. After that, damage is someone else's problem. This means that each channel member may avoid investment in that extra packaging expense if they are not going to see a benefit. This is one of the areas that channel leadership, discussed in Chapter Three, needs to address.

Carriers are generally not responsible for damage resulting from inadequate packaging. In the U.S, the Carriage of Goods by Sea Act of 1936 states that "neither the carrier nor the ship shall be responsible for loss or damage arising or resulting from insufficiency of packing." However, carriers often get stuck with cargo that has inadequate packaging, and the carrier may be liable if they cannot prove the packing was inadequate. For this reason, carriers usually set standards and will not accept cargo that is not properly packaged. One can also see this at the post office. Packages must meet some standards or the postal service will not accept them.

One of the new trends in logistics is **environmentally friendly packaging**, and cheaper packaging. Both of these have led to the increased use of returnable containers or packaging. For example, car assembly plants receive auto parts in reusable racks. These racks must be returned to the auto part manufacturer. This is reverse logistics discussed in Chapter Four. Biodegradable packing has had mixed consequences. Anything of natural origin such as straw or wood has the potential of carrying infestations, such as bugs or rodents. As an example of this issue, see the inset on Forbidden Pallets.

The product does not need to be retail packaged for transport. Some products, notably those little things in the store that get lost or stolen, are often in big, bulky showy packages. The packaging can be included, folded or otherwise condensed, and assembled at or close to the final destination.

Pallets are metal, wood or plastic platforms used for handling small batches of cargo. Pallets solve many handling problems, but they are not cheap. There is a problem for those who use pallets to ship their product. Attempts are made to get pallets back, but solutions are not always easy. One solution has been **pallet pools**, in which a group of companies share pallets among themselves. Because of the expense of pallets and the difficulty of getting them returned, they are not used as often on international shipments.

Packing refers to the process whereas packaging usually refers to the material itself. Of large industrial goods, packing is a special skill and there are companies that just offer this service. Vincent Guinto describes how major construction projects often fall behind schedule when they deliver their goods to the export packing company without prior planning. The packers need accurate cargo descriptions, weights and measures. This is needed for a variety of legal reasons. If the shipper did not provide this information, the packing process soon becomes a major source of delay.[27]

Dangerous goods have special requirements, including their packaging and marking. This is discussed in Chapter Thirteen, Security.

Forbidden Pallets

Untreated wood pallets from China were banned in the US because of wood-eating longhorn beetles, which pose a threat to live trees. China responded that this is just another attempt for the western countries to attack Chinese commerce. One alternative to pallets are slip sheets, which are fabric or plastic sheets that serve the same role as a pallet but do not work nearly as well.[28] This is not just an American concern. Brazil also has strict rules on the import of pallets, and will not accept a certificate of fumigation from the US, but requires that the pallet be fumigated there.

[1] *"They've Got Mail!"*, Brian O'Reilly, Fortune, February 7, 2000, p. 110.

[2] *"The World Economy"*, Frederick P. Stuz and Anthony R. de Souza, Prentice Hall, Upper Saddle River, NJ, 1994, p. 164.

[3] *"Transportation"* 5th, Coyle, john J, Edward J. Bardi, and Robert A. Novack, West Publishing, St. Paul, MN, 2000, p. 7.

[4] Coyle, Ibid, p. 11.

[5] *"Careers in Logistics"*, Council of Logistics Management, Oak Brook, IL.

[6] Coyle, p. 35.

[7] Coyle et al, p. 6.

[8] JOC, April 8, 1999, p. 1a.

[9] *"Micromotives and Macrobehavior"*, Thomas C. Schelling, Norton, New York, 1978, p. 68.

[10] JOC, March 2, 1999, p. 1.

[11] JOC, March 3, 1999, p.1a.

[12] Bowersox and Closs, Ibid, pp. 365-367.

[13] *"Shippers say they just want to know pickup, delivery"*, Chris Barnett, The Journal of Commerce, September 1, 1998, p. 6A.

[14] JOC, March 2, 1999, p. 14a.

[15] JOC October 8, 1999, p. 1.

[16] *"Logistics contracts need standard rules"*, Gerard Verhaar, American Shipper, March 1998, p. 28.

[17] *"Crossroads for the NIT League"*, Chris Dupin, JoC Weekley, volume 2 (47), 2001, p. 9.

[18] *"Global Logistics Management: A Competitive Advantage for the New Millennium"*, Kent N. Gourdin, Blackwell, Malden, MA, 2001, p. 95.

[19] Gourdin, Ibid, p. 105.

[20] *"Tackling forwarder identity crisis"*, Terry Brennan, The Journal of Commerce, September 1, 1998, p. 1A.

[21] *"GATT and the Environment"*, Steve Charnovitz, International Environmental Affairs, Summer 1992, p. 214.

[22] *"Will the Barriers Come Tumbling Down"*, Karen E. Thuermer, Global Trade, August 1992, pp. 10-15.

[23] *"Motor Carrier Deregulation and the Filed Rate Doctrine: Catalyst for Conflict"*, Jeffrey M. Sharp and Robert A. Novack, Transportation Journal, vol 32, no. 2, 1992, pp. 46-54.

[24] *"The Business of Shipping"*, Lane Kendall, Corness Maritime Press, Cambridge, 1980.

[25] *"International Marketing and Export Management"*, 3rd ed, Gerald Albaum, Jesper Strandskov, and Edwin Duerr, Addison Wesley, Reading, MA, 1998, p. 343.

[26] U.S. Department of Transportation, Hazardous Materials Transportation Training Modules, 2000 edition, Chapter 3.

[27] *"Avoid Export Packing Pitfalls"*, Vincent G. Guinto, JoC Week, June 3-9, 2002, p. 32.

[28] *"Beetle threat could force change in import packaging"*, Tom Baldwin, The Journal of Commerce, September 17, 1998, p. 1A.

CHAPTER 6
INTERMODALISM AND
LAND TRANSPORT

Intermodalism is one of the most complex forms of transportation. In fact, it is not a mode of transportation, but a technique of combining different modes for the most efficient movement. While the modes of transport have been around for a long time, intermodalism is one of the most important developments in the global business environment of recent decades. This chapter discusses how the intermodal network operates and how it improves overall performance. The second half of the chapter includes different modes of land transportation. While these topics are not inherently intermodal, almost every intermodal move includes a land segment.

WHAT IS INTERMODALISM?

Intermodalism is not a mode of transportation, but a **system of coordinating different modes of transport for a shipment**. It is one of the most revolutionary developments in logistics in this century. Although intermodalism is a way of putting together shipping, trucking, rail and other modes of transportation, it is more important as a system in which the sum is greater than the parts.

The literal definition is "the use of more than one mode of transportation", which we can see from the roots of the word: "inter" and "modal". A more realistic definition would be "a shipment that uses different and coordinated modes of transportation". The point is not that multiple modes are being used, but that they are coordinated so one leg of the journey is coordinated with the next. The definition is not complete, because when we think of intermodalism, the first thing that comes to mind are the cargo containers. **Intermodalism is most commonly understood as the use of intermodal containers in a coordinated shipment using multiple modes of transportation.**

Intermodalism
A shipment that uses different and coordinated modes of transportation.

In 1956, McLean Trucking Company built a prototype ship called the **Ideal X**. The ship was an ex-World War II cargo ship redesigned to hold 58 truck trailers for an experimental trip from Newark, New Jersey to Houston, Texas. The advantages of intermodalism were undeniable. Most importantly, one of the crew members of the Ideal X was a man named Al Long, who would go on to become the father of the author of an international logistics textbook. In 1958 Matson, a shipping company, started intermodal service between the US West Coast and Hawaii. The next year the first intermodal crane put into service at Alameda, CA (part of the Port of Oakland). From then on, there was an explosion of intermodal growth around the world.

It is possible that multiple modes are used, but there is no coordination between the different modes. The shipper could make arrangements with a trucking company, and separate arrangements with the shipping company. The trucking company and the shipping company would not be coordinating their activities. This is known as **multimodalism**. The important thing about intermodalism is that there is coordination between the modes to insure a smooth and efficient flow.

Containerization is the difference between just multimodalism and true intermodalism. We have all seen the large steel boxes on trucks, trains and planes. Those are intermodal containers, also sometimes referred to as ISO intermodal containers. The ISO is the same as in the ISO9000 standards, because the standards of the container are set by that organization. In the logistics worlds they are generically called **cargo containers**. Instead of moving cargo in their own boxes, all different sizes (known as **break-bulk**), it is put into the containers which all adhere to certain standards which will be described below under Intermodal Equipment. A similar term is **unitization**, which means that the cargo is all of the same size and dimensions. Cargo may be unitized but not necessarily containerized. For example, hay is made into bundles, but they are not containerized.

The important thing about containerization is that the equipment adheres to the same standards worldwide. There are several million cargo containers circulating around the world, and they all adhere to the same standards. If this were not so, they could not be handled by the carriers, or they would require special handling which defeats the entire purpose.

Intermodalism could use all modes of transport, but some are much more important than others. By the very nature of the technology, intermodalism is used primarily for medium value cargo, as we saw in the spectrum of transport options in Chapter Five. Intermodal cargo tends to be more valuable than bulk cargo but less valuable than air cargo. The most

common modes of transport used in an intermodal movement are **trucking, rail and ocean shipping**.

Gantry cranes loading a ship at the Port of Sudan.

Air cargo does have intermodalism, but the containers are air cargo intermodal containers, which are very different. Air cargo containers are much lighter and prone to damage. When we speak of air intermodal, it usually just means packing an air intermodal container at the exporter's location, trucking it to the airport, and putting the container on a truck at the other end. This is much simpler than the intermodalism that is the subject of this chapter. There are occasion use of sea and air, in which cargo is carried by ocean and then flown.[1] This is relatively rare since ocean and air, as we will see in the coming chapters, are very different in speed. Combining the two makes sense in very rare situations, and even then, there is also truck legs at the beginning, between the sea and air legs, and at the end. In other words, it is an extremely complex and costly arrangement.

Intermodalism was developed shortly after WWII to address one concern, security. However, it was soon discovered that there were some other benefits, particularly the safety of the cargo, the safety of workers, efficiency, and speed:

1. **Security.** The cargo is more concerned for few reasons. First, the cargo is in a container, which makes it harder to get into. Second, containers are only identified by a serial number on the outside. This gives enormous protection because the only way to know what is in a container is to open it

up or know the serial number. In a container yard with thousands of containers, this means that a thief would have a very poor chance of getting high value cargo just by arbitrarily opening containers. Recall the movie "Raiders of the Lost Ark", in the final scene where the Ark is rolling off in a crate, in a huge warehouse filled with identical crates. That is what any potential thief of containerized cargo is faced with.

Finally, containers have seals put on the latches. While this will not prevent a break in, it is very obvious that there has been a break-in. For security reasons, workers normally check the seal so that they do not accept cargo that has already been opened. If the seal was broken, the carrier would note this on the cargo documents so they do not become liable for any losses.

2. **Safety**. Closely related to security, containerized cargo is usually safer than in any other type of container. The metal boxes are extremely strong, and almost impossible to crush. They are more weather resistant than most other containers, though they do leak if not maintained properly.

They are also much safer for the workers who handle containerized cargo because machines do the handling and the humans stand at a safe distance. In the old style break-bulk cargo, workers were very close to the cargo. If cargo drops, or it swings, anyone nearby is risking injury or death. At modern container terminals it is rare to see humans outside the protective shield of a truck cab or crane box.

3. **Efficiency**. It is sometimes assumed that the dramatic increase in efficiency was the reason intermodalism was developed. Yet is interesting to note that this was only a byproduct. How much more efficient is intermodalism than break-bulk? By one estimate, one dockworker can handle .5 ton per hour for break-bulk, and 2.45 tons per hour for container operations.[2]

That means that when intermodalism is introduced to a port, about nine out of ten dockworkers (known as **stevedores** or **longshoremen**) were out of a job. When intermodalism was first introduced in New York, the effects on labor relations were quite bad, even to the point where labor riots could have happened. One longshoreman, "upon noting the shape of the container, referred to it as the ILA's (the labor union) coffin".[3]

Efficiency is gained in a few ways. First, because each container is of standard dimensions, specialized handling equipment is designed just for them. Second, because each container is the same, no time is wasted making adjustments. Third, given that one can now design a single, most efficient container, the boxes were made of such a size that they are much bigger than the boxes before, and that means that fewer movements are required. In other words, it is much faster to take one intermodal container off a ship than 10 small containers.

4. **Speed**. This is closely related to efficiency, but the consequences are different. If the cargo can be handled quicker, more cargo can move through one port in a given amount of time. There are some important implications to speedier loading that were not obvious at first. Because a ship can get through port faster, the ships can be larger. Imagine if a break-bulk ship was as big as the current containerships. It would take a month for that break-bulk ship to offload and load the cargo. No customers would wait that long. Therefore the speedier port times allowed the ships to be built larger, a major trend in the shipping industry, to be discussed in Chapter Seven.

Speed and Efficiency

These two concepts are *not* the same, though they are often confused. They
· are related concepts but there is an important difference. Speed is the ability
to perform a task in a short period of time. Efficiency is the ability to perform
a task with minimal resources. It may be possible to increase speed but only
by increasing the amount of resources consumed, such as when a car is
driven fast and the gas consumption *per mile* increases.

There were many consequences of this new system. Economies of scale required that a containerized port handle much more than a break-bulk port. Container ports (this can also be called an intermodal port) grew far larger than before because they could handle far more cargo. Fewer ports were needed, which meant that some grew to be huge, while many other ports shrank in terms of throughput (the amount of cargo that goes through the port) or even stopped operations. There are now in the world fewer ports, but the ones that are still operating handle much more cargo. The trade lanes that handle intermodal cargo handle much more cargo, but that also means that the number of trade lanes has decreased. Rather than shipping cargo in any one of many possible routes, intermodalism means that there are a few main routes that handle much more cargo.

The fact that the ports are larger and fewer are the result of industry-wide changes. Some individual companies choose to pick one port in a given region, and make that its **load center**. That port is used as a hub for the entire region, whereas before it would use multiple ports along the coast. This will be discussed further in Chapter Nine.

An Example of Intermodal Service

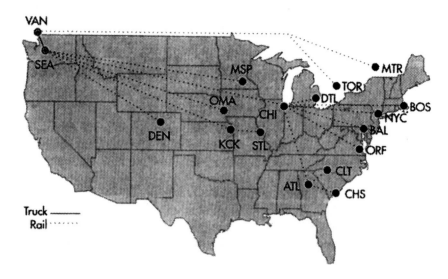

INTERMODAL EQUIPMENT

A reason intermodalism has resulted in fewer ports and fewer trade lanes is that the equipment is very specialized and operate most efficiently with large volumes of cargo. Intermodal equipment is surprisingly simple in terms of engineering, given the impact that it has had on logistics. The equipment may have some high-tech features, but the idea is quite basic. The most important aspect of intermodal equipment is that it is all standardized.

The basic element of intermodalism is the cargo container. They are 8 feet wide, 8 feet 6 inches high, and their length varies. The vast majority of the several million containers in the world are either 20 feet or 40 feet long. They also come in 45 feet, 48 feet and a few others, but these are very specialized and do not conform to the universal standards. The dimensions are not metric since the technology was developed in the US. Changing to a metric standard is not a practical option and there is no particular incentive at this point. On each corner of the container are hitches to allow them to be stacked on each other, or stacked onto a ship, truck, train or anything else with the same dimension hitches. Maximum weight limit is around 20,000 kilograms for the 20', 25,000 kilograms for the 40'.

Why the different sizes of containers? First, shippers have different needs. Those shipping small volumes prefer small containers. The second reason is the nature of the commodities being shipped. The different size containers are used for different commodities. Low value cargo tends to move in 40s where it is more important to keep costs low. Higher value cargo naturally tends to move in smaller lots, so 20s are more common. Heavy cargo moves in 20s because the weight limit would be reached before the space is filled. There has been a trend toward more use of 40s. The percentage of leased 20s has gone down from 44% to 35% over the 1990's.[4] The reason for this is not clear, but it may be that the overall volume of trade is increasing, so shippers are better able to use the more cost-efficient 40s.

Types of Containers

There are hundreds of types of containers, varying from the most common, known as a "dry container", to those that are custom designed for a particular shipper for a particular cargo. The following are the most common types. However, within each type, there are different materials and special treatments. For example, dry containers may be treated to carry certain types of food. If a container is used to carry some chemicals, it may be contaminated so it can only be used for a few types of cargos.

Dry. This is the basic container.

Open top. No top, so cargo can be laid down into the container, useful for cargo that is awkward to handle.

Flat rack. No top or sides, just the bottom and two ends. This is also for awkward cargo.

Platform. There is only the bottom. Having an awkward piece of cargo on the platform is still much easier to handle because it can be safely picked up, and placed on a surface and bolted down as if it were a container.

Refrigerated. Known as a reefer, a refrigeration unit is built into the container, running on a diesel generator or plugged into an external power source. Reefer cargo is one of the fastest growing and profitable areas of transportation.

Live animals A variety of containers have been designed for live animals. Horse carriers have windows so the horse can stick out its head while on the ship. Multiple decks are built into the container for smaller animals.

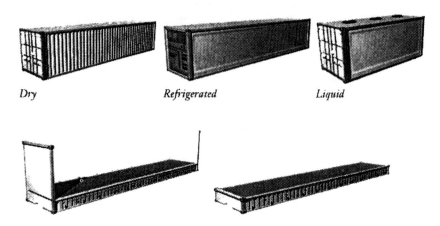

Dry   *Refrigerated*   *Liquid*

Containers have been adapted not only to carry special cargo, but for uses that have little to do with transportation. They are commonly seen as storage sheds. They are filled with building materials and brought to remote areas where they are then stacked and used as part of a pre-fabricated building. The materials stored inside are then used to make them into living quarters. Even the United Nations tribunal prosecuting war criminals are using intermodal containers because the jail has filled up its regular cells.

Intermodal **cranes** are used at ports to pick up and handle the containers. Ships traditionally had their own cranes to load and unload their own cargo. Containerships are among the very few types of ships where it has become more efficient to rely on port cranes, although there are a few containerships with their own cranes. The intermodal cranes known as **gantry cranes** are usually at an ocean port, though some rail and truck yards have them (these are the rail and truck equivalent to a port, to be discussed in Chapter Nine). These are fixed to the ground or run along tracks for short distances. Rail and truck yards are more likely to have fork-lift type machines that can move around. These types of cranes include **top-pick loaders, side-pick lifts** and others.

TEUs and FEUs

The volume of containerized trade is often measured in terms of twenty-foot equivalent units, or TEUs. One TEU is the amount of cargo that would fit in a twenty-foot intermodal container. Trade statistics, shipping companies and ports often report the amount of cargo in TEUs. This is a unit of measurement. The cargo may actually be moving in 20s, or it could be moving in 40s or other sized containers. One 40' container is equal to two 20s. A 20' container has a volume limit and a weight limit, so this measurement is not an exact indicator of the amount of containers.

Containers that are moving partially empty are still counted as a container. Another measure is forty-foot equivalent unit, or FEU, but this is not often used. Also note these measures are only used for containerized cargo. Non-containerized cargo may be compared to the equivalent number of TEUs, but trade statistics using these terms only refer to cargo that is being shipped in containers.

Conversion Table

TONS: A short ton = 2,000 pounds
A long ton = 2,240 pounds
A metric ton = 2,205 pounds

To Convert	Into	Multiply
Long Tons	Short Tons	1.12
Long Tons	Metric Tons	1.016
Metric Tons	Long Tons	0.9844
Metric Tons	Short Tons	1.1025
Short Tons	Metric Tons	0.9078
Short Tons	Long Tons	0.89287

MANAGING INTERMODAL TRANSPORT

The most important aspect of intermodalism is not the equipment, but the operating and management issues. Traditionally, a carrier operated as an independent company. The shipping company worried about shipping, the trucking company worried about trucking, and so on. Intermodalism requires the coordination of multiple carriers. Sometimes these are the same company, but usually they are independent or at least different subsidiaries in the same company.

Why is this significant? **Intermodalism requires cooperation and coordination** among a diverse group of companies on a level that is rarely seen in other industries. We know how difficult it can be to run any organization. Running an organization in close cooperation with many others, to the point where every one of hundreds of shipments needs to be coordinated, is quite a task. Also, the different companies are often very different in organization, structure, and culture. Railroad companies are very different from trucking companies, which are very different from shipping companies.

What are intermodal companies? Sometimes they are asset-based carriers, such as a shipping company that expands into providing intermodal service. This is common for containerized shipping companies, which are mostly carrying cargo that would only move intermodally. Railroad and trucking companies sometimes provide intermodal service, but it is not so critical for them. Then there are companies that do not control any assets, but only provide intermodal services. They coordinate shipments between the other carriers. Sometimes this is done by buying up space on the other carriers at a wholesale rate.

One distinctive thing about intermodal carriers is their **sophisticated information system**. They need to process a lot of information, under time pressure and in coordination with other carriers. This means that information systems tend to be exceptionally important. This will be discussed further in Chapter Sixteen.

Intermodal **traffic patterns**, from the local to the global, can be clearly seen. This is because there is a lot of required infrastructure to handle intermodal cargo. Large cargo volumes are most efficient, so there are usually a few major trade lanes, and not as many minor trade lanes. For example, across the US there are only two major lanes carrying intermodal cargo across the country. One is from Seattle across the north, and the other from Los Angeles across the south. In Europe the landscape is more crowded and complex, but intermodal trade lanes still exist.

Equipment balance was discussed at length in Chapter Five, but one can see how that issue is exceptionally important with intermodal cargo. In addition to the vehicle moving less than full from a surplus port, there is also the problem of handling empty containers. In intermodal companies, the marketing and traffic departments try to work together to address this issue. One way to deal with equipment balance issues is to lease containers instead of using one's own. It is much easier to lease a container to meet excess demand than to retrieve a container from a foreign country.

The following sections discuss aspects of overland transportation. Trucking, rail and to a lesser extent inland shipping are important to intermodal shipments. However, many international shipments go overland that are not intermodal, or they may be containerized but only use one mode of transport. The reason overland transport was included in this chapter is that they play a key role in the intermodal chain, even if they are not crossing borders themselves. The overland transport is used to link the line haul, such as the ocean ship, to the origin or destination.

The Intermodal Move

How exactly does an intermodal shipment work? The shipper decides what size container is needed, and orders it from the carrier. A container is delivered to the customer's location, and left there for a designated amount of time to be loaded (usually around 24 hours). The container is most likely brought on a truck, but very large shippers may have their own rail head where containers are picked straight off a train.

The truck (or train) comes back at the appointed time and takes the container away. It then goes on its journey. At the final destination, it is usually delivered to a building (again, usually by truck but some facilities have their own train platforms), where it is left for a designated amount of time to be unloaded. The container is later picked up and taken to its next customer.

ROADWAY TRANSPORT

Trucking is an indispensable part of almost every shipment, both domestically and internationally. It is an important link in the intermodal chain, but it is also used in some regions for international trade. Its role in intermodalism is to link the ocean port or rail yard with the local origin or destination. In other words, the cargo is loaded or unloaded at someone's facility, not at the port. Trucking is also used for some international shipments, and of course for non-intermodal shipping.

The role of trucking varies depending on the region. In large, sparsely populated areas where railroads are well developed, trucks would be used more for local delivery and defer to the rail for long distance trips. In areas where the railroad is not so well developed or the market area is heavily populated, trucks become more useful. They are the main mode for international shipments in Europe because the rail system for cargo is not well developed and the market is very dense. Russia is very large, has a decent rail system, and poor roads. Trucks in that country are more for local traffic. In Asia, the land is rugged and divided by water. Trucks are for local delivery while international shipments are probably going to use water transport.

Comparison of Vehicle Size

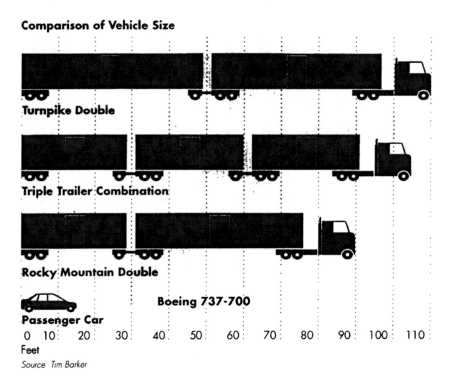

Source Tim Barker

In Chapter Five we discussed the business arrangements with transportation, as either **private, contract** or **common carriage**. For trucking, there is a fourth one, **exempt**. Trucks are normally licensed and must adhere to the rules of the road, but for those that stay on or near private property, they may be exempt. The best example of this is the trucks that operate in and around ocean ports.

The trucking industry has some distinctive features. It has low entry requirements. Anyone who can afford a truck (even if it is just leased) can offer trucking services. Of all the different modes, this has the lowest entry costs. There are relatively many players in the field, which creates a highly competitive market.

Some of the trucking fleets are owned by shippers, some are independent, and some are owned by other carriers. For example, an air cargo company may have their own trucks to provide local delivery from the airport. One of the newer developments is shipper-owned fleets offering contracts to other companies. In other words, imagine a large manufacturer that owns their own trucks to deliver their products. They decide to offer trucking services to other shippers, maybe during their slow season, or maybe to earn extra money. The other trucking companies do not like this new competition because the shipper-owned fleets do not necessarily need to make money, and thus may offer below-cost rates. Why would they do this? The shipper may only be trying to earn extra income for trucks that otherwise they would not be using, so it does not hurt them to offer below-cost trucking services.

Every transportation mode is influenced by **regulations**. Trucking is more influenced by local rules, such as speed limits, registration and so on. Other modes such as shipping do not use local roads and are not so concerned about local rules. This means that the longer the trip, the more legal jurisdictions it passes through. This creates a strong disincentive for long distance trucking.

The Eurasian landmass looks as if there are great opportunities for international trucking, yet the reality is far less optimistic. Due to political conflicts, attempts to tie together Asia and Europe overland have had little success. China and Russia have poor relations and both have poor infrastructure. Further south, the Indian subcontinent and the Middle East face serious political barriers. A UN agency, The Economic and Social Commission for Asia and the Pacific (ESCAP), has been trying to put together a 56,000 Asian Highway through 25 countries. They claim this is 85% complete, but Myanmar is the major barrier. The 10 members of ASEAN are trying to create a region of free borders by 2004, to open South East Asia to regional trucking. [5]

Safety issues are more important with road transport than any other mode. In 1997, there were 711 fatalities by trucks in the U.S, which is low considering this is a very safe environment. In other countries, truck-related accidents are much higher. By comparison, there were 746 railroad deaths and 22,000 for passenger cars in the U.S.[6] Worldwide, each year more than 700,000 people are killed and 10 million injured in accidents, costing the global economy about $500 billion. About 70% happen in developing countries, where accidents are 20 to 30 times more frequent than industrialized countries, and can cost up to 2 percent of GDP.[7]

Competition between the trucking and rail industry has affected regulations. Trucks compete mostly with railroads, but rarely with other modes. The competition between trucks and railroads can be fierce. In the US, trucks are limited in the amount of cargo (by weight) they can carry specifically because the railroad industry has lobbied the government to limit their size. In Europe, trucks may carry more cargo, which is more efficient. Another limit is the number of trailers that can be pulled at one time. Tractor-trailer productivity can be improved by 21% with 6 axles, and 48% with 7 axles. This is common practice in Europe, Canada and other countries, but the only state in the US that permits this is Michigan.[8]

Road capacity is measured by a passenger car equivalent (PCE). On level roads a truck may be only 1.2 PCE, which is to say they only use up 20% more of the roads capacity than if they were a car. But on curving, mountainous roads the same truck can be four PCEs. This is based on number of lanes, lane width, curves, grades and other engineering factors.

Truck Sizes

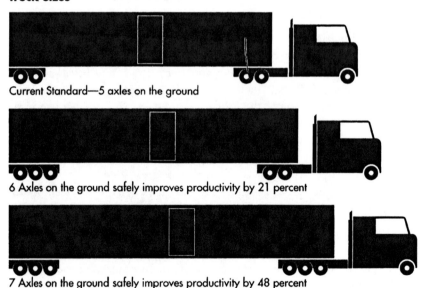

Current Standard—5 axles on the ground

6 Axles on the ground safely improves productivity by 21 percent

7 Axles on the ground safely improves productivity by 48 percent

Source: *Truck Weight Limits, Issues and Options, Special Report 225 Transportation Research Board, National Research Council, 1990*

Weight laws are partly for safety reasons, and partly because roads can only stand a certain amount of weight. Road damage is based on weight per axel, known as equivalent standard axle load, or ESAL, which is about 18,000 pounds. The damage drops off exponentially, so a damage of a car is about 1/10,000 of a truck. Pavement thickness also varies exponentially; pavement that is 11 inches thick is twice as durable as one 9 inches thick, yet it costs only a fraction more to build.[9] A US Federal Highway Administration study showed 33.5% of container loads violated truck weight laws. The biggest problem is 20' containers. Most violators were exporters, so it was not foreign importers ignorant of the rules, as some had alleged.[10]

Trucks need to be registered somewhere. This is simple when the truck is only used domestically. International trucking gives the carrier the option of where to register their trucks. This is essentially like a 'citizenship' (this is a key issue in ocean shipping). When the registration requirements of one country become too costly or bothersome, a company may reregister elsewhere. In one example, British trucks were reregistering, mostly in Benelux countries, because taxes in their own country were too high.[11]

Key Terms

Motor carrier. Another name for a trucking company.

Drayage. The trucking service. A dray is a truck shipment.

Cab. The part of the truck where the driver sits.

Chassis. The part of the truck that is pulled by the cab. The chassis is only the bed, and not the cargo that goes on top of it.

Trailer. If the chassis is for intermodal use, it would only be a bed and the container is placed on top of it. The chassis and container combined is a trailer. If it is a 'regular' type truck, the trailer includes the chassis and the part that holds the cargo.

Truck Terms

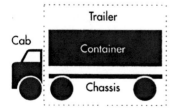

Railroads are best for large loads going long distances. To be more specific, rail is preferred for loads over 30,000 pounds over distances exceeding 300 miles.[12] If trucks have small entry barriers, the rail industry may have the highest. Train tracks would need to be laid down, which in this ever-crowded world creates immense costs for even the shortest routes. For long routes across open terrain, most places in the world already have tracks. In sum, there is not likely to be any significant change in the current system of train tracks, though China is the one big exception. They are working on some major new lines.

Railroads have been in steady decline over the past several decades, mostly due to competition with trucks. Whereas the trucking industry has been politically strong in many areas, particularly the US, the rail industry has not been as successful. Roads have not internalized all of their social costs such as pollution and congestion. In other words, tax money pays for roads that benefit trucks, but the railroad industry has not benefited as much from public support.

What is changing is the system of **operations** and **ownership**. Historically, trains in the US were privately owned and operated, while almost everywhere else in the world they were government owned and operated. Government run railroads are being privatized, especially in Latin America and Europe. Brazil was selling off last of the government's rail line leading into the center of the country, which is the most valuable of any rail in Latin America.[13] Mexico has also been trying to privatize with mixed results. Germany's rail system, like many others in Europe, is being privatized.[14]

There are high fixed costs because of the tracks and the trains, but the good news is that there are very low variable costs. There are also almost no weight or volume restrictions, which means that extremely heavy cargo that could not go on a road can go on a train. There are restrictions on dimensions. Trains go through tunnels and bridges, which means that their load cannot exceed their dimensions. There tends to be high damage (about 3% of total tonnage) due to vibrations and shock from steel wheels on steel track.[15]

Intermodalism is a large and growing part of rail cargo. Although intermodal cargo is relatively small in comparison to the other cargo carried by rail, as a part of the intermodal network railroads are an important link for long distance overland travel. Containers may be placed directly on the back of the flatcar, known **as container-on-flatcar (COFC)**. They may also be left on the chassis, and the whole trailer placed on the flatcar, known as **trailer-on-flatcar (TOFC)**. TOFC can be faster because the trailer is pulled off the train and is instantly ready to leave, whereas the COFC requires that the container be picked up and placed on a chassis, which takes more time. Both of these arrangements are known as piggybacking, and this process is the second largest market in US after coal.[16]

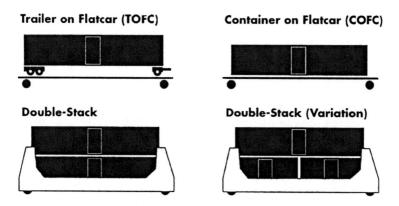

Trailer on Flatcar (TOFC)

Container on Flatcar (COFC)

Double-Stack

Double-Stack (Variation)

Railroad service is not nearly as flexible for obvious reasons. They can only go where there is track, and they only stop at **railheads** (the railroad equivalent of a port or a station). Trains are much less dependable in keeping to a schedule, which is an important part of intermodalism. While truckers are credited with 95% on-time performance, railroads achieve 70% in a good year. This general lack of performance has been hampering growth in intermodalism despite the efforts of companies to promote intermodalism. Intermodal shipments are 10-20% cheaper than all-highway moves.[17]

Trains are quite **slow**. Canadian Pacific Railway's average speed was the industry's fastest for North America, at 25.2mph. Trains are slow because tracks may be of poor quality, but mostly because of stops. A truck can approach an intersection and execute just about any turn in a matter of seconds. They can drop off or hook up trailers in minutes. Trains need to play a complex and slow game of switching tracks, waiting for other trains to get out of the way before they can proceed, and so on.

Shippers that use rail are either moving very large volumes, or they are intermodal shippers working through an intermodal carrier. In other words, it is highly unlikely small shippers would contact railroads directly. Competition is low in this industry, even in those rare cases where more than one railroad is offering services on the same track. This has been an ongoing problem in improving rail service. One trade group proposed that the following information be provided to compare the performance of different train companies to shippers: switching efficiency, freight bill accuracy, and car inventory.[18] UPS is the largest consumer of rail services in the US, but this does not mean that they would use it for time-sensitive shipments. Sometimes a company will have shipments large enough to charter an entire train, known as a **unit train**.

Railroads do not have the standardization that is seen in other modes, probably because it resides in one geographical area and there has been little incentive historically to have the different railroads connect. The **rail gauge** is the width of the track. There are different gauges in use, and this is the main problem for international connections. Many borders require that passengers and cargo get off one train, and get on another for the next country, all because of differences in gauge. Australia was and an extreme example, in which each province had a different standard.

The **engine** of a train is that car which has the engine and pulls the train. There are diesel engines, and there are trains with overhead wires. The overhead wire system is more energy efficient but requires more infrastructure investment. Some shippers have tracks leading directly to their facility so a train can be loaded or unloaded right there. This is a major investment and the shipper is captive to the railroad company operating that rail.

Governments have traditionally kept close control over the industry for military and social reasons. Rail was often the only viable link for some communities or industries to the rest of the country. The US railroad industry is made up of private companies, and until the 1970s they were heavily regulated. Rates in the US were strongly controlled until the Railroad Revitalization and Regulator Reform Act of 1976, introduced zone of rate flexibility (ZORF). Complete freedom in pricing was not given because some companies are captive to the local railroads. For example, the

1980 Staggers Rail Act deregulated railroad industry, liberalized rate making, allowed rail lines to abandon poor performing rail lines, and liberalized the rules regarding mergers. There has since been a major consolidation of the railroads, and they are even allowed to merge with motor carrier.

Types of Train Cars

Car. The train refers to the entire chain of cars, and a car is the smallest unit of the train.

Engine. This is the car that has the engine.

Boxcars. Essentially a box, what people usually think of when they think of a train.

Hoppers. A big bowl for pouring in cargo such as coal.

Flatcar. A flat platform, mostly used for intermodal containers but also for large pieces such as vehicles or machinery.

Refrigerated. Like a boxcar only refrigerated.

Tank. A tank is built onto the car, much like an intermodal tank container.

The Interstate Commerce Commission was created in 1887 to ensure a normal rate of return for railroads, with provisions for investment, pricing, operations, conditions of entry and exit. Railroads could not abandon lines just because they were not profitable. As a result of deregulation, many unprofitable lines were abandoned and active rail line mileage decreased from 179,000 miles in 1980 to 148,000 in 1989.[19] The number of major railroads was reduced from 45 in 1979 to 13 in 1991.[20] It has been estimated that in the trucking and rail industry, rates were excessive by $1 billion annually because of government regulation that limited competition. Preventing railroads from abandoning unprofitable routes and other operational inefficiencies cost an estimated $2.5 billion.[21]

The European railroad system is extremely well developed, but that does not always translate into efficiency. Passenger train service is among the best in the world, but cargo trains are not nearly as good. Cargo trains are too slow to move during the day, so night operations are required. During the day trains need to be able to maintain a top speed of 120 kilometers per hour (75 miles per hour), but cargo trains are usually around 80-100 kph. Commercial speed (point to point) is about half that speed.[22]

The Trans-Asian Railway is an example of an international project to join the rail systems of many countries to provide a single rail link from

Singapore to Istanbul, Turkey. The UN has been active bringing to together the many countries to establish standards and rules to allow shipments along this entire corridor. Challenges to be overcome include politics, economics and engineering. Burma has been politically isolated for years, and yet it holds a blocking position between Southeast Asia and the West. Similar issues are found with lines going through Iran and Iraq. The engineering of the various systems also pose a problem. There are least five gauges in use throughout the region.

INLAND SHIPPING

Inland shipping, which includes rivers, lakes and canals, is discussed here because it is essentially overland service even if it moves over water. The ships and their service are very different from maritime (ocean) shipping. There is a gray area between coastal shipping, which is in some ways like inland shipping but is in an ocean. The fundamental difference is that the service is mostly domestic for coastal shipping, so it is similar to inland shipping.

Where is inland shipping mostly done? By the limits of geography, there are a few areas that have the vast majority of inland shipping:

- US Great Lakes
- Mississippi River
- Russia's great rivers
- China's great rivers
- Europe's river system

One of the biggest limitations is geography. Even in large lakes and rivers there are navigability issues. Rivers may be too shallow, or the depth changes with the shifting mud and sand. Seasonal variation can create problems. If the water level is too high from rain, the vessels may not be able to get under bridges. Northern waterways are blocked with ice, because small bodies of water freeze even if ocean ports do not.

One example of how geography affects inland shipping has been the case of the US Great Lakes. The water level dropped by a foot, resulting in serious consequences for the shipping industry. The major commodities in this area are iron ore, coal and limestone, which move in large volumes and rely on low transportation costs. The one-foot drop in water level meant that shipments were reduced by 5%, or 2,500 to 3,000 tons.[23]

Inland shipping companies have low fixed costs since they do not need to pay for the right-of-way, and government usually pays for dredging and navigational aids. The vessels are much more expensive than trucks, but not very different from trains and less than ocean ships. The biggest competition is railroads and pipelines.[24] The types of ships are similar to some ocean ships, but for inland shipping they are usually much smaller. This will be covered in Chapter Seven.

European Waterways

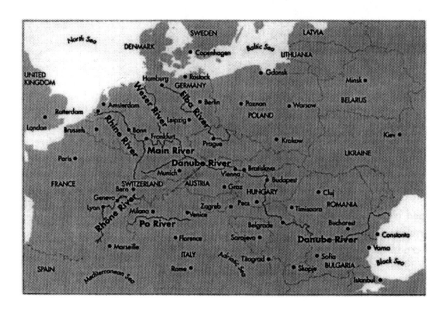

PIPELINES

Of all the modes of transportation, pipelines are the last thing people think of. They move comparable amounts of cargo as rail and motor, but they are largely invisible. They require very high investment, but with very low variable costs. One unique thing about pipelines is that the cargo only moves in one direction, with rare exceptions. Cargo safety is a strong point, with low loss and damage. Pipelines are very dependable, requiring low maintenance and are not affected by weather.

We already mentioned that there is a storage function in transportation, and this is especially true in the pipeline. Hence the term 'it is in the pipeline'. One reason for this is that pipeline cargo is slow. There is very limited accessibility since the cargo can only be delivered and accessed at certain places.

Of the different types of cargo (containerized, break-bulk, bulk), piped commodities are considered bulk. The commodities carried in pipes are quite limited. Petroleum dominates the list. There are at least 15 grades of crude oil plus other petroleum products. Different grades can be put in the pipeline, one following the other. Mixing is not much of a problem because of their different grades that keep them separated. If there is any mixing, it just goes with the lower grade. There is one coal slurry pipeline, in which crushed coal is mixed with water and later separated at the destination. This is in the US at Black Mesa in the Southwest. A coal slurry is quite specialized and requires a lot of water.

Larger pipes are much more cost effective than smaller ones. There is slight friction between the liquid and the inside of the pipe, so the larger the pipe, the lower percent of liquid is in contact with the pipe. Energy is needed to move the cargo through the pipes, though in some limited cases gravity is used. There is also an efficiency gain from larger pipes because more cargo is being moved but there is much less additional cost with a larger pipe. A 12-inch pipeline can transport three times as much 8-inch.[25]

The companies that operate pipelines are sometimes the shippers, or they may be independent. In the US, most operate as common carriers. They are very oligopolistic, with very few players (worldwide that is, not just in the US). They tend to be natural monopolies because parallel pipes would be grossly inefficient. The main competition is water carriers, the only other mode that has the same cost characteristics. Most are domestic, because they require long-term stability and are sensitive to political disruption. In one case, during the cold war, there was a major dispute over a Russian oil pipeline to Europe. The Europeans wanted to buy Russian oil but the US claimed it would lead to dependency.

[1] *"Sea-air Cheap and Fast"*, Global Trade, February 1992, pp. 16-8.

[2] *"Containerization and the Labor Equation"*, Thomas W. Gleason, VIA Port of NY-NJ, April 1986, p. 27.

[3] Gleason, Ibid, p. 1.

[4] JOC, June 8, 1999, p. 1

[5] JOC, October 7, 1999, p. 5.

[6] JOC, August 12, 1998, p.11A.

[7] *"Toll Roads"*, Antonio Estache, Manuel Romero, and John Strong, in Estache and de Rus, ibid, p. 240.

[8] Transportation Research Board, Special Report 225, 1990.

[9] *"Toll Roads"*, Antonio Estache, Manuel Romero, and John Strong, in Estache and de Rus, ibid, p. 283.

[10] American Shipper, June 1989, p. 72.

[11] JOC, March 17, 1999, p. 14a.

[12] *"Transportation"*, 4th edition, John J. Coyle, Edward J. Bardi, Robert A. Novack, West, St Paul, MN, 1994, p. 169.

[13] *"Brazil puts rail link to heartland up for bids"*, Thierry Ogier, The Journal of Commerce, August 17, 1998, p. 1.

[14] The Economist, June 16, 2001, p. 66.

[15] Coyle et al, *"Transportation"*, p. 170.

[16] *"Railroad Facts"*, American Railroad Association, 1991, p. 6.

[17] *"Intermodal sector warned it's flirting with flat-line growth"*, Rip Watson, The Journal of Commerce, September 25, 1998, p. 14A.

[18] *"Railroad data show little change in average train speed"*, JOC, February 12, 1999, p. 12a.

[19] Eno Transportation Foundation, *"Transportation in America"*, 9th ed, 1991, p. 64.

[20] Moody's Transportation Manual, Moody's Investors Service, New York, 1980.

[21] *"Conceptual Developments in the Economics of Transportation: An Interpretive Survey", Clifford Winston, Journal of Economic Literature, volume 23, 1985, p. 83.*

[22] *"Which intermodal technique for Europe?", Henrik Baasch, American Shipper, February 1999, p. 34.*

[23] *JOC, March 10, 1999, p. 1B.*

[24] *Coyle et al, "Transportation", p. 224.*

[25] *Coyle et al, "Transportation", p. 243.*

CHAPTER 7
MARITIME SHIPPING

No industry represents international transportation as well as maritime shipping. By its very nature, crossing oceans, it is the mode of transport that has played a central role in history tying together markets across the globe. That may be why the term ship has the double meaning, a boat and the act of shipping something.

The shipping industry is the backbone of international shipments. Overland travel in historical times was much more difficult and hazardous, which is why ocean-borne cargo was often the only viable option. Even today, ships carry the great bulk in terms of the volume of cargo shipped internationally. This chapter describes the industry, including the types of ships used and the equipment related to ships. In Chapter Four we introduced economic and regulatory issues, but we now consider those aspects unique to this industry.

SHIPS AND SHIPPING EQUIPMENT

The maritime shipping industry refers to that part of water transport that operates on open ocean, as contrasted to inland shipping. The companies are called **ocean carriers**, also sometimes known as **shipping lines**. While there is a lot of dynamism in this industry, compared to the air carriers, ocean carriers tend to be older and more conservative.

By the nature of the industry, ocean carriers need to be big because ships are big, and the carrier needs to have a fleet in order to be competitive. There are very few ocean carriers with only one or a few ships. The industry is dominated by fleets of 10 to 40 ships. For comparison, a typical dry-bulk freighter holds about 65,000 tons, compared to 50-80 tons for a railcar.

The ocean carriers tend to be well-established companies with long histories. The industry is very old, the technology has not changed dramatically in the past century, and barriers to entry are high. This may be why ocean carriers have a conservative corporate culture. A ship is vastly more expensive than a truck or a train, but similar to the cost of a plane. Unlike planes, a ship's design is specialized for a certain role, and cannot be easily changed. Planes are relatively generic in design.

Ships play a military and economic role that has greatly influenced the regulatory and business environment. National defense requires that a country have the transportation assets to move around, especially for major powers that try to exert their influence around the world. That is why they go to great lengths to insure that there is a substantial fleet under their control. This does not mean that the ships are owned or operated by the government, but that they are subject to that country's regulations. This will be a major topic in this chapter.

There are several types of ships based on engineering and function, which also determines type of operations and the business arrangement. Recall that when we look at carriers, try not to think of the vehicle (the ship, plane, train etc), but think of the market demand for transportation, and then look how that demand is met by the transportation services. The major types of ships are as follows:

- **Containerized**. Also known as container liner. Because the large containerships carry as much as 7,000 containers, this industry has a large number of relatively small customers. One ship may carry a hundred containers for a large company as well as a single container of household belongings for a person moving overseas.
- **Bulk**. Bulk covers a wide variety of ships but the thing they have in common is that the ship is like a big bowl into which the cargo is poured. The distinctive feature of bulk cargo is that it can be poured, either as liquid like oil or dry like wheat.
- **Bulk: Tankers**. Most are oil tankers, but there are also tankers carrying chemicals, liquid food products and other commodities. Instead of thinking of the ship like a bowl, a tanker is like one big tank, except that the tank is divided into compartments.
- **Bulk: Dry bulk**. The ship deck has large hatches that can be removed and the ship is like a big bowl.
- **Bulk: Roll-on/Roll-off**. More commonly called a **roro** or **car carriers**, this is technically considered a bulk carrier only the cargo clearly cannot be poured. This is for anything that can be rolled on or off, such as cars and trucks, but also machinery and sometimes containers. Containers are put on a trolley and rolled in.
- **Breakbulk**. This is what one thinks of as the classic freighter, a cargo ship with compartments below that can carry just about any sort of cargo. The cargo is in its own container, not adhering to any particular standard like a container ship, but not loose like a bulk ship.

- **Refrigerated**. This is usually a breakbulk ship, but the cargo holds are refrigerated. This is an expensive new service that is taking advantage of the growing market in fresh foods. Containerships also carry containers that are refrigerated.
- **Barges**. A large floating bowl or platform that holds cargo, but has no engine. It is pulled by a tugboat.
- **Mixed**. There are mixed-use ships such as part containerized and part breakbulk, or part tanker and part break-bulk. There is a wide variety of such ships, but they are becoming less common since the economies of scale greatly favor ships with a specific design.

Some types of ships are becoming more common, while others are fading out. Container ships are growing in numbers, while general cargo ships are being reduced in numbers. For general cargo ships, as of the late 1990's there was around 17 million deadweight tons of capacity, about 1,300 ships, and the numbers were going down by a hundred or so every year.

Types of Ships

Type	Number of Ships	GTM
Oil	6,900	147
Bulk Dry	5,000	140
General Cargo	17,400	56
Container	2,100	48
RoRo	1,700	22
Liquefied Gas	1,000	16
Chemical	2,200	13
Passenger/RoRo	2,400	12
Bulk Dry/Oil	242	11
Refrigerated	1,400	7
Passenger	2,800	7
Other Bulk Dry	1,000	7

GTM: Gross Tons, Metric
Source: Lloyd's Maritime Information Services, 1997

Some ships are **self-loading**, which means they have the means of loading or unloading themselves. This generally means a crane, but a tanker may also have the pumping system. This is important if the ship is going to ports that do not have their own loading equipment. Most of the modern ships only call on ports that have equipment, so self-loading ships are limited to the less-modern trades. The big exception is military ships, which need to have the option of loading themselves if in armed conflict the port equipment is damaged.

Shipbuilding is undergoing a transition. Western countries have historically been the major builders, but now the countries that are building the most ships are Korea and Japan. These two countries each produce about 35% of the new ships, while China has a 5% share that is growing fast.[1] As already mentioned, ships tend to be associated with a given country. In order to promote jobs and also for national defense reasons, some countries have subsidized their local shipbuilding industry. A ship built in the US costs twice as much as a Korean built ship, yet the US government pays the difference.

Ships eventually retire and are scrapped for their metal, a process known as **ship breaking**. The ship is typically run aground and an army of workers start picking it apart. This creates an environmental nightmare because of the large amount of toxic materials in a ship. This is only done in some very poor countries, particularly Bangladesh and the famous Alang Shipyard in India.[2] An alternative is to sink it at sea. There has been an unfortunate habit of abandoning ships in whatever port they are in when the company that owns them decides that they are no longer profitable. These issues are not so much of a concern with other modes of transport because ships are so large.

FastShip

Traditionally, there has been a very large gap between the service capabilities of ocean ships and air cargo planes. That gap may be reduced by a radically new ship design. FastShip, Inc is developing a ship that can move at 40 knots, twice the speed of a normal container ship, and cross the Atlantic in four days.

The key to this new ship is its design. As a ship pushes through water, it creates a 'captive wave'. The faster a ship moves, the larger the captive wave, which requires yet more force. The ship then drops into the trough behind the wave. In 1975, a British engineer solved the problem with what he called a 'semi-planed monohull'. The bow has a deep v-shaped cut, a wide, shallow stern, and a concave bottom. Passenger ships normally use catamaran hulls or hydrofoils to lift the boat out of the water for more speed. That would not work for heavy cargo ships because their weight makes such a design unstable.

FastShip is spending $1.5 billion for the venture. Their parent company, a British engineering firm called Vickers, was bought by Rolls Royce, who is making the initial 25 jet engines. The French port of Cherbourg has agreed to be the port of entry in Europe, while in the US the Port of Philadelphia will be used.[3]

Rates are planned to be at a 60% premium. It will offer six-day door-to-door service, as opposed to three weeks for regular intermodal service. This has many implications for the shipping industry as well as for the business opportunities created by filling the gap between slow/cheap ocean shipping and fast/expensive air cargo.

SHIPPING COMPANY OPERATIONS

The type of ship determines to a large degree the business operations, and vice versa. For a shipper, the decision of what type of ocean carrier to use depends on the cargo characteristics and to a lesser extent the service needs. Low value cargo would most likely move in bulk carriers. Medium and high value cargo goes with a containership. Specialized cargo may take advantage of some of the specialized ships like chemical carriers or refrigerated ships.

Top Shipping Companies

The following are the rankings of the largest ocean carriers based on imports into the U.S.:

1. Maersk Sealand (Denmark)
2. Evergreen (Taiwan)
3. APL (Owned by NOL, Singapore)
4. Hanjin (Korea)
5. Cosco (PRC)
6. P&O Nedlloyd (Great Britain)
7. Hyundai (Korea)
8. OOCL (Hong Kong)
9. K Line (Japan)
10. Yangming (Taiwan)

Source: JoC Week

There is a limited amount of **competition among the different types of ships**. Breakbulk carriers have shrunk dramatically since the introduction of intermodalism. Roro carriers have also lost a lot of business to containerships, but also because of politics. Many countries did not like importing many cars because they felt it was harming local workers. This led to import restrictions, so now most cars are being manufactured in their market area rather than being shipped across oceans.

There are three basic business arrangements, liners, tramps, and private carriers.

- **Liners.** The ship follows a regular route and a regular schedule, which are publicized. The best way to think of trampers and liners is a taxi service (trampers), and a bus service (liners).
- **Tramps.** The ship has no set schedule, but operates on a voyage-by-voyage basis. Trampers are those types of ships where there is one type of cargo on board, or at least one shipper.

- **Private**. The ship is operated as a private carrier. The ship itself may be owned by the user or chartered.

A **liner has special legal benefits and obligations**, discussed in the section on shipping regulations. Liners are almost exclusively containerships or roros. This is because a containership carries thousands of containers, every one of which could be from a different shipper. That means that every voyage is carrying cargo from many shippers, so the only feasible arrangement is to operate like a bus. Bulk ships have one, or maybe a few, shippers with their cargo on a voyage, so they can make special arrangements and schedules.

Port Rotation

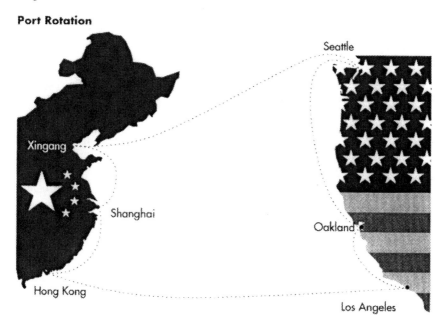

Liners follow fixed routes, known as **port rotations**. Arranging the port rotation is one of the very sophisticated skills of this business. They need to look at trade flows, locally, regionally and even global, and decide what service to provide. One option is to cross a landmass as if it were the opposite of a bridge, known as **mini-land bridge**, or **MLB**. For example, cargo from Asia to Europe often goes to the US West Coast, is put on a train across the continent, and shipped across the Atlantic.

The following is just one example of the planning considerations that go into routing. When a ship arrives at a port, it does not necessarily pick up or drop off cargo. Some ports it will only do one but not the other. For example, imagine a ship with the following route:

Remember that other ships are offering competing services, and the goal for the shipper is to find the fastest and cheapest service across the ocean. In this example, cargo may be picked up in Los Angeles destined for Xingang, but the carrier would probably not pick up cargo for Hong Kong. Why? Imagine you are on a bus. You may be willing to wait while it makes a few stops, but if there are a lot of stops ahead, you would start wondering if there was a more direct bus. Likewise, if the Los Angeles cargo were destined to Hong Kong, the shipper would probably start looking for a ship that is either going to bypass Oakland, or at least call on Hong Kong first. Recall from the discussion of cabotage that foreign ships are restricted from dropping off cargo that they picked up in that country, but this is a different issue than the customer service effect of port rotations.

There are some unique **labor** issues as a result of the fact that ships are off by themselves for the vast majority of the time, traveling the globe as if they had no home. Many ships have poor work conditions. It is not uncommon for a ship to drop off workers in foreign countries, stranded without any money. This also happens when a ship is abandoned. Even in relatively progressive ocean carriers, there is often a division of labor similar to a caste system. The officers are from developed countries, particularly Japan, Hong Kong, Britain or Norway. The crew is usually from poor countries. One nation stands out for providing an exceptionally large number of ship workers, the Philippines. This is apparently because maritime work is part of their history and culture.

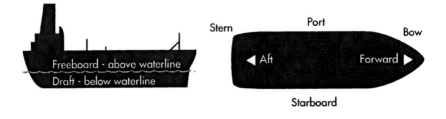

Ship Terms

This industry has a lot of tradition. As you can see, some of these terms are used not because there is anything special to the term, but because that is what has historically been used. Since the industry, on shore and at see, use these terms, one wants to know them to understand what mariners are talking about.

Port and Starboard: Left and right, respectively, as one faces the bow from on board the ship.

Bow and stern: the front and back of the ship, respectively.

Draft: the measurement of how deep a ship goes underwater. This determines if a port is deep enough to accommodate a ship.

Beam: the width of a ship.

Freeboard: the height of a ship above the water

Bridge: the "command post" or headquarters of the ship, typically on the highest part, from where the ship is run.

HOW BIG IS BIG?

There are different ways of measuring ship size:

Deadweight tonnage (DWT): weight it can carry including fuel and supplies

Gross revenue tonnage (GRT): space available for cargo.

Displacement: total weight of the ship. This could theoretically be measured by looking at the weight of the water that a ship displaces. In other words, imagine if one could freeze an ocean, take the ship out, and then weigh the amount of water that would fit in that hole left in the ocean by the ship.

The largest roros carry about 6,000 cars. This can vary because the decks on many ships can move up and down. If they have small cars, the decks come down, making more space for bulky cargo on top. The largest containerships claim to carry 7,000 TEUs, but their theoretical limit is not the same as what they can actually carry if the containers were full. Some ships would need to cut their capacity by as much as 25% if all the containers were loaded. The weight would make the ship unstable. The limit of a containership with one engine is about 9,000 TEU, but if and when ships are made with two engines, experts believe that instead of just being a little more than 9,000, they will go all the way up to 12,000 TEU.

The Panama Canal, 110 feet wide, was a major limiting factor on the size of ships. Now there is a special term, panamax, for the ships that are too large to fit in the canal, and thus are restricted in the routes that they serve. There is also the limit of bridges in port areas, known as "air draft". The proposed 9,000 TEU ships would need 166 feet of clearance, which is more than some major ports.

Tankers have been coming down in size. The Jahre Viking was the largest, at 564,000 metric tons with a 24.6m draft and 458m long. Currently the largest ships being built are around 450,000 tons, though the majority are around 310,000 metric tons and 333m in length.

MARITIME ECONOMICS

The economics of the maritime shipping is affected by many variables, but a few of the most important issues will be discussed here. Other chapters included general issues, such as cargo types, regulations, and so on. However, there are some issues unique to maritime shipping.

Ship size is a key variable in the industry, mostly because ships may be extremely big or small. Planes, trains and trucks are much more limited in how much they can change their size because of technology, infrastructure, and so on. There is almost nothing limiting the size of ships, which is why this is an area that the industry has been taking advantage of.

There are **economies of scale** in larger ships, though many industry analysts are wondering where the limit is. Tankers are now so large that some of them cannot come into any port. They go to a pumping station located several kilometers offshore, not even in sight of land. This is done at the origin and destination, so the ship never sees land during its operating life. Containerships have been getting larger, and are now over 7,000 TEUs. There is talk of building 12,000 TEU ships. The global routes used by these big ships are more limited, and thus less ports are being used.

Shipbuilding: 2003, from the American Shipbuilding Association
• Korea 34% • Japan 32% • China 11% • EU 15% • US 1% • Other 7%

Business needs are **limiting the growth** in the size of ships, not engineering. By using a larger ship, the number of ports becomes more limited. There are only a few ports in every region that can accommodate the largest containerships. Yet the real limit comes from the fact that with larger ships, the frequency of service is reduced and there is more stress on land-based operations. Smaller ships mean that departures are more frequent. If a large ship can take one-week's worth of cargo, which means that there will only be one departure a week. Miss the one sailing and you need to wait an entire week. The ports are struggling to handle larger ships, to be discussed in Chapter Nine.

In addition to the size of ships, there is a **consolidation of the shipping companies**. Shippers like to work with as few carriers as possible. If a carrier offers service to anywhere in the world, that would probably be preferred to one that only services a few regions. This means that the carriers are building up their fleets to operate globally. The containership industry is very large so there is still competition in each market. In the roro market, this is not the case. A merger between Wallenius and Wilhelmsen Line created the largest roro fleet in the world, with 20% of the world's 351 roro ships, and they are still seeking to merge with other carriers. It is not clear what effect this will have on the market.

There are many different types of alliances being used in the shipping industry. In some cases, a few carriers will work together on a trade lane. The actual containerships may only be provided by one or two members of the alliance. Others may operate no ships at all. They would simply pay the other carriers to carry their cargo. In other cases, one carrier will service one route, and other carrier will service a different route. In this case, they can offer more routes, not more departures. The spread of alliances and use of centralized hubs also encourages the development of larger ships. Rationalization of assets, described in Chapter Five, is very common in maritime shipping.

The pricing function varies depending on the type of ship and the way it operates. Liners publish rates to the public to see, and thus there is little room for negotiation. Service contracts are being used instead. There are legal reasons for this, which will be discussed shortly. All other carriers may or may not fix rates. It is difficult to generalize because of the different operational methods.

There are a few basic methods of developing rates. Containerized cargo is charged per container or by weight, and different rates are charged depending on the commodity. Bulk cargo is charges per weight. Breakbulk cargo is charged in a variety of ways since the types of cargo varies. If a company charters a ship, then there is no shipping charge as such. They are paying for the entire ship to use as they wish. Some cargo is priced in specialized ways. Coal is unique in that it is based on BTUs. For example, in a coal shipment, an independent party takes a sample and determines the btu's per unit of weight, which may vary depending on the moisture.

Pricing is more important for some commodities than others. Containerized cargo is often medium to high value, with many additional services provided. In this case, price is one issue but there are also service issues. For low value cargo, such as bulk cargo, price is the issue. For these commodities, the cost of transport is much greater as a percentage of the cargo's total price.

Sometimes shipping lines are able to increase their rates significantly due to market conditions. This often makes them a victim of their own success, because the rest of the industry sees that and can enter the market, adding capacity to the trade lanes and driving prices down. There is some limit to a ship's ability to shift from one trade lane to another. First, it needs to be the same type of ship. A bulk ship or tanker cannot go after intermodal cargo with rare exceptions. Even among the intermodal ships, their design make them better suited for particular markets. For example, the Atlantic sees a lot of high-value, low volume cargo that would use 20' containers. The Pacific sees more low-value, high volume cargo such as raw materials and food, that would go in 40' containers. Ships are built with more or less 20 or 40' slots depending on what trade lanes they would most likely serve.

SHIPPING REGULATIONS

As we discussed in Chapter Five, regulation of the ocean shipping is unique because governments can only control one side of the shipment. Once a ship leaves national territory, the exporting government has little control over what happens. Furthermore, the regulations of all countries in which the ships visit need to be roughly aligned. Regulation of the maritime shipping industry has a few main priorities:
- National fleet for national defense
- Access to international markets
- Preservation of competition

How far does a country's control extend? In 1930 there was an attempt at a treaty over how far state's rights extended, and they could not agree on three miles or further. The debate had nothing to do with shipping, but over coastal fisheries.[4] In 1982 the UN Convention on the Law of the Sea established a 12 mile territorial limit. In some passages there is not enough space for a ship to stay in international waters. In this case the Law of the Sea affirmed the right of **innocent passage**, called **transit passage**.

Governments often try to influence or control the behavior of parties outside of its territory, known as **extraterritoriality**. This is often resisted, as one would expect. Governments have considerable choice in how it regulates trade coming into and out of its territory. However, there is a gray area where national regulations affect foreign companies in a manner that they find unacceptable. For example, governments have tried to enforce their own standard of safety, health and environmental protection by applying those standards to any ship that enters its territory. The Law of the Sea denied countries the right to use ship construction and equipment standards on ships at sea.[5]

There are international organizations that seek to develop consensus on regulations. The **International Maritime Organization** (IMO) is a UN agency dedicated to the ocean shipping industry.[6] As an example of their work, the IMO has become very active since the 9-11 attacks in developing security standards against terrorism. The **Comité Maritime International** (CMI) is an organization of lawyers, based in Belgium, founded in 1897, to promote standardization of maritime law.

Monopoly regulation has always been a major issue with maritime shipping, and has influenced the pricing process. Liners are one of the very few industries exempt from some anti-trust regulations. In other words, the rules designed to prevent monopolies from forming do not apply in some cases. To understand why, a little history is helpful. Modern shipping began in the 1870s when the steamship was invented and the Suez Canal opened. This resulted in overcapacity and rate wars. The Calcutta Conference was organized by a prominent British ship owner to divide the market in the British-India trade 1875. This was the beginning of **shipping conferences**. Members (all of whom were carriers, not individuals) were allowed to talk with their competitors to discuss rates and seek agreement on rates. In other words, they were allowed to fix rates. By 1900 there were 100 conferences and by the 1970s about 360.[7]

Included with the anti-trust exemption was the requirement to file tariffs with the government. This was done because the tariffs needed to be available for the public to see. The **Ocean Shipping Reform Act** (OSRA) was a landmark piece of legislation that represented a compromise between the interests of shippers, which wanted conferences eliminated, and the carriers, that wanted to retain anti-trust immunity.

What market forces caused this deregulation? One of the last major conference attempts was the Trans-Atlantic Agreement (TAA) described as "the last gasp of an old era".[8] Conferences lost a large percentage of the members and the ability to control rates because containerization created competition among the different modes of transport (between intermodal and break-bulk), and carriers were offering door-to-door rates in which the ocean portion was not clear and thus could not be compared among the carriers. In the North Pacific, the trade went from 42 liners in 1985 to 10 joint ventures and consortia in 1993. In the break bulk era there was a considerable difference in service quality between conference and non-conference carriers, but containerization was the great equalizer.[9] Also, the Asian liners did not share any sense of solidarity with the Western companies and did not actively participate in conferences.

The OSRA allows confidential service contracts. Previously, if a carrier were in a conference, all of the members needed to know what prices were being offered. That was the only way they could fix prices and pressure their members not to undercut each other. Yet the large shippers wanted more complex service arrangements, which was the result of companies' sophisticated supply chain management. The OSRA eliminated the need to file tariffs with the government, but they would only need to post them on the Internet. More importantly, confidential service contracts were allowed.

Conferences disappeared because they could not determine what rates were being offered and thus could not enforce their own agreements. Talking agreements, which allow discussion of rates and "non-binding voluntary guidelines", are still permitted, although there is pressure from various shipper groups to eliminate these as well.

Currently the vast majority of ocean shipping is based on service contracts. Even when small shippers often move under these contracts because when they use a freight forwarder, the forwarder is operating under a service contract with the carriers.

Whereas air transport deregulation in the US resulted in a wave of deregulation around the world, the same is not happening with maritime shipping. The OECD has been discussing deregulation, and while some people are pushing for a worldwide equivalent to the OSRA, others are strongly opposed. It does not look like there will be major changes in the near future.

Another major area of shipping regulation is **registration**. Ships would not appear to have any home given the way they travel around the world. Yet they must be registered in a country, essentially like a citizenship. Ship registration has become a controversial issue. The ship must adhere to the regulations of that country in which it is registered. To be registered in the US, a ship must be built in a US shipyard and the crew must be mostly American. This adds an enormous cost. Therefore, other countries offer to register ships, charge very low fees, but do not check their ships for safety. These are known as **flags of convenience** (FOC).

FOCs began as far back as the 16th century when the English used them to get around Spanish blockades. In the 1920s when alcohol was forbidden in the US, ships used a Panama registration to operate cruise boats where the passengers could drink as much as they liked. Not all countries approved of the use of FOCs. The US wanted to allow FOCs and pushed for non-discrimination in which a state must allow ships regardless of the flag. European states disagreed, and in the 1958 Convention on the High Seas they were able to include a provision in which there must be a "genuine link" between the owners of the ship and the nation of registry. By the 1970s the European states were finding it hard to abide by their own rules and their shipowners transferred ownership to FOCs. Ironically, it was India and Brazil leading the objection to FOCs since the 1970s.[10]

The Global Mariner

The Global Mariner sailed around the world bringing publicity to the working conditions of sailors. Few are aware that in many parts of the world sailors are working under dangerous and harsh conditions, often left unpaid and even abandoned in foreign countries. Shipping companies find it possible to do this because they are difficult to control in the international community. Ironically, the Global Mariner was hit by another ship and sank in the mouth of the Orinoco River in Venezuela in 2000.

The Global Mariner was operated by the International Transport Workers' Federation (ITF), representing 500 unions in 125 countries, with over 5 million members. This includes all aspects of the transport industry, including railways, road, aviation, ports, fishing, and others. The ITF is one of 15 'international trade secretariats', international bodies covering a given industry. This becomes necessary as businesses become global, and thus labor issues need to be discussed in a global context.

The European Commission calculated that tax breaks and lower labor costs can save a ship owner about $1 million a year per ship. The Organization for Economic Cooperation and Development estimates that the worst owners spend about $3,000 a day on a ship's maintenance, while the best owners spend around $10,000 a day. About 55% of the shipping tonnage flies FOCs. Of the top 35 maritime countries, only China and Korea and a few others have more home flags than foreign ones.[11]

The concern is that there are some important safety, environmental protection, and labor standard issues that are being evaded with FOCs. The countries that offer themselves as FOCs are often not at all interested in the maritime industry other than to earn some money. There is nothing that says any country may or may not offer themselves as a place to register ships. As the box below shows, some of the most common flags on ships are countries that are very small and not involved with maritime shipping.

Liberia is a common example because it is the second largest fleet, and the country has experienced some dramatic political problems, leading some to question whether they should be allowed to offer a registry. Liberia's registry costs about $1000 per ship. The company that operates the registry, a law firm named by International Registries, Inc,[12] near Washington DC, also offers Liberia as a tax haven for companies to avoid taxes. This corporate registry service brings in about $40 million per year ($1000 per company), as compared to the $2 million a year from ship registries.[13] Yet the law firm claims that out of this money, the registering countries are able to monitor their ships and insure full compliance with the law. Yet Liberian, who is supposed to enforce standards, has been in the midst of civil war for years. At times the law firm could not even be sure whom to send the money because there was no apparent government in control. In response, IRI offered to change registries to Marshall Islands at no cost and same day.[14]

Piracy

Piracy is not just a tale from history. It is still quite common in some parts of the world. The normal practice is for a handful of men in a small high-speed boat to board a ship at night to rob the crew. Sometimes the ship is taken and the cargo sold. Often the crew are injured or killed. The Paris-based International Chamber of Commerce reported that there were 198 attacks on ships in 1998, and 247 in 1997.[15] The most common place for this is Southeast Asia, particularly among the Indonesian islands. This is where there is heavy traffic, multiple national jurisdictions, and weak international cooperation or policing.

Ships need to be inspected regularly to insure they are safe. The inspections are done by private companies called **classification societies**. There are a few good ones, and many that do not adhere to such strict standards. Among the reputable ones are Bureau Veritas and Lloyd's. Note that Lloyd's of London is a major insurance company, and Lloyd's Registration is an affiliate. If one were to insure a ship, it would be best to have good information about its condition, and that would come from its own classification society. Inspections are normally done every five years. Problems start to appear with older ships, and these are the ones that often reregister with less reputable FOCs.

Using a FOC does not necessarily mean that the ship is trying to evade quality standards. A recent study commissioned by the ITF found Singapore and Hong Kong, which they consider FOCs, to be near the top of their Flag State Conformance Index (Flasci).[16] Also, the ship's registration applies only to the ship, while the company that owns it is based in its home country. There does not need to be any particular relationship between the two countries. See the list below of ship ownership compared to the list of registries.

Why would a carrier pay extra to have a given flag? The most common reason is protection. A US flagship is treated as part of the US. This means that a country that cannot protect its fleet should not try to offer such protection. There were cases in which Kuwaiti ships were registered in the US so that anyone thinking of attacking the ships would need to consider the chance of the US retaliating. Another advantage is cabotage. To offer shipping services within a country requires that the ship be registered in that country.

The US has one of the costliest flags in the world. To be registered in the US, the ship needs to be built there, the crew must be US citizens, and repairs may only be done in the US. The last major containership built in the US cost an estimated $129 million.[17] A similar ship built in a non-US country such as Korea would have cost $25 to $30 million.[18] No company would do this if it were not for the fact that the government offers subsidies to make up the difference. Also, the government guarantees all of its own cargo is only shipped with US flag carriers, a practice known as **preference cargo**. Only if there is no service provided may the US cargo go with another carrier.

Top Fleets In The World

Country	Gross Tonnage ('000)
1. Panama	124,000
2. Liberia	36,000
3. Greece	28,500
4. Bahamas	27,200
5. Malta	25,000
6. Cyprus	22,000
7. Singapore	22,000
8. Hong Kong	17,000
9. Norway Int'l	17,000
10. Mashall Islands	16,000

Source: Drewery, 2003, for vessels 100 gross tons and over.

Every transportation mode has some **environmental issues** of concern. In the air transport industry, the concern is mostly noise. In trucking, the main issue is air pollution. In the ocean shipping industry the primary concern comes from spills of toxic substances. Spillage, particularly oil spills from tankers, is one of the single greatest influences on the tanker industry. The Chairman of Chevron Shipping noted that he spends more time on safety issues than any other subject. In fact, regulations in the US are such that he could go to jail for failure to take proper precautions. This is something that few other business leaders deal with.

There is a new rule, first developed in the US and extending internationally, that requires double-hulls on tankers. This was a direct result of the Exxon Valdez spill. Interestingly enough, non-US tankers spill the equivalent of several times the amount that was spilled by the Valdez. What makes the difference is where the oil is spilled. Some parts of the world are very sensitive to oil spillage, while others are remarkably resilient. There was a tanker that ran onto rocks on the northern coast of Scotland and spilled quantities similar to the Valdez, yet the rough conditions and rocky coast meant that the vast majority of the oil was dissipated into the ocean.

Ironically, the problem with oil spills is not so much the large spills as all the small spills, with results from either **bilge** or **ballast**. Tankers carry water when they do not have oil, to maintain stability. This water is referred to as ballast. Previously, the ballast would be pumped out, even though it was mixed with oil. The IMO now requires segregated ballast.

Bilge refers to water that collects at the inside bottom of the ship. This is also mixed with oil, fuel and other toxic substances. Traditionally it was pumped into the ocean. Regulations attempting to control this practice have been very difficult to enforce.

Ballast can also result in the migration of marine life, with negative effects. A large ship can take on as much as 15,000 tons of ballast. This ballast carries marine life that are then left wherever the ballast is pumped out. There has been discussion of the issue for years with little effect. To treat ballast as pollution would result in major changes to maritime operations. Oily bilge and oily ballast is normally pumped onto shoreside tanks for treatment. Yet water ballast is discharged in massive volumes, 2.4 million gallons per hour in the US alone by one estimate. The zebra mussel, native to the Black Sea, has become established in the US Great Lakes and has caused an estimated $5 billion worth of damage. One solution is to have ships exchange ballast on the high seas instead of in ports, and report on their activities to the authorities. A primary concern for the maritime and port authorities is a clear definition of what is clean enough for discharge.[19]

Double Bottom vs. Single Bottom Tanker

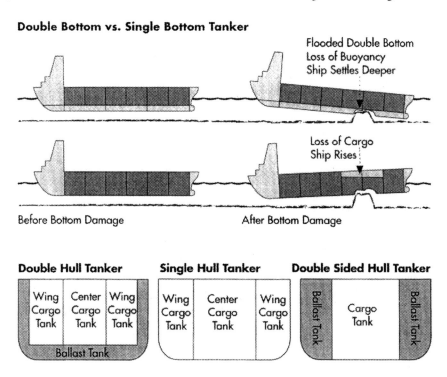

Flooded Double Bottom
Loss of Buoyancy
Ship Settles Deeper

Loss of Cargo
Ship Rises

Before Bottom Damage   After Bottom Damage

Double Hull Tanker

Wing Cargo Tank	Center Cargo Tank	Wing Cargo Tank
Ballast Tank		

Single Hull Tanker

Wing Cargo Tank	Center Cargo Tank	Wing Cargo Tank

Double Sided Hull Tanker

Ballast Tank	Cargo Tank	Ballast Tank

A Crude Oil Carrier Using Load On Top System for Anti Pollution

Stage 1
Vessel at Sea in Dirty Ballast Condition and Cleaning Tanks

Slop From Tank Cleaning to Slop Tank

Stage 2
Vessel at Sea when Tank Cleaning Complete and With Clean Ballast Shipped Disposing of Dirty Ballast

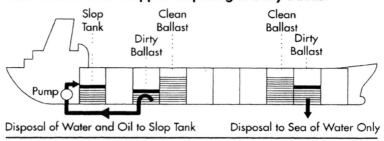

Disposal of Water and Oil to Slop Tank   Disposal to Sea of Water Only

Stage 3
Vessel at Sea in Clean Ballast Condition. All Polluted Water Concentrated in

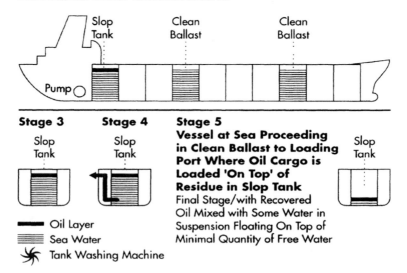

Stage 3   **Stage 4**   **Stage 5**
Vessel at Sea Proceeding in Clean Ballast to Loading Port Where Oil Cargo is Loaded 'On Top' of Residue in Slop Tank
Final Stage/with Recovered Oil Mixed with Some Water in Suspension Floating On Top of Minimal Quantity of Free Water

— Oil Layer
≡ Sea Water
✳ Tank Washing Machine

The Following is the ITF's Policy on FOCs:

In determining whether to designate a register as an FOC the ITF continues to take into account the degree to which foreign owned vessels predominate on the registry. If the majority are foreign owned and there is no "genuine link" between the ship-owner and the flag of the country concerned, then the registry will automatically qualify to be designated by the Fair Practices Committee (FPC) as an FOC. In addition, the ITF now also scrutinises the performance of ship registers measured against the following criteria which could lead to national registers which do not meet the ownership criteria nevertheless being added to the FOC target list if they exhibit serious deficiencies with regard to their:

Ability and/or willingness to enforce international minimum social standards on its vessels, including respect for basic human and trade union rights, freedom of association and the right to collective bargaining with bona fide trade unions;

Social record as determined by the degree of ratification and enforcement of ILO Conventions and Recommendations;

Safety and environmental record as revealed by the ratification and enforcement of IMO Conventions and revealed by port state control inspections, deficiencies and detentions.

Based on the above criteria, the ITF considered the following FOCs:

Antigua and Barbuda	Cayman Islands	Malta
Aruba	Cook Islands	Marshall Islands
Bahamas	Cyprus	Mauritius
Barbados	Equatorial Guinea	Netherlands Antilles
Belize	Gibraltar	Panama
Bermuda	Honduras	Sri Lanka
Bolivia	Lebanon	St. Vincent & The Grenadines
Burma	Liberia	Tuvalu
Cambodia	Luxembourg	Vanuatu
Canary Islands		

Top Ship Owning Countries

Country	Tonnage (000GMT)
Greece	8,500
Japan	7,700
PRC	4,000
Germany	3,000
USA	2,700
Norway	2,600
United Kingdom	2,100
Singapore	1,800
Taiwan	1,600
Korea	1,600

Source: Drewery, 2003.

MANAGING OCEAN CARRIERS

Shippers have a different relationship with ocean carriers depending on the type of carrier. Tankers and bulk carriers are usually chartered. This may be a **trip charter** (also known as a **voyage charter**), which only lasts for the trip, or a **time charter**, which lasts a specified amount of time and may involve multiple trips. There is also the choice of **bareboat**, which means the shipper gets just the boat and must find a crew, or a **demise charter**, in which the ship comes fully crewed. This is all specified in the **Contract of Affreightment**, an agreement between the ship owner and the shipper. Almost 90% of bulk ships are on long-term charter, and of those about a third are owned by multinational firms involved in the trade of bulk goods.[20] The **Baltic and International Maritime Council** (BIMCO), based in Denmark, is an independent organization whose motto is "working for the benefit of the shipping community". As one part of their many activities, they offer model contracts for bulk goods.

Container lines deal with many shippers simultaneously, and the arrangement can be much more complex. Container lines are synonymous with intermodalism. Sometimes the shipper only wants the ocean carriage. Often they want a point-to-point shipment, which means that the ocean carrier is actually providing intermodal service, of which the ocean trip is one part.

The effects of deregulation and the OSRA are still ongoing. Small shippers were not disadvantaged by big shippers (we are referring here to the amount of cargo shipped, not necessarily the size of the company). But with private service contracts, big shippers can keep their shipping costs a secret. Recall that transportation costs are a major factor in being competitive in foreign markets, and it is more important to look at one's transportation costs *in comparison to one's competitors.* The result is that small shippers are banding together in shipping associations, discussed in Chapter Fourteen. Shipping companies have already indicated that, in the face of deregulation, they are diversifying into other logistics services.

Top 20 Container Carriers

The following are the 1998 rankings for container lines. These rankings change every year, but it shows who are the main players, and what country they come from. Note that the 20th place goes to a company that is the transportation arm of a food company.

Volume in TEUs

1. Sea-Land Service	1,376,887
2. Evergreen Line	1,298,143
3. Maersk	1,064,916
4. Hanjin Shipping Company	902,305
5. American President Lines	895,518
6. China Ocean Shipping	657,531
7. Hyundai Merchant Marine	637,103
8. P&O Nedlloyd	551,830
9. NYK Line	527,027
10. Yang Ming Line	525,422
11. Orient Overseas Container Line	498,227
12. K Line	480,906
13. Mitsui OSK Lines	435,045
14. Hapag Lloyd	406,307
15. DSR Senator Line	370,100
16. Crowley American Transport	369,264
17. Zim Container	365,738
18. Mediterranean Shipping Co.	318,791
19. Cho Yang Line	285,543
20. Dole Fresh Fruit Co	212,352

Source: Journal of Commerce

[1] JOC, September 9, 1999, p. 12.

[2] "Toxic Exports: The Transfer of Hazardous Wastes from Rich to Poor Countries", Jennifer Clapp, Cornell University Press, Ithaca, NY, 2001, p. 101, see also "Ships for Scrap: Steel and Toxic Wastes for Asia", Judit Kanthak, Andreas Bernstorff, and Nityanand Jayaraman, Greenpeace, Hamburg, 1999.

[3] "FastShip selects French harbor", Jack Lucentini, The Journal of Commerce, October 15, 1998, p. 10A, and "Joining the jet-set", The Economist, September 25, 1999, p. 77.

[4] "The three-mile limit: Preserving the freedom of the seas", Bernard G. Heinzen, Stanford Law Review, volume 11, May 1959.

[5] "Governing Global Networks: International Regimes for Transportation and Communications", Mark W. Zacher with Brent A. Sutton", Cambridge University Press, New York, 1996, p. 48.

[6] www.imo.org

[7] "Liner Conferences in the Container Age", Gunnar Sletmo, and Williams Jr, Ernest W, Macmillan, New York, 1981.

[8] Zacher with Sutton, Ibid, p. 72.

[9] Ibid, p. 73.

[10] Zacher with Sutton, Ibid, p. 64.

[11] "Following the Flag of Convenience", The Economist, February 22, 1997, p. 75-76.

[12] www.register-iri.com

[13] "Changing of guard at Liberian registry", Robert Mottley, American Shipper, February 1999, p. 55.

[14] "Liberia fights for control of registry", Aviva Freudmann, The Journal of Commerce, October 28, 1998, p. 1B.

[15] JOC, June 18, 1999, p. 16.

[16] "Hong Kong Beats Singapore in Ship Register League", David Hughes, Business Times Singapore, July 4, 2001.

[17] The ship was called the R.J. Pfeiffer, built for Matson Navigation, who mainly services mainland US and Hawaii

[18] JOC, March 31, 1999, p. 5a.

[19] JOC, September 27, 1999, p. 9.

[20] "Review of Maritime Transport", UNCTAD, 1989.

CHAPTER 8
AIR TRANSPORTATION

If the maritime industry is noted for its long history and tradition, the air transport industry is recognized for its newness and innovation. This chapter begins with a brief history of air transport because this industry, more than any of the other modes of transport, is shaped by events that have been quite recent. Anther aspect unique to air transport is the fact many carriers combine cargo and passenger traffic. This creates some interesting comparisons between the operations of people and cargo. We also discuss the regulations and economics of this industry, which are still changing rapidly.

AIR TRANSPORTATION

Air transportation has been the single most significant development in logistics of this century. It is often forgotten, given the more recent development of information technologies, that this is a mode of transport that was unknown just a couple generations ago. Thirty four percent of world trade, by value, is moving by air.[1] This is the high value, low density cargo that needs to get somewhere fast. While ocean shipping carries the bulk of world trade, air cargo carries the cream of the crop.

1903 The Wright brothers invent the first plane at Kitty Hawk.

Global air cargo value passed $2 trillion in 1998, with a tonnage of 52.4 billion pounds. The value of air cargo has been increasing at about 2.5% a year, which is less than the volume or tonnage. Tonnage within Asia rose a stunning 40% in 1998.[2] This is because as the market expands, it is pushing into the less valuable cargo as air transport becomes cheaper.

1909 First international flight- across the English Channel

While other modes of transport are seeing growth resulting from increased international trade, the air cargo is experiencing exceptionally high growth. This is a result of increased trade, but also the increased wealth and resulting demand for premium service. According to the premier air cargo professional association, **The International Air Cargo**

Association (TIACA), the following are the primary reasons for the growth in air cargo:[3]

- Deregulation and liberalization of the air cargo industry.
- Global interdependence helped by world trade agreements.
- International production and sales of goods and services.
- New inventory management concepts such as "JIT" and "Zero" stocks.
- New air-eligible commodities.
- The vast development of high value and limited time-consumable commodities.

Air transportation for business was first developed for mail service. Mail is very small and light, and very time sensitive. The industry then evolved into passenger service with air cargo as only a sideline. Toward the end of the century, air cargo grew to become a significant industry in its own right.

1918 First regular airmail route- New York City/Washington DC

Unlike ocean shipping, there is considerable mixing between passenger transport and cargo transport. Airplanes carrying passengers on regularly scheduled flights, known as airliners, may also carry significant amounts of cargo. That is why this chapter will look at the air transportation industry as a whole, even if a lot of it is more relevant to passenger movement.

1944 Chicago Convention: International Civil Aviation Organization Established

The air cargo industry has evolved into an air logistics industry. The difference is that while air cargo only refers to the air portion of the journey, air logistics manages the door to door delivery, much of which is land-based transportation. In fact, when looking at door to door shipments, the air leg is often the simplest and easiest.

1946 Bermuda Conference: International Air Transportation Association Established

AIRPLANES AND AVIATION EQUIPMENT

The great majority of aviation is for passenger traffic. Because this book looks at cargo logistics, we will only include a general introduction to aviation, but concentrate on air cargo.

The Airbus Beluga. Illustration courtesy of Airbus.

Airplanes used in air cargo are essentially the same as for passenger traffic. Ocean ships are relatively cheap to design and build, which is why there are passenger ferries and cargo ships. Planes, on the other hand, cost billions of dollars to research, design and build. The technical demands of flight also limit the possibilities of what a plane can do, whereas ships come in a wide variety of designs. What we call "air cargo" is carried in a few ways, not always by air:

- **Bellyspace.** This is the space in the belly of a passenger plane, normally used for luggage. About 60% of air cargo moves in luggage compartments.[4] Cargo, unlike luggage, is not accompanied by the sender, more handlers, different and more documentation.
- **Flexbelly.** Some planes are basically passenger planes, but the interior can be adjusted to include more or less space for cargo or passengers.
- **Freighter.** This is a plane used just for cargo. Often a passenger plane toward the end of its useful life is converted into a freighter. Freighter planes built just for this purpose normally would not include windows, since there is no need for windows along the body of the plane.
- **Truck.** What many do not realize is that a lot of 'air cargo' actually goes by truck. Major airlines have been replacing their larger planes (B747s, DC10s and L1011s) with smaller planes for shorter flights. This reduces cargo space available, and thus means air cargo companies need to find a quick alternative.[5]

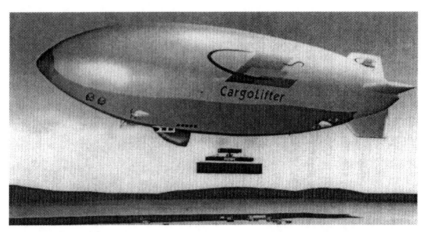

Source: CargoLifter

CargoLifter

It is not often that the transportation industry sees radical change in the modes of transport available to a shipper, and German-based CargoLifter is a good example of the difficulties associated with being too original. Aircraft face a much more difficult environment than ocean shipping, both in engineering and business. CargoLifter tried to build a dirigible (also known as a blimp or zeppelin) that would have been the first of its kind in 60 years, and the largest airship on earth by volume. The CL-160 was designed for cargo, and particularly heavy cargo. It would have offered a unique ability to pick-up or drop-off cargo anywhere in the world without a runway. With a crane capable of holding 160 tons and a loading bay 165 feet long and 26 feet wide, it could pick up the weight equivalent to a diesel locomotive engine (120 tons) and a humpback whale (40 tons). Flying speed would be between 50 and 75 mph loaded, or up to 84 mph unloaded.

After several years of progress and setbacks, it seems that they are changing their plans to a smaller model. The CL-75 looks more like a round balloon and would carry up to 75 tons for short distances. In fact, the CL-75 is considered by the Guinness Book of World Records® as the World's Largest Balloon.

Air transport has some unique characteristics that affect its operations, cargo type and services. Among the obvious is speed. Less obvious is the fact that transport-related stresses are much less on air travel than other modes, partly because the trip is quicker and partly because a flight is smoother than land or sea travel. For this reason, less packaging is needed, which also saves expense, bulk and weight. Insurance premiums on air cargo tend to be less. We will see later in this chapter how that affects service and the types of cargo carried.

In the chapter on maritime transport, we saw how industry demand led to different types of ships. In the air cargo market, there is much less diversity because a plane is so expensive to design from scratch. Furthermore, those designing a plane have much less choice given the demands of physics and engineering. Boeing's jumbo (747) was originally designed as a military aircraft, with a large nose to allow for tanks. They lost that contract to a rival firm, and only then began thinking of the plane for passenger use.[6]

For these reasons, the aircraft manufacturing industry is concentrating on improving efficiency and distance. The Concorde was introduced in 1976, but has never been profitable. There is now little incentive for faster planes. Instead, airlines have been asking for planes that are more fuel-efficient, would reduce labor costs, and operate more quietly. One of the newest planes, the Boeing 777, is designed for long-range flights with high passenger loads. Passengers prefer direct flights, especially on the longer routes. This could mean less belly space for cargo if it filled with passenger luggage. The result may be more cargo going on freighters, putting upward pressure on freighter rates.[7]

Freighter Fleet Doubles by 2013

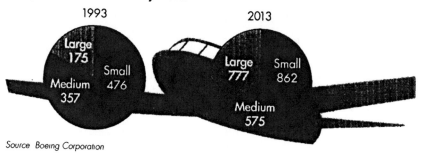

1993

Large 175
Medium 357
Small 476

2013

Large 777
Medium 575
Small 862

Source Boeing Corporation

Airplane manufacturing is currently one of the most concentrated industries in the world. There are only two major companies for large planes, and only a couple more for smaller planes. Boeing, who recently moved from Seattle to Chicago, and Airbus, in Europe, essentially control the market for large planes. About two thirds of the large planes are built by Boeing, and the remaining third from Airbus, but the trend is toward a 50/50 split of the market. The reason there are only two companies is because the cost of designing and manufacturing large planes are so great, there is only room in the market for two companies. Airbus is a consortium of European companies, almost all controlled or owned by national governments. For this reason the company is highly political in its operations.

The current competition between Boeing and Airbus is centering on their new jumbos. Airbus plans for the introduction in 2004 of the A380 which will carry 555 passengers. Boeing is taking a different tact, with a 250-seat Sonic Cruiser that will fly much faster, 98% of the speed of sound (about 20% faster than current planes, cutting an hour off a transatlantic flight). This suggests that Boeing expects future growth to be in point-to-point routes, while Airbus expects the growth in cheaper hub routes. They do not even seem to agree on the amount of growth. While both estimate a tripling of the air-travel market in the next 20 years, Boeing expects that of the total demand of 18,120 new planes, only a third will be twin-aisled (ie, large). Airbus forecasts 14,670 planes, but thinks that half of those will be large planes.[8]

The market for medium size planes includes a couple of other companies, such as McDonnell Douglas and Fokker, but these planes are rarely used for international and long flights. Indonesia has attempted to create an airplane industry from scratch as part of their push for high technology development. The project was started by Mr. Habibi (who would later become President), and many critics have noted that the government spent large sums of money with little commercial success.

Two-Thirds of New Jets Will Be Delivered Outside the United States

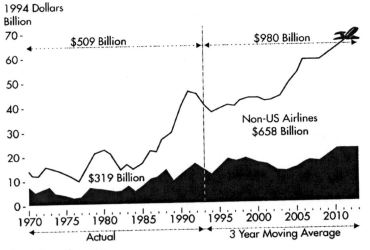

Source Boeing Corporation

The Total Market for New Airplanes Is $980 Billion

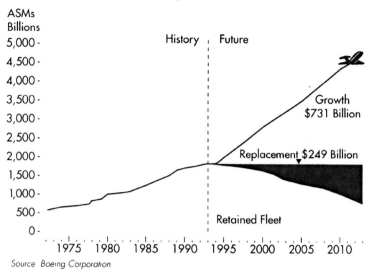

Source Boeing Corporation

AIR CARGO OPERATIONS

Air cargo has traditionally been a side-business for airlines to fill up extra bellyspace. Mail now accounts for less than 3% of airlines' revenue.[9] While US airlines only get 5 to 10 percent of their revenue from cargo, other airlines, such as Lufthansa, Japan Airlines and Air France, get as much as one-third of their revenues from cargo.[10] This limited the amount of innovation and investment going into cargo operations, even though there has been a dramatic increase in the airline industry. In 1975, the **International Air Transport Association** (IATA) found that the average delivery time for air cargo from point to point (not just the flight time but pick-up and delivery) was 6.5 days. Twenty-one years later, another study found that that time was still about six days.[11] While air transportation has been growing at 4% per year, air cargo has been growing at 7 to 10% rate.[12]

There are two types of carriers, airlines that carry cargo as a sideline, and dedicated air cargo carriers. Up until 1970 there were a growing number of freighters. When the jumbo jet, the Boeing 747, was first introduced in 1970, they had the ironic effect of expanding the air cargo industry while running many all-cargo carriers bankrupt. A 747 could carry much more cargo in its belly even with a full load of passengers. Yet because there was so much space in the passenger airlines, all-cargo carriers lost much of the business. Flying Tigers, acquired by FedEx in 1989, was the only survivor of the original three.

In the 1970's, there was a major deregulation of the aviation industry. This is discussed later in the chapter, but at this point it is only important to note that with deregulation, prices for aviation (cargo and passengers) dropped and the overall demand increased significantly.

Currently air cargo is a small part of airlines' revenue, but the air cargo industry as a whole is growing. Most important is an increase in time-sensitive cargo and luxury goods being traded worldwide. As overall wealth increases, there is a shift in demand from more goods to higher quality goods and more service. That is why the wealthiest regions in the world are seeing much more demand for those goods that are shipped air cargo: exotic foods, medicine, and so forth. There are four major air cargo markets, North Atlantic, transpacific, Europe-Far East and US domestic, make up nearly three quarters of the total market, with each of these four holding roughly equal volumes.[13] Yet the greatest growth is in the smaller markets where incomes are rising.

Air cargo carriers have now evolved into three forms. There are the **integrated carriers**, such as FedEx and UPS. **Scheduled carriers** are typically airlines offering cargo services. **Charter carriers** are the specialized all-cargo carriers.

Types of Passenger Routes[14]
We have been discussing air cargo, while making the point that there are important similarities and differences between moving people and cargo. The following are the major types of passenger routes. Note that some of these parallel cargo trade lanes, yet other route types would see very little cargo.

- Business travel
- Personal travel
- Gateway
- Diplomatic
- Poor surface accessibility
- Strategic policy
- Colonial
- Prestige
- Military

One reason that growth of the air cargo market has been limited is its **connection with passenger airlines**. Cargo and passengers have different characteristics, which sometimes forces an airline to favor their passenger customers over cargo customers.

First, as a general rule for all modes of transportation, people do not follow the same **paths** as cargo. People travel between major population centers, vacation spots, between home and work, and so on. Cargo travels from where raw materials are obtained, to manufacturing centers, and then to the market, where they finally meet up with people. FedEx's use of a national hub is a good example of the difference that can be made when all-cargo aircraft are used, free from the constraints of passenger routes.

Second, passenger traffic is also **seasonal** on some routes, which have very little correlation to cargo traffic seasons. People travel to vacation spots in the summer, or to visit family during holidays. Cargo has some seasonality such as food harvests, fish seasons, flowers, and so on, but these have little correlation to passenger seasons.

Finally, traffic balance is entirely different for passengers as cargo. People generally return from wherever they fly. Cargo rarely does. All-cargo carriers thus have a difficult time maintaining a balanced trade when their customers and commodities are different for each direction of a round trip.

Cargo containers used for air cargo are very different from that used for other modes of transport. The intermodal container is tough, designed for bad weather and rough handling. Air cargo containers, on the other hand, need to be lighter. Technically, many of the containers used for air cargo are intermodal in that they can be put on the bed of a truck, but they are only used from the airport into a nearby town.

Intermodal-size containers (8X8X20) are used for air cargo but only on freighters. They are built to more exacting specifications, designed to be an integral part of the plane, and their tare is much less. Other air containers are shaped to fit into the plane. These containers are designed for specific aircraft, and thus their compatibility with other modes of transport is very limited.

Tare
The tare is the weight of a container. This is important because there are strict weight limits on planes, so there is a strong incentive to have lightweight containers. One must also be clear in shipping goods to distinguish between gross weight (which includes the tare) and net weight (which does not include the tare)

Air cargo can be divided into **four distinct markets.** This division is based mostly on the service provided, and indirectly what kind of commodity is being moved. These are general classifications and not completely distinct. As companies try to expand, their services will sometimes cross from one market into the others:

• **Mail** - This is normally the monopoly market of the official postal service in a given country, defined by the size of the letter. This is an area where there are major changes as some governments privatize their postal markets. See Chapter Seventeen for a discussion of postal systems.

- **Express transportation**- this market was defined by FedEx, and is the fastest growing of the three air cargo segments. The difference with mail is obvious. The difference with freight is that their emphasis on speed. Air express also stress their ability to off delivery that is time definite, which means that the package is promised to be delivered at a certain time.
- **Courier**- courier services are an extension of air express. Instead of second day delivery, courier service means either same day service or next flight out. The package requires that someone fly with it. While courier services can be ten times as expensive as air express, the information provided is not nearly as good. Air express carriers have an information system that can track a package. Yet courier service is so fast that the carriers have not bothered to develop such tracking ability. One of the reasons that courier service is so expensive is that it requires someone from the company to be at the airport 24 hours to get the packages through customs.
- **Freight**- this market covers everything else that moves air cargo, and is mostly the larger packages that would be too expensive to be sent air express. The difference between air express and air freight has become increasingly vague as express carriers have all but eliminated weight limits on packages, and freighters are increasingly schedule-conscious.

U.S. Domestic Expedited Freight Market Competitor Segmentation

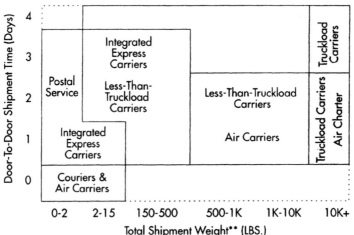

Excluded from Analysis and Forecast
* *Same-Day Service*
** *Regardless of the Number of Pieces*   Source. MergeGlobal

The above three services are based on the marketing strategy of the carrier, which brings up an issue of customer expectations. An air cargo company needs to set up its operations one way for international service, yet customers on both sides can have very different expectations. An air express carrier can only provide international service by organizing its operations to meet the needs of one market or the other. Tailoring its service to both can be difficult and inefficient.

Perishables: The Golden Market

Perishables are one of the fastest growing and lucrative markets in air cargo. There has been exponential growth in the number of people worldwide who can afford luxury goods such as imported perishable food products. Perishable cargo requires high standards of service and is rewarded by large profit margins.

Perishable Imports: the following is a breakdown of the total perishable imports worldwide.

Meat and animal products	.241
Pharmaceuticals	.17
Dairy products	.139
Nuts	.115
Fish and seafood	.114
Miscellaneous	.062
Vegetables	.055
Flowers and plants	.040
Fruits	.033
Chocolates and candy	.03

Source: International Trade Center, Geneva, April 1998.

The following is an approximate distribution of costs for a shipment of perishables moving by air cargo.

Value-added services	.082
Local pickup	.089
Transportation in refrigerated truck	.144
Export service and customs	.015
Air cargo	.49
Import service and customs	.091
Warehousing	.09

Source: HPL Hellmann, Ltd.

We can also see a difference between the B2C and B2B markets. Mail is mostly direct marketing, and of the private letters sent, 40% are Christmas cards (in the US market).[15] The courier market is overwhelmingly B2B, as is freight. The air express market, on the other hand, has been experiencing rapid growth thanks to e-commerce as people buy online products that are delivered by air express carriers. B2C deliveries are expected to grow exponentially, but deliveries to homes are less lucrative because homes are more scattered.

During the 1990's, the integrated carriers have grown dramatically and taken a large portion of the airline and forwarder's business. This trend is mostly the result of increased demand for the high-end and more costly services of the integrators. Business travelers are on the decline given improved telecommunications, which means that cargo operations are increasingly important. One way that airlines are meeting the challenge of the integrators is to work more closely with air freight forwarders to get more business from them.

The **commodities carried by air cargo** are different from that carried by other modes of transport. The fact that cargo can be moved such long distances in such a short time has created entirely new markets. For example, fresh cut flowers are shipped around the world, which would not be an option with other modes of transport. This also applies to some fresh fruits. Shipping live animals is stressful for them, which is why the speed of a plane makes it more likely they will survive the trip. There is also less stress from noise and movement than overland transport. The interior of the plane needs special protection so the animals' hoofs do not penetrate the floor, and they are given mixture of food to adjust to new area.

Security, discussed in more detail in Chapter Thirteen, is a major concern for obvious reasons. Whereas luggage can be associated with a specific passenger, cargo flies without anybody. There is also concept of the Known Shipper, a person that can be held responsible for their shipment. However, this does not mean there has been a problem with bombs. Edward Emmett, head of the NIT League, notes, "there's never been a terrorist incident involving commercial air freight".[16] Recommended practices and procedures (RPP) on air safety were developed from ICAO's Annex 17 of the Chicago convention, revised and reinforced by the IATA's recommendation 1630.[17]

Dangers come not just from sabotage, but from hazardous cargo, which are highly restricted if it is to move by air. Rules concerning the transport of hazardous goods are developed by the ICAO, but need to be controlled and enforced by national law. Many types of hazardous goods cannot be sent in the belly, but need to go on an all-cargo carrier. There has been concern with the number of plane crashes and bombings that cargo is a threat. Yet of the major crashes, such as Lockerbie and Valuejet, the cause has not been cargo. The threat comes primarily from luggage or other non-cargo sources.

Air cargo companies, unlike airlines, are not just in the business of running planes. A large part of the operation is **trucking**. Just as with ocean carriers, trucks are at least as important to the air carrier as the planes. Air cargo carriers, in choosing an airport to use, normally look at the trucking facilities at least as much as what the airport has to offer.

Trucks play two main roles for an air cargo carrier. First, they bring the cargo from the shipper to the airport. On the other end they bring the cargo from the plane to the ultimate destination. This job is often done by local trucking companies, but can also be done by the air carrier. Second, air carriers may choose to place cargo on a truck for trips under 500 miles. Most common is to use trucks to deliver cargo to the consolidation airport. In the US, trucking is used more by Asian and European carriers. This is because aviation law and the bilaterals restrict them to a few major airports, which means they are not allowed to fly the cargo to other areas. US air carriers, however, can fly their cargo anywhere in the country. In Asia and Europe the opposite does not occur. US carriers are not trucking cargo in Asia and Europe for lack of flying rights because the countries are so much smaller and thus they have bilaterals with all the different countries.

Just as in other modes of transportation, air cargo companies can provide service into a market it does not fly to. **Interline agreements** allow this even if the carrier does not fly its planes there, or it does not own any planes. For example, Emirates Airlines entered the US market by "taking a weekly allocation on a Boeing 747 freighter operated by KLM Royal Dutch Airlines between New York and Amsterdam". The sales agent said "initially we're taking two upper-deck pallet positions on the KLM freighter once a week."[18]

A major legal case between Britain and the US is a useful example of equal access on very different terms. If a British carrier is given permission to fly to a US port, they then have access to the largest economy in the world. Yet if a US carrier has access to a British port, they only have access to a market that is a small fraction of what the British counterparts gained.

However, in order to service that US market, the British carrier would need more than just one port of entry. It would either need multiple ports of entry, or extensive domestic (in the US) distribution system.

That would probably mean a partnership with a US firm. It might seem that this 'equal' trade was not so equal, but there is one other thing to consider. The US carrier gained access to 'all' of Britain. Yet that US carrier would need to draw from the entire US market, which means that it too would need to be part of a nation-wide (in the US) network. However, it would be far easier for a US carrier to create that network than for any foreign carrier. In fact most US carriers already have a nation-wide system in place. Thus what looks like a biased agreement in favor of Britain may be biased in favor of the U.S [19]

Forces and Constraints for Air Cargo Growth

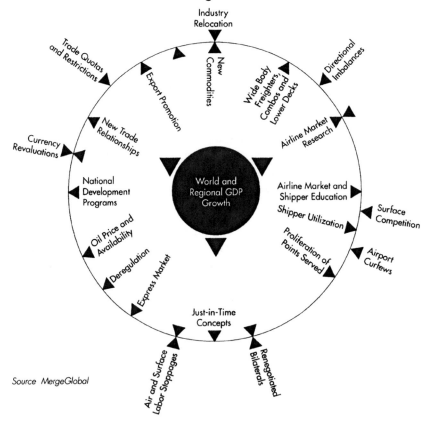

Source MergeGlobal

Characteristics of Air Cargo Commodities

The following are characteristics of cargo that is likely to use air transport:

1. When the commodity is:
   A. Perishable
   B. Subject to quick obsolescence
   C. Required on short notice
   D. Valuable relative to weight
   E. Expensive to handle or store

2. When demand is:
   A. Unpredictable
   B. Infrequent
   C. In excess of local supply
   D. Seasonal

3. When distribution problems include:
   A. Risk of pilferage, breakage, or deterioration
   B. High insurance costs for long in-transit periods
   C. Heavy or expensive packaging required for surface transportation
   D. Special handling or care needed
   E. Warehousing or stocks in excess of what would be needed if air freight were used

Source: "Air Transportation", Alexander T. Wells.

AVIATION ECONOMICS

Two major factors influence the **growth** of the air cargo market, **economic conditions** and **rate levels**.[20] Economic conditions are out of the control of carriers. Rate levels, on the other hand, are controlled by carriers. As we will see, carriers have sophisticated pricing strategies, but overall rates have been falling as the market has grown.

Airfare for cargo and passengers were originally developed by IATA, but that is now forbidden. With fixed prices, only the most efficient (typically US carriers) were profitable. Since most non-US carriers were government owned, they required subsidies to keep themselves solvent. Shortly after deregulation, there was a massive change in which the industry as a whole lost billions of dollars, as much as they had earned in the previous 40 years. Rates are now based on market conditions.

Yield management is the way an air carrier gets the maximum amount of revenue from each unit (passenger or cargo) carried. You may recognize this when you as a passenger get widely different airfare depending on what time you want to fly, what days you fly, and how far in advance you buy your ticket. It is an important business concept that applies to many sectors, but is especially relevant to aviation. Effective yield management is a critical part of an airline's profitability. While passenger yield management is a fascinating art and science, the important thing to note here is that cargo behaves very differently. We as passengers are highly sensitive to changes in spot rates. Cargo is not. There are much more volume contracts with cargo while passengers generally buy one ticket at a time. Air cargo rates are determined by the following factors:[21]

- **Volume of traffic.** The higher the traffic volume, the cheaper it is to provide service, and thus the service is cheaper.
- **Direction of traffic.** While passengers normally travel round trip, cargo does not. Many trade lanes have more traffic going one way than the reverse.
- **Characteristics of the traffic.** This includes cargo density, size per piece, and average weight of the shipment. Note dimensional weight discussion above.
- **Value of the service.** This is determined not just by the airline, but the needs of the customer.
- **Competition.** Carriers cannot avoid the influence of other carriers in determining their rates.

Types of Rates

Specific commodity rates, usually lower than standard rates, are available for certain commodities that move in large volumes, such as fish from Alaska or flowers from Colombia.

Exception rates, usually higher than standard rates, are higher for cargo that requires special handling, such as furniture not in crates or live animals.

Priority reserved rates, also higher than standard rates, allow a company to reserve space on a specific flight so they know exactly when the cargo will arrive at its destination.

Speed package rates are designed for small packages moving airport to airport. This is also a higher priced premium service.

Container rates apply to the container, which may or may not be adjusted depending on the commodity.

For cargo that is unusually lightweight, the rate is based on **dimensional weight.** For every 194 cubic inches, one pound of weight is assessed. Air carriers need to consider both weight and density in order to get the optimal load. In other words, if cargo is bulky but light, the plane would fill up but not earn as much revenue as cargo that is more compact and heavier. Therefore, the optimal cargo is typically at least 12 pounds per square foot. There is also the problem of irregular shapes, which means that space is being wasted. This is known as **stacking losses**.

Top 20 Air Freight Carriers
(Based on air line-haul traffic carried, 1993)

Carrier	Tons (000s)	Carrier	Tons (000s)
1. FedEx	2,796	11. Singapore	467
2. UPS Airlines	997	12. Delta	459
3. Lufthansa	794	13. Emery Worldwide	454
4. Japan Air Lines	719	14. British Airways	442
5. Korean	649	15. Burlington Air	430
6. American	604	16. Airborne	397
7. Air France	571	17. Cathay Pacific	395
8. United	536	18. All Nippon	390
9. Northwest	495	19. Air Canada	344
10. KLM	478	20. DHL	301

Source: MergeGlobal

AVIATION REGULATION

The aviation industry has been one of the most heavily regulated and controlled of any industry, mostly because of its military value. Managing international aviation requires regulation at the international level as well as the national level. No one nation could be expected to organize the regulations for the world, yet there is clearly a need for global agreement on many matters. That is why there are some organizations that play important roles in the airline industry, among which are the following:

International Air Transport Association. Represents the airlines. Its members are only carriers, including resellers such as NVOCCs. IATA used to be able to fix prices up until US deregulation. It then lost the power to do that for the US market, and has lost the power for so many other countries (due to the spread of deregulation) that it is no longer involved in price management.

International Civil Aeronautics Organization. The UN agency that deals with civil aviation issues. This is the agency that brings government representatives together to organize a common set of regulation. These mostly include technical standards, safety standards, and to a lesser degree business standards. The fact that all international aviation uses the English language, for example, was set by the ICAO.

As we have seen in Chapter Five, national regulations are still far more important because they can be enforced. Air transport regulations tend to be developed at the international level, for the sake of uniformity, but applied and enforced at the national level.

Every scheduled flight (passenger and cargo) that moves between two countries needs to be agreed upon in advance in a treaty between the two countries known as a **bilateral**. This is a treaty between two countries, not between any airlines (treaties are always between countries, not private parties). A country needs a bilateral agreement between it and every country that its airlines have any contact with. That means for a country like the US, there are many bilaterals. For a small country, there may be only a handful.

A bilateral is a very long and detailed agreement on what rights each country gives each other in regards to airline service. These rights are divided into what is known as the Five Freedoms:

1. **First freedom, Overflight**. The right to fly in a country's airspace. While this may seem minor, remember that Flight KA007 was shot down by the Soviet Union for flying into their airspace. Airspace is part of the nation's territory and is protected as such. A nation's airspace extends as far as the atmosphere, which is why outer space is considered open territory.
2. **Second freedom, Service stops**. The right to stop for the purposes of servicing the plane or its passengers. No revenue passengers may get off or on. Services include such things as refueling, repairs etc.
3. **Third freedom, Passenger deliver**. The right to bring revenue passengers to a foreign country.
4. **Fourth freedom, Repatriation**. The right to bring revenue passengers from a foreign country to the home country.
5. **Fifth freedom, Oncarriage**. The right to bring revenue passengers from one foreign country to another. Note this requires bilateral agreements with both of the foreign countries.

The above five freedoms are the basic elements of a bilateral, but there are also two other areas that are informally known as sixth and seventh freedom rights:

6. **Sixth freedom, cabotage**. The right to carry passengers within a country. This, just as in ocean transport, is rarely granted.

7. Seventh freedom, cargo. The above freedoms only apply to passengers. The seventh freedom includes all of the above arrangements but for cargo.

How does the bilaterals turn into actual air service? When two countries decide they should have air service between them (usually because airlines in one or both countries pressure their governments for it), they come together and agree on the treaty. The rights are usually reciprocal. Whatever rights are granted apply to both countries. The bilateral is usually VERY detailed. For example, between India and South Africa, the bilateral would not just say "10 flights a week, five for each national airline". Instead they would specify exactly which airport to which airport. Ten flights a week between Mumbai and Johannesburg. Once all the flights for that year have been used, they cannot add more flights without amendments to the treaty. If the Johannesburg Airport (remember, this is only an example) were closed due to a labor strike, they could not use another airport.

Once the bilateral is agreed and signed, each country turns to its own national airlines and offers them the new routes. Policies in each country vary, but most use the following procedure. The national airlines apply for any or all of the new routes, and the national airline authority decides which airline deserves it. The decision is usually based on a combination of fairness (distributing the routes among the airlines) and the airlines' ability to service the route. For example, imagine an Indian airline was purely domestic, and they wanted just this route to Johannesburg. Then there was another airline that already served other cities in southern Africa and wanted to add this service. The Indian national authority may decide that the latter airline was in a better position to service the route, or it may decide the former airline needs a chance to start up international service.

There tends to be a **mirror effect** when new routes are agreed upon. That route goes from having no service to two airlines at the same time. This means there is a big jump from no capacity to two airlines.

The above description of the five freedoms still exists, but there has been a dramatic change in the regulatory environment beginning in the US in the 1970s. At that time, airlines worldwide were notoriously inefficient and heavily regulated. Air travel was restricted to the wealthy and business travelers, and routes were limited. **Charter flights**, however, were not subject to rate controls. A charter is a one-time flight and not part of a routine schedule. It needed to be for an **affinity group**, a group of people with some significant connection such as members of a club. The Chicago Conference made no mention of charters. Yet because this was a legal loophole, charter operators were creating all kinds of ridiculous clubs in order to provide reduced-fare service.

In one case, the **Left Handed Club** offered flights for anyone in that club. The federal aviation authorities showed up just before the flight was going to take off, and checked to see who was left handed, and who was not. The scene was quite amusing except to those getting kicked off the plane for not being left-handed. Then there was the **Sandwich Wars**. If prices are fixed, the other way to attract customers was better service. One airline started offering better meals in flight. Other airlines complained, and resorted to offering better food themselves. They then complained to the authorities about their competitor's sandwiches. High level meetings were held to discuss the quality and standards of sandwiches.

The US then passed the **Airline Deregulation Act** of 1978 which began allowing airlines to operate whatever routes (domestically) they wanted, and charge whatever they wanted. The result was a large upset in the industry, new airlines coming into the market, and prices dropped dramatically. Between 1978 and 1998, fares in the US dropped 40 percent and flights increased by 50 percent, and the percentage of seats occupied on each flight is much higher.[22] Air travel became possible for the average person, planes that used to fly only a quarter full were now full. There were also some major problems. Airlines did not know how to compete in the new business environment. In one year the US airline industry lost more money than had been earned in the 40 years of its existence. Many airlines went bankrupt.

The most famous of these casualties was **Pan American Airways (Pan Am)**, once the flagship of America around the world. Prior to deregulation, Pan Am was the only airline given the right to service overseas routes. Yet they were not allowed any domestic routes. When one goes on an international flight, they rarely start at the international port, such as Los Angeles or New York. They mostly start with a domestic flight to one of those international hubs. Since Pan Am did not have domestic routes, after deregulation, people felt more comfortable with their domestic carrier that was now flying international routes. Thus Pan Am was not able to survive the changes.

After deregulating the domestic airline industry, the US government started working on international flights by pushing for **open skies agreements**, bilaterals that were very liberal to promote competition in international flights. Since this was a revolutionary idea, they started by pressuring a relatively small market, Bermuda. After that was agreed, they went on to other countries. Since then, open skies agreements have been spreading around the world. There are now still many airline markets that suffer from heavy regulation and uncompetitive environment. According to the Journal of Commerce, the worst cases for liberalization is China, which heavily controls its air industry and has the most dangerous airlines in the world, and Britain, that refuses to open landing slots at Heathrow (this will be discussed further in Chapter Nine).[23]

European airlines have been heavily regulated, and that region has seen considerable debate over competition. The US tried to sign bilaterals with some countries, but the European Union stepped in and said this was a European issue, and individual countries were not allowed to sign their own agreements. There was some fierce debate between the EU and some member nations, which still continues.

Asia has been remarkably liberal in their competition policies. There are few countries that have not embraced the open skies approach, notably China and North Korea. One reason for this difference between Asia and Europe is colonialism. European governments were active in the ownership and control of their airlines, so they had a vested interest in protecting them no matter how inefficient they became. Asian countries were under colonial rule and later did not invest in airlines as much. Therefore Asian airlines tended to be more independent. The terrain for Asia is better for air travel, while Europe has an easy alternative, trains.

Elsewhere in the world airlines have been deregulating at various speeds. **Latin America** has seen major changes in their business environment, and has been supportive of open skies. Brazil is the key market because it is by far the largest and most powerful of the Latin nations. For this reason their efforts at deregulation were seen as an omen for the other countries.[24]

Charter flights still exist in the deregulated environment, but they are not nearly as important. When there is no bilateral in effect, flights go based on vague 'continuous charter' arrangement. This happened when China began allowing all-cargo flights from the US, but the political system at that time was too sensitive to agree on a bilateral.

How does this deregulation affect air cargo? The rapid increase in the passenger traffic caused a general increase in the entire aviation industry, thus improving cargo services. New airports have been built, and obscure ones brought into heavy usage. This means that the number of routes, both passenger and cargo, have increased. However, as planes are now full of passengers, the luggage compartment is often full. Since cargo used to travel in the belly, there are times when there is not enough space.

Worldwide Air Freight Market
Largest Regional Air Trade Markets

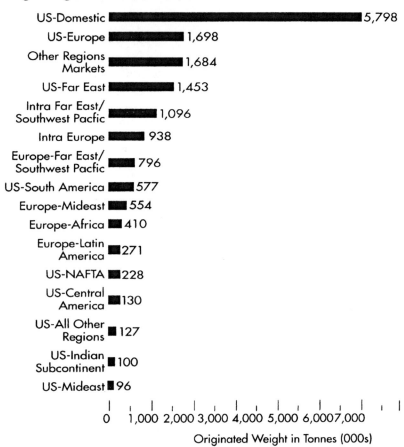

US International data include estimated low-value shipments
Source MergeGlobal primary research

The all-cargo airplanes were increasing gradually up until the introduction of the Boeing 747s, known as the widebody. This plane's introduction was a large step forward in the size of a plane, and it had an odd effect on cargo. An all-cargo 747 could carry greater amounts of cargo more efficiently. Yet the passenger 747s had so much capacity in their belly that it almost ruined the cargo industry by taking all of its business. Since then, the all-cargo industry slowly recovered. When FedEx introduced their service, one of the major issues facing it was that passenger service took substantial airport space. Deregulation has meant that cargo operations need to compete for airport space, and often need to find alternate airports.

In developing bilaterals, the US has been negotiating air cargo issues along with passenger services to gain leverage. Air cargo carriers have complained that they are being held hostage to benefit the passenger airlines. There are cases where the cargo industry would like to expand service, yet cannot do so until the overall bilateral is agreed.

There has been debate whether deregulation of the air cargo industry should happen in the WTO or some other organization. Normally the WTO has handled free trade issues in a wide variety of industries. Yet air industry leaders have suggested that other organizations are better suited. First priority goes to the ICAO. The predecessors to the WTO have dealt in so-called soft rights such as computer reservation systems, aircraft repair and maintenance, and selling of air transport services. But the most important parts were excluded, such as landing slots.

Dead Cargo

Their frequent flyer miles are not going to help them now, but that does not stop 120,000 dead people from getting on airplanes every year. That is how many times a cadaver is shipped air-cargo, according to funeral industry estimates. The increase in the number of people who die away from their homes reflects the general increase in mobility for people.

It costs about $225-500 to ship a body within the US, which makes it very lucrative cargo. Some airlines even offer free tickets to the funeral homes to attract their business. No, the bereaved do not get the tickets. If the body is cremated, the US Postal Service may be used, at a cost of around $40.[25]

MANAGING AIR CARRIERS

We now look at air cargo from the shipper's perspective. How do the changes in the industry affect shippers who want to use air cargo services? This has been a rapidly growing industry for a few reasons:

- **Increased demand for premium service**- there is an increasing number of people in the world who can afford the benefits of premium service and want the luxury goods that can best be provided by air cargo. For example, cut flowers from Holland and fresh fruit from Latin America, delivered daily to the US, require a large market that can afford these goods. This trend is part of an overall increase in international trade.

- **Increased service**- there has been an overall increase in the aviation industry, including many new routes and increased services on routes.
- **More competitive service**- the business environment is less regulated and more competitive, which translates into reducing costs.
- **Carrier integration**- air cargo carriers, rather than operating independently as airlines used to do, are now part of integrated carriers that increase the value of air cargo services. Instead of just providing air carriage, these companies can provide door to door service.

Carriers are becoming much more sophisticated and the range of services has been increasing. The variety of services range from next-flight-out to next day delivery to the cheapest and slowest service available. In Hong Kong, one air express company is now using convenience markets to drop off urgent packages where they have pickup three times daily. Door to door delivery is now commonplace. The territory served is increasing worldwide so places that used to be considered 'remote' now enjoy service close to that of the largest cities in the world.

Shippers make a choice on how to transport cargo, so it is helpful to see how air cargo compares to other modes of transport. Competition among the different modes of transport has caused them to change their operating methods. One reason ships are more schedule conscious is because of competition from air cargo service. Still, this competition between air and sea is not very strong for the simple reason that there is a very large difference in the cost and service provided by the two. In other words, small differences in air cargo rates are not going to have much of an impact on ocean cargo that is only a fraction of the cost. Likewise the service provided by ocean carriers could not come close to the speed of air cargo.

There is more competition between air service and trucking service over short distances. As already mentioned, air cargo is sometimes trucked if the distance is not too great, and there are no natural barriers. This is common in the US and Europe. Trains are rarely a viable alternative because their service is too slow and inflexible. Air cargo is seeing some competition from non-transportation industries, particularly the fax and Internet. These new technologies are the main threat to air express, mail and courier services.

There has been a trend in the passenger airlines toward alliances, the biggest being the Star Alliance. These alliances improve service by increasing the number of destinations while dealing with only one carrier. Yet air cargo does not seem to have benefited from these alliances, which are based on the passenger market. Computer systems are far less compatible on the cargo side of airlines than on the passenger side. Shippers have not indicated much interest in whether a carrier is an alliance member or not. Part of this may be that cargo moves mostly point to point, so the interline agreements that passengers enjoy are of no value to the cargo shipper. This is one reason why all-cargo carriers have been gaining over the passenger/cargo carriers.

Air cargo carriers can save money in the planes used, and the fleet selected. Older planes can be acquired more cheaply, but use up to 40% more fuel than new models. Having a standardized fleet makes maintenance and personnel cheaper. Cargo planes may either be retired passenger planes, or custom-designed cargo planes.

How many airlines are required to service a given market? The markets are almost always divided by national boundaries rather than economic boundaries. Analysts ask how many airlines a given market can support. Recall that cabotage is almost never granted, which means that even the smallest nations tend to have their own airlines. That means that for international travel, routes face that 'double or nothing' problem. When a nation establishes a bilateral treaty with another nation, each route that is previously not serviced is now served by two airlines. It is very difficult to arrange an agreement in which only one airline gets a given route.

Armored Air Cargo

Armored transport companies are now expanding into air cargo, including Brinks and Dunbar. This is an example of the specialized air cargo services being provided in a rapidly expanding market. Brinks does not contract out any of its operations, and keeps control over the cargo for the entire trip. Two major solutions are offered to shippers of gold, diamonds, financial instruments, cash etc: absolute security and hassle-free insurance. They will pay for the entire value of the cargo, which is not common in the shipping industry. That means there is a strong incentive not to lose any cargo in the first place. The 'high val' market is expanding, and new commodities such as computer chips and pharmaceuticals are also being added to the list.[26]

What may be the largest movement of high valued goods in history took place in the months before January 1, 2000. In anticipation of Y2K hysteria, the US Federal Reserve shipped out $267 billion in cash to the 37 field offices, enough to allow every person in the country to withdraw $1,000. This was in case people decide to withdraw lots of cash for fear of Y2K computer problems with their bank.[27]

By Eric Bernhardt

The County of Sacramento, California has long prepared for the challenges and opportunities associated with growth in aviation activity and the need for economic generation. The County owns and operates Sacramento International Airport, Mather Airport, Executive Airport, and Franklin Field. Sacramento International provides scheduled passenger service for the entire Sacramento region. Executive and Franklin accommodate smaller general aviation aircraft. However, the role of Mather as a dedicated air cargo facility has been the subject of extensive debate since the Air Force Base was officially re-opened by Sacramento County as a civilian airport in May 1995. Such debate and controversy is typical at air cargo airports located in urban environments.

Background

Mather Airport is located approximately 12 miles east of downtown Sacramento in unincorporated Sacramento County. Mather served as a U.S. Air Force Base from 1918 until the decision to close the Base was announced by the Department of Defense in 1988. In 1993, facilities were transitioned to Sacramento County; and on May 5, 1995, Mather was officially re-opened as a civilian airport.

Until 1995, all cargo in the Sacramento region was accommodated at International Airport. To preserve capacity for passenger operations at International, Sacramento County decided that Mather would serve as a dedicated air cargo airport. To facilitate this role, the County set out to (1) construct airfield improvements consistent with air cargo needs, (2) improve and develop facilities to attract a wide range of tenants, and (3) attract quality aviation and non-aviation industries and commercial tenants. Since most of the facilities were constructed in the 1950's to military criteria, the County implemented numerous improvements to enhance safety and satisfy Federal Aviation Administration (FAA) design standards. Improvements included installation of a new instrument landing system for all weather operations, new runway lighting, and a new airport traffic control tower (ATCT).

Following Mather's re-opening, the majority of all-cargo carriers operating at International relocated to Mather because of (1) limited space and opportunities for expansion at International, (2) the need to develop new and independent air cargo transfer and sort facilities, and (3) Mather's location relative to growing commercial/industrial markets. By 2000, Mather was accommodating approximately 150 million pounds of airfreight. In addition to air cargo, Mather was also accommodating some of the region's general aviation demand, including corporate aviation, government uses (such as for the U.S. Forest Service), and flight training.

With the development of Mather as a cargo airport, there was a net increase in regional cargo activity due, in no small part, to the efforts of the County to attract air cargo service to Sacramento from the San Francisco Bay area. However, community acceptance issues, primarily related to further airport development and the environmental impact of air cargo operations on the surrounding communities, became significant.

Issues

From the community's perspective, aircraft noise and over-flights associated with the ultimate disposition of Mather's runways was of particular concern.

Mather has two parallel northeast-southwest runways. The primary runway is 11,300 feet long, which is sufficient to accommodate the existing and future mix of air cargo aircraft, including the Airbus A-300, Boeing B-727, B-757, B-767, B-747, and McDonald Douglas MD-11. The extensive length of the primary runway allows users to reach most national and some international markets at maximum takeoff weight, which is an essential financial consideration for air cargo airlines given the use of large, heavy aircraft. The secondary runway is only 3,700 feet long, which is sufficient to accommodate smaller general aviation type aircraft, but not long enough for use by the larger types of aircraft used by air cargo airlines. In effect, Mather was a one-runway airport for air cargo operations.

By 2000, many of Mather's air cargo airlines were expressing the need for a backup runway to be used during periods when the primary runway is not available due to routine maintenance, inclement weather, reconstruction or rehabilitation projects, or aircraft accidents/incidents. There was logic to support this argument. Backup runways are common at airports served by air cargo airlines. For instance, 48 of the top 50 air cargo airports in the United States have two or more runways, with 15 having four or more runways. In addition, the air cargo airlines pointed out that when the primary runway was closed at Mather temporary relocations to other airports required (1) the transfer of employees and equipment, (2) increased operating and labor costs, (3) lower levels of service as a result of disrupted time schedules, (4) and potential customer refund costs if delivery times are not met. From the County's perspective, the provision of a backup runway would allow the County to market Mather as a potential West Coast air cargo hub; and enable Mather to compete with airports outside of the Sacramento area, such as San Francisco, Oakland, Stockton, Fresno, and Reno.

Nevertheless, providing a backup runway at Mather was extremely challenging because of the considerable off-airport development constraints. Mather is surrounded by a mix of residential and open land uses. Based on FAA established criteria for acceptable noise exposure levels, changing aircraft flight patterns, upgrading Mather's secondary runway, or constructing a new runway would cause significant noise impacts to existing residential areas and/or require changes to planned land uses—neither alternative was acceptable to Sacramento County.

Benefit/Cost Analysis

Given the potential constraints, Mather's air cargo carriers began to express a desire to relocate to other airports outside the Sacramento area that possess two or more air carrier-capable runways, such as Oakland, Stockton, Fresno, or Reno. In response to these issues, Sacramento County initiated a study to determine whether a backup runway should be constructed at Mather to allow the Airport to continue to serve as the region's primary air cargo airport. It was assumed that if a backup runway were not provided, a small percentage of air cargo airlines would relocate to airports outside the Sacramento region. The relocation of air cargo airlines would result in a loss, or "leakage", of air cargo-generated economic benefits provided to the Sacramento area.

Air cargo is a significant generator of economic activity in the Sacramento area. Economic impacts are measured as follows:
- Direct impacts—Jobs, sales revenue resulting from air cargo operations.
- Indirect impacts—"Spin-off" effects of air cargo operations, such as business-to-business activity and expenditures.
- Induced impacts—Economic activity of employees spending wages.

To quantify the economic impact of air cargo operations, constrained and unconstrained forecasts were prepared to estimate the aviation activity and resulting air cargo tonnage that could be accommodated in the future with and without a backup runway at Mather. Economic impact studies verified that the potential "leakage" of air cargo to other airports equated to approximately 200 direct jobs, $30 million in direct revenue, and $20 million in aggregate income in a 20-year period.

However, this was only half the equation. Sacramento County also assessed the financial costs and environmental impacts of various backup runway development alternatives at Mather. Development alternatives ranged from converting an existing taxiway into a limited-use runway, to extending and strengthening the secondary runway to air carrier standards, to constructing a new full-length parallel backup runway. Extensive technical analyses were conducted to quantify the (1) noise impacts on existing and planned residences; (2) effect on natural environmental resources such as wetlands; (3) effect on air traffic control and aircraft operations; and (4) economic impacts resulting from the loss of developable land due to FAA-required safety and runway protection criteria.

Results

General order-of-magnitude development and environmental mitigation costs ranged from $50 million to $100 million, depending on the degree to which a backup runway would accommodate user needs. However, potential noise impacts on planned residential areas and loss of developable land due to runway protection areas were significant for those alternatives that considered extending the existing secondary runway. In one of these alternatives, approximately 400 acres of land planned for residential development would be limited to light industrial uses given the altitude of landing air cargo aircraft and resulting noise exposure levels. Such actions would preclude construction of 1,200 planned residential units in a fast growing community in need of residential development to facilitate recent growth in high-tech, and other professional industries.

Such factors were given careful consideration by the Sacramento County Board of Supervisors—who ultimately had to decide upon a course of action for Mather. On the one hand, continued development of Mather as an air cargo airport would provide economic benefits and jobs for the region. On the other hand, runway development alternatives were either extremely costly, or resulted in adverse environmental and socioeconomic impacts.

In the end, the potential positive economic impacts provided to the region by accommodating air cargo at Mather were given considerable weight. In October 2001, the Sacramento County Board of Supervisors decided to continue developing Mather as an air cargo airport, but within the limits established by pre-approved land use decisions/policies. In other words, if a backup runway were to be provided, it had to be constructed in a location and to a length that did not affect land uses established to benefit the growing communities surrounding the Airport. Since locating, designing, and funding such a runway would require substantial resources, as well as detailed analysis, planning, and stakeholder coordination, the Board of Supervisors also decided to prepare a detailed master plan study to consider the balance of interests and analyze the options and consequences in greater detail.

[1] "Global Voice of Air Logistics: Manifesto", The International Air Cargo Association (IACA), p. 13.

[2] "World Air Cargo Traffic Flow Models", The Colorography Group, 1999.

[3] IACA.

[4] IACA, Ibid, p. 14.

[5] "Road feeders exploit opportunity", Chris Gillis and Brian Reyes, American Shipper, September 1998, p. 28.

[6] "Towards the Wild Blue Yonder", The Economist, April 27, 2002, p. 67.

[7] JOC, June 16, 1999.

[8] "Towards the Wild Blue Yonder", The Economist, April 27, 2002, p. 67.

[9] "Air Transportation", Alexander T. Wells, 3rd ed, Wadsworth, Belmont, CA, 1994, p. 357.

[10] Wells, Ibid, p. 363.

[11] "Rates vs Service", Chris Gillis, American Shipper, September 1998, p. 16.

[12] JOC, September 15, 1999, p. 21.

[13] Wells, Ibid,, p. 364.

[14] "The Politics of International Air Transport", Betsy Gidwitz, Lexington, MA: D.C. Heath, 1980, pp. 161-167.

[15] O'Reilly, Ibid, p. 110.

[16] "NIT League looks skyward', American Shipper, March 1998, p. 20.

[17] See ICAO's International Standards and Recommended Practices titled "Safeguarding International Civil Aviation Against Acts of Unlawful Interference."

[18] JOC, September 14, 1999, p. 6.

[19] This is only offered as an example of the issues that arise with alliances, not as a definitive judgment on this particular agreement.

[20] "Wells, Ibid, p. 364.

[21] Wells, Ibid, pp. 376-7.

[22] "Fear of Monopolies in the Sky", San Francisco Chronicle, July 5, 1998.

[23] JOC, March 10, 1999, p. 14a.

[24] "As Free as a Bird", The Economist, May 9, 1998, p. 64.

[25] JOC, September 1, 1999, p. 20.

[26] JOC, June 8, 1999, p. 9.

[27] JOC, June 8, 1999, p. 12.

CHAPTER 9
PORTS AND FACILITIES

This chapter discusses the many functions of the port. As we see from the example of Rotterdam, the functions are often not those that we traditionally associate with ports. No longer are ports the dirty and dangerous place for pirates. They are an integral part of the global business network. There are in this chapter sections on ocean and air ports, as well as materials management. Some aspects of port management are the same for different modes of transport, but other aspects are quite different. In the final section we introduce free trade zones, which are closely associated with ports and play an important role in logistics.

PORT AND FACILITY FUNCTIONS

Ports play a vital role in logistics as the intersections of transport. They play a wide variety of important roles, operating at multiple dimensions. The largest ports are like cities unto themselves, with as much complexity. Not only do they serve the practical role of handing cargo and passengers, they are the link between a country and the rest of the world. The first view we all have of a foreign country is upon our arrival at a port.

What is a port? This can be a little hard to see at first because they look very different when one looks at the various modes of transport. A port may be defined at the **intersection of different modes of transport**. The fact that there are more than one mode of transportation is to distinguish a port from facilities and those places where cargo is relayed between vehicles of the same mode. For example, a truck arrives at a warehouse and the cargo is put on another truck to continue the journey. This would not be considered a port.

A port is the intersection of different modes of transport.

A **facility** is a general term used for the fixed locations where logistics activities are carried out, particularly manufacturing locations and warehouses. There is a gray area between what is a port and what is a facility. Non-port facilities are discussed here because ports and facilities play some very similar roles, though there are some differences.

There are several important **reasons for distinguishing between ports and facilities. First**, at a port there are parties other than the shipper or consignee handling the cargo. We do not mean just moving the cargo, but loading, offloading and other activities. This creates a risk which is not present when the shipper hands the cargo to a carrier, and that same carrier takes it directly to the consignee. **Second**, different modes of transportation come together at a port. If there was one important message from intermodalism, it is that each mode of transportation has unique characteristics, and making them work together is no easy task. The port is exactly where they must be made to work. **Third**, as an intersection, there are different directions in which the cargo can go, and that means the risk of misrouting. The important aspects of a port include:

- Location in relation to markets
- Location in relation to its competitors
- Inland connections
- Infrastructure and technology
- Accessibility to the trade lane
- Management

When we refer to the location in relation to its market, that includes the cargo's origin and destination, and to a much more limited extent, the shipper's and consignee's home office. In other words, a shipment would go through those ports that are along the ideal route. Ports need to be positions along trade lanes where they are needed. What is not so important is that the shippers or consignees be located anywhere near these ports. The organizations directing the shipment are not commonly located anywhere near these ports, so the movement is being handled by third parties. This has some interesting consequences.

This chapter discusses port management, but there are two levels of organization. There is the Port Authority that manages the port, and the carriers that operate within this port. The carriers are the customers of the Port Authority, while shippers are the customers of the carriers. Each has their own priorities. The Port Authority wants to increase overall trade volume. The carriers are competing with each other. Shippers want the best service from the carrier and have no direct connection with the port operators. A study that asked shippers what they look for in a port showed that the most important factors are equipment availability, low loss/damage, and convenient pickup times.[1] Some of what shippers want is under the control of the carriers, and some under the control of the port.

Ports are often natural monopolies when there are no viable competitors. In other cases there is active **competition**. That is what we are referring to by 'location in relation to its competitors'. As with all monopolies, they can be inefficient and corrupt, which is why there is are ongoing efforts to introduce competition. In Chapter Seventeen there is a discussion of privatization and how to best structure competition in a natural monopoly. At this point it needs only be mentioned that ports do in fact compete both within and with other ports.

The primary way to introduce competition is to disaggregate the port into those parts that need for economic reasons only one provider, and those parts that may be privatized and competitive. Within a port, the different terminals may be under different management. The largest geographic ports may have different ports within the region. They can then specialize or compete based on price. Ports that specialize may concentrate on passengers, break-bulk cargo, intermodal, and so on. Or they may offer the same services as nearby ports but compete on quality of service or price.

This implies modern ports are highly sophisticated corporations offering a wide array of services to many customers. As we just noted, ports operate on many dimensions. Shippers and carriers are dropping off or picking up cargo, which is the most basic service. Yet there is extensive cargo handling, packing and repacking, light manufacturing, and so on, within a port. There are also commercial activities that may have little to do with transportation, such as office space on port lands.

Rotterdam

What does it take to make an excellent port? Perhaps the best people to ask are the people that have made the Port of Rotterdam the largest port in the world. It is not just large; many consider it one of the best managed and foresighted port organizations in the world. The port is not actually on the ocean, but runs along a river thirty miles inland. It offers feedership service to 110 European ports and inland shuttle trains that can offer nonstop delivery to most cities on the continent.

Yet this is not what distinguishes the Port of Rotterdam. What sets them apart from the rest is the way they have strategically developed their port not just for the traditional tasks of loading and unloading ships, but to meet the challenges of global production systems. The management talked to industry to ask what they needed from a port. Instead of just being a stop along the route, the port plays important roles in the overall production and distribution systems of many companies.[2]

Ports have many options to keep them competitive. The German ports at Bremerhaven and Hamburg decided to coordinate operations using the same train heading into the European interior.[3] The Port of Corpus Christi, Texas, is trying to reduce their reliance on petroleum. They are adding refrigeration facilities, a rail link to Mexico, and a cruise ship berth.[4] The Port of Mumbai (Bombay) used to be India's main port, but it has steadily lost cargo to the newer, more modern Jawaharlal Nehru Port, only a short distance south. Mumbai suffers from a variety of problems, such as inferior equipment, shallower draft, and smaller berths. However, the most important issue is labor. Mumbai has a massive staff of 25,000, a very strong union, and a reputation for poor service.[5]

Ports now have a new source of competition, **non-port industry and tourists**. Non-port industries generate demand for limited space. If these other industries can pay more, they may take over port lands and reduce the port's competitiveness. While this may seem appropriate for a competitive economy, the port services not just that area but the larger region. This will be discussed later in this chapter. The waterfront of many cities is considered a good location for museums, tourist attractions and other forms of development that is forcing out port users.

Cargo does not end its journey at the port. That is why the definition of a port does not include the origin or destination, because these areas are not acting as an intersection. This means that the port needs to consider where the cargo is going on to. International trade makes up a global network, so a port's location is relevant for how it fits in relative to ports around the region and around the world. For example, cargo from Asia to the US has a choice of Seattle, Oakland or Los Angeles. These ports look at how much of their cargo is destined for that city, and how much is continuing inland. This determines how they should arrange their infrastructure. For Seattle, 80% of its cargo goes inland, 20% for Oakland and 50% for Los Angeles.

Port infrastructure, unlike location, is something that a port can change. Ports can invest in equipment or it can neglect it, which some do. There are some things that are absolutely necessary for it to qualify as a port of a given type. Container ports need the gantry cranes, airports need air traffic control equipment, and so on. However, there are many things that can be done to improve the infrastructure. The most common areas are the facilities located at or near the port, such as warehouses or processing facilities. The technology driving port operations varies widely and has become a distinguishing factor for some ports.

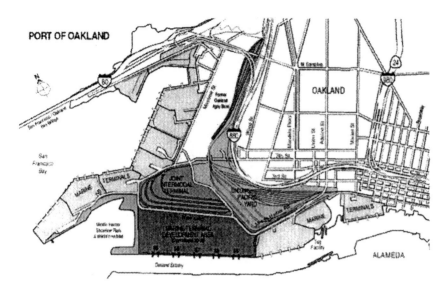

PORT OF OAKLAND

Source: Port of Oakland

Inland connections refer to the way the ports can connect the different modes of transportation. Some ports are designed so that the cargo flows through with minimal handling or delay. In other cases, there are connections that take time. For example, an ocean port would ideally have rail tracks leading to the side of the ship, so cargo can be taken off the ship and placed directly on the train. Often, crowded ports mean that the cargo is taken off, trucked to a storage location, and later taken to a rail yard that is away from the port area. Many ports do not have any connection with other modes. There may be no rail connection in the area.

Access to the trade lane is the opposite of inland connections. The shortest distance between two points is a straight line. In finding the ideal shipment route, one first looks to the shortest route to the destination, and will probably use the ports that are along the route. Some ports are not in line with the trade lane. For example, ports are needed in the Middle East for cargo going between Asia and Europe. Ideally there would be one right along the route that goes from the Southern tip of India, through the Suez Canal. However, the major port in this region is Dubai. Other ports that are more in line with the trade lane, like Yemen, suffer from management, political or equipment problems. In other cases, ocean ports are located up a river far from the open ocean.

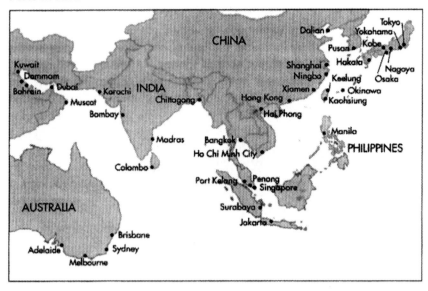

Taking a more local view, there is a tradeoff between the need to be near the market, and the need for land. Ports take up a lot of space. Large metropolitan areas or industrial areas have higher land values. This means that a port needs to find a balance between locating close to the market, but not so close that it cannot afford the land. Many ports are experiencing competition for land with other users that can afford to pay more. In many cities, the industrial waterfront is being redeveloped into housing, retail and office space. Not only does the port find itself squeezed out, but the business community as a whole finds that their ability to trade is being hurt.

Port management can make the difference between success or failure of a port. A good location and modern equipment cannot make up for poor management. The vast majority of ports are run by government entities, usually by a commission assigned by a local or regional government. These are called **port authorities**. Ports benefit not just the immediate area, but the entire region and often an entire country or more. Therefore the mission of the port is not necessarily to make a profit, but to provide its services as efficiently as possible to promote trade. In other words, they should not necessarily be trying to maximize their own profit.

Ports in Western Europe

As with all public organizations, ports are prone to bureaucracy and inefficiency. Some ports are famous for employing hundreds, even thousands, of people more than an efficient port should need because they are seen as a way to create jobs. Japan's ocean ports are managed by a labor/management organization noted for its secrecy and has total control over port regulations. Foreign carriers are not permitted to operate their own terminals.

The primary way to increase efficiency is the **structure of competition**. There are four types of port authorities.[6] The **landlord port** is one in which the port owns and manages infrastructure, and private parties manage everything else, such as cranes and vehicles. This is the most common model. With the **tool port**, the port also owns the superstructures such as cranes and facilities, but private parties rent assets through concessions or licenses. Examples include Antwerp (Belgium) and Seattle (US). The **services port** is where the port has complete ownership and management. The Port of Singapore used to be the prime example, but they are moving toward private participation. Finally, there are a few **privately owned ports**, but these are almost all very small ports.

New Zealand privatized their ports in all modes in 1984, and it has been reported that they are now as competitive as the best run private businesses. Railways are handling twice as much cargo as 10 years ago, and Auckland's ocean port is handling six times as much cargo. The number of dockworkers fell from 3,300 to 1,800 almost immediately, and cargo charges were reduced by 50%.[7]

Example of an Airport Commission Organization Chart

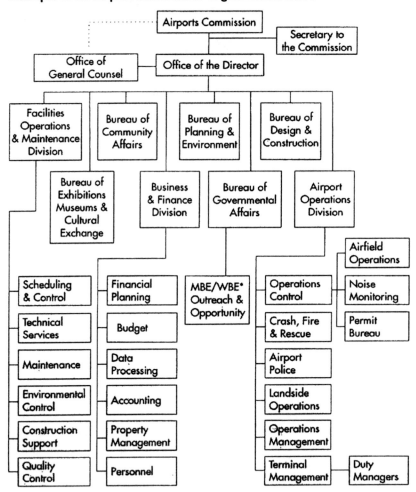

*Minority Business Enterprise/Women Business Enterprise

Privatization has created a new type of corporation, exemplified by the Port of Singapore. The port authorities offer themselves as management agencies elsewhere. The Port of Singapore Authority (PSA) is famous for managing other ports worldwide, including ports in Beirut and India. While PSA is a public entity, there are a growing number of private corporations offering similar services.

Comparative Landing Fee Rates
Major U.S. Airports As of July 1, 1994

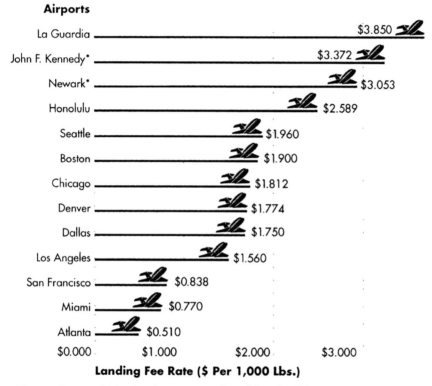

Airports

La Guardia — $3.850
John F. Kennedy* — $3.372
Newark* — $3.053
Honolulu — $2.589
Seattle — $1.960
Boston — $1.900
Chicago — $1.812
Denver — $1.774
Dallas — $1.750
Los Angeles — $1.560
San Francisco — $0.838
Miami — $0.770
Atlanta — $0.510

$0.000 $1.000 $2.000 $3.000

Landing Fee Rate ($ Per 1,000 Lbs.)

*Amount shown is the landing fee rate equivalent of the takeoff fee rate. (Takeoff rate for JFK is $2 65, Takeoff rate for Newark is $2.40)
Source San Francisco International Airport

Port privatization is not popular with everyone, and particularly the labor force. One of the most common changes made by privatized ports is the reduction in labor force and increased work standards. **Labor issues** are one of the main issues for ports. Unlike the merchant marine, port workers are usually unionized, and these tend to be very strong unions. One reason is that they are a specialized organization with no close competitors nearby. In other words, if a port union went on strike, the port authority would have a hard time hiring other dockworkers given that ports are often hundreds of kilometers apart. Another reason for the solidarity is that ports, like the marine industry, are very old and well established. This union solidarity is not quite as common with airports, which are much newer. In the US there has been turmoil in the industry since deregulation, which also led to an increase in non-union workers.

Airports and ocean ports are sometimes combined under one port authority. This is not so unusual even though there is little if any cargo going between these two. In other words, it is rare for ocean cargo to be transferred to air cargo, though it does happen. There are still benefits from having one port authority. Both airports and ocean ports connect with trucks. Both use the same kind of facilities like warehouses. Both face the same mission of supporting trade and not profit.

Ports **earn their money** from a variety of ways, but the most common source is user fees. An airport receives landing fees from every plan that lands. Ocean ports receive money from every ship that comes to its dock. There are also a wide variety of other user fees. Virtually everything that happens at a port is charged in some way. The ocean port rents out space to the carriers. Airports rent out space in the terminal to concessions. Even taxi drivers who want to pick up a passenger pays for the privilege.

The fact that ports are officially run for the benefit of the users raises some issues, such as the fact that they can earn a profit as we generally understand it. If the mission of the port is to operate as efficiently as possible and not earn a profit (or as they would phrase it, create a **budgetary excess**), what if there are profits? The extra money is usually reinvested into the port, or at least reduces fees. The Port of Oakland, for example, can remit up to 10% of their budgetary excess to the city, but there have been attempts by the city to take more than this amount.

In the process of charging users, ports often get stuck in the middle of international politics. There is a international legal norm that says the federal authorities have the final say on anything that regulated international relations, including trade. That means that taxes and other charges on trade are subject to the final authority of the federal government. Ports need to charge someone, and that means the users, many of whom are foreign entities. In other words, the port tax is actually a tax on foreigners.

To make the situation more complex, local companies often complain that port charges are preventing them from being competitive on international markets, a charge to which politicians are very sensitive.

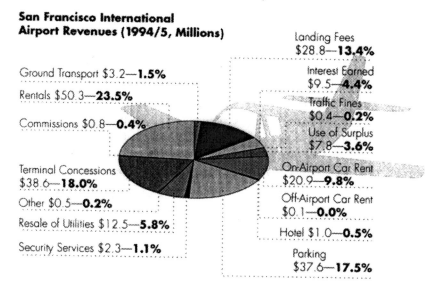

San Francisco International Airport Revenues (1994/5, Millions)

Ground Transport $3.2—**1.5%**

Rentals $50.3—**23.5%**

Commissions $0.8—**0.4%**

Terminal Concessions $38.6—**18.0%**

Other $0.5—**0.2%**

Resale of Utilities $12.5—**5.8%**

Security Services $2.3—**1.1%**

Landing Fees $28.8—**13.4%**

Interest Earned $9.5—**4.4%**

Traffic Fines $0.4—**0.2%**

Use of Surplus $7.8—**3.6%**

On-Airport Car Rent $20.9—**9.8%**

Off-Airport Car Rent $0.1—**0.0%**

Hotel $1.0—**0.5%**

Parking $37.6—**17.5%**

For these reasons, port charges and harbor taxes are an ongoing political debate. Anytime a new tax is proposed, port users often complain to the federal authorities that the port is interfering with federal jurisdiction or hindering trade. Ports try to make a clear distinction between a **port tax** and **user fees**. Taxes can only be charged by governments, and usually only to their own citizens or companies. What is charged to international carriers or shippers should only be a user fee. In Europe the situation is more complex because there are the national authorities and the European Union. Charges on imports and not exports are commonly challenged, because that is a clear bias against foreign companies. In other cases the discrimination, if any, is less clear. A harbor maintenance tax proposal was contested because it included channel maintenance. The harbor was used by international ships, but the channel, it was alleged, was only for domestic shipments.[8]

Ports have enjoyed anti-trust immunity in some cases, much like the ocean shipping industry. This means they can meet with other ports to discuss prices and other matters. Among the justifications for port anti-trust immunity are:[9]

• Ports are costly.

- Ports provide a public service.
- Ports are capital intensive.
- Ports need countervailing power against shippers.
- Ports cannot be moved.
- Ports need to exchange ideas.

The above justifications were listed by a US commission that was also reviewing ocean shipping anti-trust immunity. Since the ocean shipping has lost most of its immunity, it remains to be seen if the same will happen to ports. A key difference for ports, however, is that they provide a public service.

Information systems have become one of the distinguishing factors of world class ports. There is a limit to how good the other factors (location, inland connections, infrastructure etc) can be. The information provided by the port to its users, especially combined with good management, is one area that can dramatically enhance the value offered by a port. For example, users want to know where the cargo is and what is happening to it, if anything.

Technology used at ports is designed to manage information for the users, and also to optimize operations. It is most important for the port to optimize the use of limited space. One way to do this is to know exactly where a container can be located to reduce searching for containers. Another way that information can help is by forecasting when a container needs to be moved in relation to other containers next to it.

Data entry is an area that benefits from technology. Every piece of cargo that enters the port needs to be input in some manner. One system currently in use photographs containers as they enter or exit the front gates, and reads the numbers and letters in the picture to get the serial number of the container and chassis. If the serial number is illegible, an alarm goes off and a worker must write down the serial number by hand, but users have reported that this is not a common problem.[10]

The challenge for the port is to tie together the many carriers and shippers, all of whom have their own information systems with different technology and standards. Imagine a ship arrives in port, with thousands of containers being offloaded. Some are going onto any one of a few trains. Some are being trucked away. Others are to be put in storage while others are immediately picked up. Still others are going onto another ship. Data on all of this is coming and going between all the carriers and all the shippers.

Port development comes from either increasing the capacity of an existing port or building entirely new ports. New port development, known as greenfields (this term may also apply to any development on land that has not been previously developed), is not common, but it does happen, notably in China. Trade lanes are well established and there are ports pretty much anywhere in the world one would need one. Port development is popular because the increased trade offers a lot of benefits for the entire country. However, there is a tendency for countries to develop their ports in competition with each other, resulting in overcapacity.

Port consolidation is one way that unproductive competition is controlled. Houston wants to merge with nearby Texas City. San Francisco has talked of merging with Oakland. The Port of New York and New Jersey is one of the largest ports authorities and have a complex agreement. There has been conflict because New York (the city, not the port) believes that most of the money is going to New Jersey. Jersey sees the port more as an opportunity to create jobs.[11]

International border crossings are not ports, but they are often closely associated with them and have many of the facilities we have been discussing. Many countries have legal restrictions on foreign carriers (particularly trucks and trains) operating in their countries. This means that a carrier must pass their cargo onto another at the border.

Another way that border crossings resemble a port is that they decide how the shipment is routed. A ship arriving at a shore must pick a port. Likewise a truck entering a country must pick a border crossing, which is often very limited. The port of entry may be further limited for certain cargo. For example, some ports of entry are designated for hazardous cargo, for live animals and so on. The idea is that some commodities require special attention. Hazardous cargo may need to be inspected by a specialist in that area. Live animals may need to be inspected by a veterinarian. Some free trade arrangements require that cargo moving under that agreement move through designated ports. For example, there are agreements between the US and Mexico, separate from NAFTA, for some specific commodities, and these can only go through certain ports of entry.

Those ports that are also ports of entry need to work with customs and regulatory controls. These agencies usually operate in the port area. An efficient port will also consider the regulatory process to insure a smooth flow. The Port of Dalian (China), operated by a Singapore company, has been trying to reduce turnaround time. They now conduct immigration checks on officers and crew at the front gates instead of at the wharves, so that when the ship arrives, they can go straight to work.[12]

The Major Canals

There are a few cases in which trade lanes have been created were none existed before, thanks to some remarkable feats of engineering. There are many canals in the world, notably in the European waterways, but the two discussed here are notable because they were far larger than anything else ever built. There has been discussion of canals across Nicaragua, and the Kra Canal across Thailand. Neither of these appear viable, even though the Kra project is still being actively researched.

The Suez Canal

The Suez Canal was begun in 1858 with a French engineering company, owned by French and Egyptian interests, to revert to Egyptian government after 99 years. By the Convention of Constantinople in 1888, the canal was to be an international passage available to ships of all nations, in war or peace. In reality, politics have followed the canal. It required 10 years of construction and 1.5 million workers, of whom 125,000 died in the process. It finally opened 1869. In 1963 President Nasser declared it property of the Egyptian government.

The Canal revolutionized global shipping. Instead of 12,400 miles (19,950 kilometers) from London to Mumbai, the trip became 7,250 miles (11,670 kilometer). Distance between Tokyo and Rotterdam shortened by 23%. For shipments from Asia to US East Coast, Suez is faster than crossing the Pacific and then rail across the US. Singapore to New York, for example, is 22 days via the Suez, but 36 days via US West Coast. However, modern ships are too deep for the canal, and an increasing number pipelines delivering oil directly to the Mediterranean.

Each year 25,000 ships annually passed through, Egypt's third largest source of revenue at $2 billion. It is the longest canal in the world without locks, which means it can be widened or deepened anytime with little problem. It is 101 miles (163 KM) long, about twice the length of the Panama Canal. Part of it is a channel going through lakes, while in other stretches a canal was dug. It is only one lane with occasional passing bays. Ships move in convoys at fixed speeds, each day two convoys south to north (mostly oil) and one north to south (mostly manufactured goods).

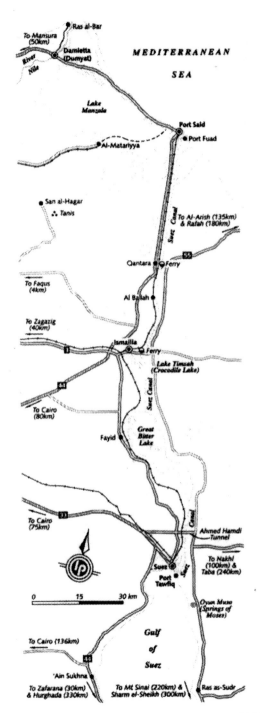

The Panama Canal

Despite the increasing number of post-Panamax ships, the Panama Canal still accommodates four out of five of the world's merchant ships. It is 80 km (51 miles) long, and the locks are 300 meters long and 30 meters wide. There is also the bottleneck at the Gaillard Cut, which required one of the largest excavations in engineering history.

The French originally planned and began construction on the Canal, but it was finished only after the US engineered a political coup resulting in the independence of Panama from Columbia. The US ran it as a non-profit utility, while the Panamanian government will operate it as a semi-independent commercial business.

Charges are based on ship size, and the largest ships are charged about $110,000 for a single passage. Annual revenue from the canal is around $605 million (1999). Rates are determined by complex economic models that estimate the alternatives that shippers have, and then charge as much as possible without diverting traffic while running the canal at capacity. For example, one analysis estimated that optimal pricing would charge $2 a ton for corn ships, $1 a ton for soybean ships. The current rate is around $1.50 a ton regardless of the commodity.

There has been concern that the Panamanian government, known to be less than stable, may use the canal as a cash cow. A Canal Authority has been set up that keeps canal operations completely separate from government control. An international advisory board is also in place, made up essentially of the canal's customers. There is also the question of how such a closure would affect each industry. Some sectors are more or less sensitive, based on what alternatives are available. For example, it was estimated that a closure would have a minimal impact on the US agricultural trade. Less than 2% of market share in corn and soybean from the US would be lost. "The numbers indicate that U.S. grain traffic is more important to the canal than the canal is to grain exports".[13]

The challenge for the near future will be upgrading the canal and its equipment. In less than 20 years more than half of the world's ships will be too big for the canal. Wider locks would cost an estimated $6 billion.

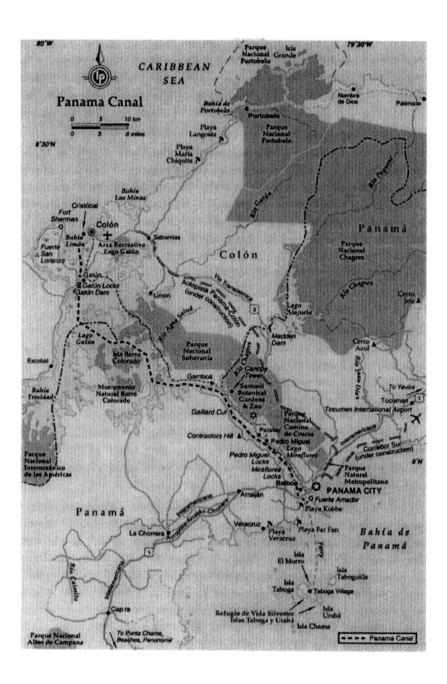

Cross-docking has become an operational practice that saves time and space in the logistics chain.[14] It is when trucks come to a warehouse loading dock and the cargo is unloaded, sorted and reloaded without being stored. This simple system requires good coordination, but it can eliminate the need for storage space, and speeds up delivery time. This is one way that carriers are improving the efficiency of port and facility operations.

Environmental issues are becoming a major issue in ports, though the specific issues vary between ocean and air ports. There is now talk of the "sustainable port", one that considers all issues relating to itself and society instead of just their short term profitability. Not surprisingly, ports that take a broader view of their operations also tend to be more professionally run, and thus more competitive in the long run. Others take a different view. In Murmansk, there is an oil slick across the water, resulting in a thick black layer in the shores tidal zone. This is not from any recent spill, but from their ongoing inability to control small oil spills.

OCEAN PORTS

Historically, an ocean port was simply any place where geography allowed a ship near land where it could safely load cargo. We tend to forget that geography is still by far the primary determinant of where ocean ports exist. This is unlike airports, which can use any flat area of land. Some ports are distinctive for their equipment, such as the container port. Others are distinctive because of the type of cargo that is handled. A port handling live animals may also be used as a general cargo port, but it is special for its facilities that hold live animals. Some of these ports are very limited in their use, like a tanker port, while others are more adaptable.

A typical ocean port is made up of maritime access infrastructure, port infrastructure, terminals, land access, and other facilities. Maritime access infrastructure refers to channels and navigational aids in the water. These are usually maintained by someone other than the port. In the U.S, the Army Corps of Engineers and Coast Guard perform some of these tasks. In the port itself, there is the infrastructure such as docks. At each terminal, which is where ships are loaded, there are the terminals, which are the cranes and buildings. Land access refers to roads and rail connections. Other facilities refer to the wide array of other commercial activities taking place around a port. Ocean ports used to be labor intensive, but as we see from the extensive infrastructural needs, they are now increasing capital intensive.

Parts of an Ocean Port
- Maritime Access Infrastructure
- Port Infrastructure
- Terminals
- Land Access
- Other Facilities

There has been a trend toward fewer but larger ports, and this is particularly so with ocean ports. One effect is that larger ships are forcing ports to improve their operations because of the huge amount of cargo that must be sent through the same amount of space without losing time. For example, a 7,000 TEU ship with 75% of its cargo intermodal (the other 25% being delivered locally) would need 9.4 double-stack trains and hundreds of truck movements.[15]

The major types of ocean ports
- Bulk
- Ro-ro
- Tanker
- Container
- General
- Hazardous cargo
- Live animals
- Company owned

Part of the reason for larger ports is not necessarily the increased size of ships, but the need for economies of scale. New York City (not the Port of New York, which is already combined with New Jersey) is pushing for "superport" in Brooklyn to counter New Jersey. Another reason for the increased size is the growth in alliances among the carriers. This means that the alliance looks for one port in a given region, instead of a few. The port they choose would need to be big enough already to accommodate the additional shipments.

Stopford describes the levels of port development in his classic book on maritime economics.[16] The **small local port** handles all types of cargo, and serves only the local community. The **large local port** serves only the local area, but now they can invest in specialized equipment. The **large regional port** invests in specialized cargo, and serves a region. Finally, the **regional distribution center**, also known as a load center, is a collection of highly specialized terminals.

These distinctions are important for a variety of reasons, particularly the level and type of competition. We already noted the importance of competition for driving efficiency. Kent and Hochstein identified the threshold values for competition.[17] For ports with less than 30,000 TEUs of throughput, it makes sense for only one entity to manage the ports. When volume rises to 100,000, there is the possibility of multiple independent terminals within the port, thus creating competition. When volume reaches 300,000, there may be multiple ports, each operating independently.

Ports need to understand how they fit in with the regional and global network of ports. For ocean ports, this includes other ports of the same type, but there is also some ship-to-ship transfers. There was a discussion of inland cargo, and that also affects maritime cargo. Smaller ships and barges often bring cargo down the river to a maritime port. Hong Kong built a terminal, Tuen Mun, designed specifically for the river cargo associated with the Pearl River Delta.[18]

Labor relations are important for ports because the unions are very strong. In the US, the operators of gantry cranes for intermodal containers may earn over $140,000 a year. One way this is justified is as a percentage of cargo that is handled. This is an extreme example, but wages are very good. In other countries, wages vary widely. Job roles tend to be strictly divided and limited, in which each worker does only what is part of his or her job. The crane operator cannot drive the trucks and so on. There has traditionally been a difference between those who move cargo from the ship's side onto the ship and securing it inside, and those working on the docks and move cargo around the port.[19] Port strikes are very disruptive, which is one reason why wages are good and unions strong. The goal for both unions and port operators, however, is for cooperation and teamwork.

In assessing the performance of labor, a few different methods are used. They all generally try to measure the amount of cargo moved, divided by the amount of labor used. In other words, how much cargo did the average stevedore move through the port. This is an average number since cargo is handled by many stevedores as it moves through a port. For example, the Port of Los Angeles/Long Beach, largest container port in the US, 'weighted tons per man-hour' was 7.63 in 1998. Another way to measure productivity is lifts per hour; the number of containers lifted per hour. Some ports reported as much as 38 container moves per hour.[20]

Governments often regulate the port operations to protect certain interests. One of the most common is to prevent foreign companies from taking control. By the nature of the port operations, it would not be difficult for a foreign company to try to buy up some businesses. Many governments limit this. Governments have also tried to make money on foreign carriers. Japan Harbor Transportation Association requires ocean carriers to respect "traditional business rights", and "prior consultation". In other words, the

Japanese business style is very different and some would say bureaucratic. Vested business interests are hard to dislocate. When a carrier signs an agreement, such as to use a certain company's services, it is very difficult to get out of the obligation. When American President Lines (APL) merged with Neptune Orient Line (NOL), it required extensive operational changes. Yet APL already had signed five agreements on who was going to handle their cargo from prior mergers at its Yokohama terminal.[21]

Maritime ports face the controversial and expensive job of **dredging**, digging up the mud on the sea floor to insure room for ships to come in. Ports have a natural depth, which is usually much shallower than what modern large ships require. For example, the largest container ships need about 35 feet of draft. Therefore, every once in a while the port needs to be dredged. The frequency of dredging varies from yearly to once every several years, depending on the natural conditions that lead mud to accumulate.

Dredging creates major environmental problems. The mud is stirred up, which harms the marine environment. Pollution and toxins that had settled into the mud (and have remained relatively harmless) are now dug up. The mud must then be dumped somewhere, causing more problems. There are a few alternatives for dumps. It may be dumped in deeper water, off to the side out of traffic lanes, or on land.

Of all the aspects of port management, the single most fundamental job is to load the ship. While this may seem simple, there is a science to it. Failure to follow procedures can result in disaster. Ships have capsized at the port, broken apart or damaged as a result of improper loading. The cargo needs to be offloaded and loaded in a sequence that will leave the weight evenly distributed. On containerships, the containers must fit in slots. Even minor mistakes in weight distribution can bend the ship enough that a container cannot be placed in its slot. Tankers use ballast, as already discussed, to keep weight in certain compartments while others are being loaded.

Loading a containership is an excellent example of the complexity involved with port and transportation management. While it may seem like the crane is just picking up containers at random, there are sophisticated computer programs that determine the sequence of offloading and onloading, according to a long list of variables, including:
• Minimize overall movements.
• Heavy cargo is placed low, light cargo high.
• Dangerous goes on top, where it can be easily accessed and handled.

- Awkward cargo on top.
- Reefer containers can only be placed where there are power outlets.
- 20' and 40' in their respective slots.
- The cargo that is getting off at the next port should be on top.

Tugboats

Tugboats play a vital but often unrecognized role in shipping. The largest ships, which are almost all the international ships, need tugboats to help them maneuver in a port area. Some modern ships have a 'bow-thruster', a propeller in the bow, perpendicular to the boat (in other words, instead of pushing a boat forward or backward, it pushes the bow crossways), which reduce their need for tugboats.

Tugboats are becoming more common for business and safety reasons. They help a ship get into and out of a port quicker, thus speeding up turnaround time. They are often required as a safety measure to prevent collisions or grounding.

Associated with tugboats are the Pilots. This is not the pilot as in the captain of a boat, but an individual who is intimately familiar with a port. She or he is brought out to the ship by the tugboat, boards the ship and guides it to the dock. Much like the tugboat, a pilot can speed up the process and improves safety.

A tugboat in Port Sudan.

Top 10 US Ports

The following ranking, based on data by the Census Bureau, compares the value of cargo that goes through the port regardless of mode (air, sea, rail or road).

	In billions of dollars		
	Exports	Imports	Total Trade
Port of Long Beach, CA	23.3	63.7	87
Port of Detroit, MI	44.2	41.1	85.3
JFK International Airport, NY	38.5	42.1	80.6
Port of Los Angeles, CA	15.8	57	72.8
San Francisco Int'l Airport, CA	33.8	37.2	71
Port of New York and New Jersey	22.2	44.8	67
Los Angeles Int'l Airport, CA	32.6	29.1	61.7
Port of Buffalo-Niagara Falls, NY	31.1	27.6	58.7
Port of Laredo, TX	18.2	20.7	38.9
Port of Seattle, WA	11.6	22.5	34.1

Source Bureau of Transportation Statistics and Census Bureau

AIRPORTS

The number of airports is much greater than most people would guess because passengers are accustomed to going to their local passenger airport. Yet when one looks at cargo airports, military airports, and local air strips, airports are a major industry in themselves. The number of airports in the world varies dramatically depending on what we consider a 'true' airport. In the US, for example, there are around 417 airports with control towers and scheduled air service, but there are 670 airports certified by the FAA. If you include private airports and those not paved and lighted, the number swells to over 18,000.

Almost all major airports are run by a public agency, usually a local government. This is because airports are vital to a community and tend to be a natural monopoly. The many small airports, though, tend to be privately run. There has been talk of privatizing airports to make them operate more efficiently. The best known example is the United Kingdom, where London's airports are run by a private company.

Airport administrations tend to look like a small town. They have all the departments of a company, as well as their own fire and police departments. Airports are usually divided into two areas, called **groundside operations** and the **airside operations**. The groundside is what the passengers see, including airline arrival and departure areas, public parking, the restaurants and so on. But there is another area, often bigger, the airside. This is where the airlines operate, the baggage is handled and the cargo moved. The airside tends to be invisible to those who do not work in aviation.

Money earned by an airport comes from a variety of sources. The two most important sources are **concessions** and **carrier fees**. Of the carrier fees, the primary source is landing fees. Although a major income area for airports, it is a small part of a carrier's costs, about 5 percent according to Doganis.[22]

Larger airports can better exploit commercial activities, whereas smaller airports are almost entirely dependent on aeronautical revenues. Concessions are the companies that rent space in and around the airport to provide goods and services, such as restaurants and gift stores. Even a small booth in a major airport can earn huge amounts of money because of its good location. These concessions then pay the airport a percentage of their revenue. Airport real estate is not an efficient market, and rates vary widely. By one estimate, rentals at gateway airports are $18.50 per square foot, compared to metro airports at $12.50 and medium metro at $9.50.[23]

Carrier fees can be calculated in two ways. First, there is the **residual** method. An airport calculates its overall expenses, minus non-airline revenue (such as concessions), and charges the airlines the remaining costs. The second way to calculate carrier fees is the **compensatory** method. Airlines pay for the parts of the airport they use. They would pay rent for all the space used. Most importantly, landing fees are charged to the airlines every time a plane lands at the airport.

Capital improvements to airports can be paid for either by selling bonds or special taxes. Passengers are often charged a fee for arrival or departure. This is usually supervised by the federal authorities because they do not want airports to create special costs for entering or exiting a country that would affect international relations. The US government has a special program in which it charges taxes and uses the money to improve air traffic safety. What happened instead was that the federal authorities held onto the money for political reasons, even though there was serious safety risks in some of the older computers used in airports.

Airport capacity refers to the amount of passengers and cargo that can move through an airport. It is determined by the runway, terminals and facilities, and is one of the most critical issues for airports today. Groundside capacity refers to the number of passengers or cargo that can get to the airport, including parking space. Airside capacity refers to the number of planes that can be handled, including the number of planes that can arrive and take off in a certain period of time. Runway capacity is based on air traffic control, demand, weather, design and configuration of runways.[24]

Air traffic control is usually done by a federal authority, to act as an air traffic cop. Commercial airports are normally certified by the authorities that then set up the air tower. These towers handle take-offs and landings. In the US and other countries, there is a different facility that handles air traffic control while a plane is approaching or leaving an airport. In the US for example, there are over 400 air control towers in various airports, but only 184 'terminal radar approach control' facilities (TRACONs). This is because a TRACON can handle more than one airport. There is yet another office that tracks planes enroute, 21 in the US. Air traffic control is more than safety. Delays at airports can be very costly. It was estimated that $3 billion a year is spent on fuel and other costs for the delayed flights in the US.[25]

There has been a shift to the use of hub-and-spoke arrangements by the airlines, in which one airport is used as a hub. Flights in a region, instead of going direct to their end destination, go to a hub. This increases the likelihood that a passenger (or cargo) may not get a direct flight, but it increases overall efficiency. Small airports in particular can be served much more efficiently when they feed into a hub. The hub-and-spoke arrangement has had one side-effect of making one airline dominant in many airports, which can result in increased prices. This is known as 'fortress hubs'. For example, there were demands for British Airways to give up 276 landing slots at Heathrow and Gatwick without compensation, valued at around $800 million.[26]

There has been some limited progress in opening up all-cargo airports. This requires a region with a very high volume of air cargo, and the current airports being at capacity. In the US there have been plans for a few of the abandoned air force bases turned into all-cargo airports. Having a complete airport that is not in use is clearly the ideal situation. Clarke Air Force Base in the Philippines was a US base scheduled for closure when a volcanic eruption sped up that process. Now FedEx has taken it over for their cargo operations.

There are two basic functions in a warehouse, movement and storage of the cargo. This includes the following functions:

- Movement
  Receiving
  In-storage handling
  Shipping

- Storage
  Planned storage
  Extended storage

This may seem like a simple job, but it is not. Warehouse management has become a sophisticated, technical and competitive field. To say that warehouses only move or store things misses out on why they are being moved and stored. Modern warehouses are being called on to perform more and more logistics, manufacturing and marketing tasks.

Warehouse design is based on a few **operating principles**. There should be only one story to aid the flow through. Going up and down between a multi-storied building is a way to save space, but it slows the movement. Ceilings should be as high as possible, in order to avoid multiple floors. There should be one general direction of flow, in one side and out the other. High volume products are placed up front, the opposite of what one sees as a customer in a grocery store.

The storage plan is part of the overall logistics plan, which in turn is based on the overall operations strategy. Storage should not be done as an unplanned event too often. If it is happening frequently, that is a signal that there is a logistics problem. Just like carriers, warehouses are organized as either private, public or contract warehouses. One major difference is that public warehouses are not like common carriers and do not have any obligation to provide space.

Materials handling refers to the way things are physically handled. This is important when one looks at a micro view of logistics, such as the logistics of a single facility. It is less relevant for a discussion of international logistics, but it does deserve a brief description. The best method of material handling is no handling at all. In other words, avoid handling to begin with, and then minimize it as much as possible. Any handling entails costs and the risk of damage. There are some basic handling considerations:

- Standardized equipment- for example, do not have different brand forklifts in the same facility.

- Continuous flow - systems should be designed for maximum continuous product flow. A material flow that stops and starts is disruptive. In other words, try to create an assembly line.
- Investment should be in handling and not stationary equipment.
- In handling equipment, the ratio of deadweight to payload should be minimized. The deadweight refers to the handling equipment itself plus packaging etc. In other words, the heavier the handling equipment, the more effort it takes just to move the equipment, never mind the material being moved.
- Gravity flow should be used where possible.

A basic tradeoff needs to be made between paying for additional space, and the labor costs associated with more confined space. In southern California, an acre of port property costs $120,000 to rent. Yet if they confined space, they would end up paying more in labor to have containers stacked up or moved more than once. Port operators prefer to leave containers on a chassis rather than keep them stacked, up to four high, and then need to sift through them when the trucker comes to pick up a load. One port expert stated that 65% of a terminal's costs are connected to delivering import containers. Larger terminals tend to be more efficient and more flexible. One of the reasons is that with larger ports, support activities, such as office space, are relatively less than small terminals.[27]

Value-added services of a warehouse

- Orders shipped within guaranteed time
- After market sales support
- Freight unloading
- Forklift service
- In-handling
- Storage
- Out-handling
- Documentation handling
- Dispatch
- Computerized inventory controls
- Regular inventory or status reports
- Inventory rotation
- Communications services
- Customer warehouse pick-up
- Local delivery

FREE TRADE ZONES

Free trade zones (FTZ) are not a port and not a warehouse. They are a truly unique sort of facility that is taking on a much more important role in international logistics. An FTZ is a place in a country that is technically not "in" the country, at least as far as the customs authorities are concerned. These are areas, rarely larger than a few square kilometers but some towns are designated as FTZs. They are physically sealed off, and they operate with the specific approval of the national government.

Operation of a Foreign Trade Zone

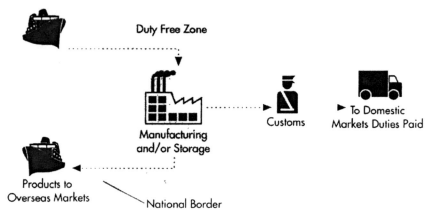

The purpose of an FTZ is to encourage international trade by giving some flexibility in the import rules. An FTZ allows someone to bring something into the country, and as long as it remains in the FTZ, it is not officially imported. There are many things that can be done to the cargo while in an FTZ (see inset). Some of these things are intended to assist the importer to get the cargo ready for legal import. In this way it helps the importer do his job. Other times the FTZ is a way to create local jobs. An **Export Processing Zones** (EPZ) is similar to an FTZ but intended for light manufacturing. In this chapter we will refer only to FTZs, which may include EPZs. The following are some of the main reasons for using and FTZ:

- **Delay tariffs**. As long as the cargo sits in the FTZ, it does not pay the import duties. The FTZ is physically just a warehouse, usually, but it is more expensive than an average warehouse.
- **Avoid tariffs before onshipment**. The cargo may not be destined for that country at all. It may sit in an FTZ and then be exported.
- **Processing**. This is more like an EPZ, but even FTZs do some processing.
- **Correct mistakes**. The customs authority may have told the importer that the cargo was not up to the local law, so it can sit in an FTZ to be fixed.
- **Sell**. The cargo may be bought and sold while sitting in the FTZ.

An FTZ competes with the local businesses, which is why port labor unions are often hostile to the idea. The money that would be going to the government is being held back. Jobs that would be created in the community are being created in the FTZ, which are often lower paid. Why would a government chose to create an FTZ? First, they provide the carriers and shippers flexibility. Without an FTZ, any mistake would mean that the cargo be sent out of the country, an expensive solution. Another benefit of FTZs is that it is a good way for unstable countries to get some free and fair enterprise. Foreign companies are often shy of entering an unstable country, but an FTZ is usually much safer.

A **Foreign Trade Zone Board** administers the operations. These boards are usually assigned by the government. The World Association of Free Trade Zones and Export Processing Zones, and independent non-profit organization founded by the United Nations, is in Phoenix, Arizona and is designed to promote EPZs as a method of economic development.

Export Processing Zones

⊛ Multiple Sites
● Single Sites

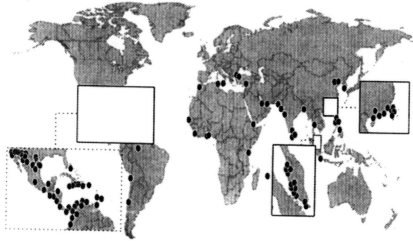

Despite the name, an FTZs are not entirely free. It needs to make money since it is offering space to its users. Therefore they collect storage fees as a regular warehouse. Other activities also entail user fees. Customs typically collects a fee for processing imports and/or exports from within the FTZ.

The US has 236 general purpose FTZs and 375 subzones as of 1999. These FTZs are used by over 3,550 firms and employed 370,000 people. In 1998 more than $168 billion in cargo moved through these zones.[28] The US distinguishes between two kinds of zones:

- **General purpose zones**- for multiple users, which must be 60 statute miles or 90 minutes driving time from a US Customs port of entry
- **Sub-zones**- that are sites owned or controlled by specific firms.

For example, New United Motors (NUMMI), a joint venture between General Motors and Toyota, assembles cars and small trucks near San Francisco. They are designated as a sub-zone. Auto parts that enter the plant are not 'imported' until they are put into a car and that car leaves the plant as a finished product. This way NUMMI delays paying duties on the parts until the finished car is shipped out, and it pays no duties on returned parts.

FTZs are often created by local businesses looking for a tax break on work that is being imported but then re-exported. Kodak has a facility that receives disposable cameras, re-uses the parts and then ships out 'new' disposable cameras. They joined with other local companies to have a FTZ established in Rochester, New York, and had their local plants designated at sub-zones.[29]

In-bond storage is a similar arrangement. The cargo is in a bonded warehouse and has not been officially imported. The warehouse is controlled by the customs authorities or contractor acting under their direction. The difference is that there may be no processing or sale of the cargo. This is used when something arrives and there is some problem clearing customs and the cargo needs to be stored somewhere.

Recall from earlier in this chapter that issues arise when ports seek to charge users a tax or user fees. There has been some debate whether those fees apply to FTZs. If the idea of an FTZ is to exempt them from normal charges, some would say they should not even pay for harbor maintenance taxes. At least one court case decided that FTZs must still pay a harbor maintenance fee since it is not a tax.[30]

Merchandise in a FTZ may be:

• Stored	• Tested	• Mixed	• Cleaned
• Sampled	• Relabeled	• Assembled	• Manufactured
• Repackaged	• Displayed	• Salvaged	• Destroyed
• Repaired	• Manipulated	• Processed	

Table II World Port Ranking – 2000

TOTAL CARGO VOLUME—Metric Tons (000s), except where noted   CONTAINER TRAFFIC—Twenty Foot Equivalent Units (TEUs)

RANK	PORT	COUNTRY	MEASURE	TONS	RANK	PORT	COUNTRY	TEUs
1	Singapore	Singapore	FT	325,591	1	Hong Kong	China	18,098,000
2	Rotterdam	Netherlands	MT	319,969	2	Singapore	Singapore	17,090,000
3	South Louisiana	U.S.A.	MT	197,680	3	Pusan	South Korea	7,540,387
4	Shanghai	China	MT	186,287	4	Kaohsiung	Taiwan	7,425,832
5	Hong Kong	China	MT	174,642	5	Rotterdam	Netherlands	6,274,000
6	Houston	U.S.A.	MT	173,770	6	Shanghai	China	5,613,000
7	Chiba	Japan	FT	169,043	7	Los Angeles	U.S.A.	4,879,429
8	Nagoya	Japan	FT	153,370	8	Long Beach	U.S.A.	4,600,787
9	Ulsan	South Korea	RT	151,067	9	Hamburg	Germany	4,248,247
10	Kwangyang	South Korea	RT	139,476	10	Antwerp	Belgium	4,082,334
11	Antwerp	Belgium	MT	130,531	11	Tanjung Priok	Indonesia	3,368,629
12	New York/ New Jersey	U.S.A.	MT	125,885	12	Port Kelang	Malaysia	3,206,753
13	Inchon	South Korea	RT	120,398	13	Dubai	U.A.E.	3,058,866
14	Pusan	South Korea	RT	117,229	14	New York/ New Jersey	U.S.A.	3,050,036
15	Yokohama	Japan	FT	116,994	15	Tokyo	Japan	2,898,724
16	Kaohsiung	Taiwan	RT	115,287	16	Felixstowe	U.K.	2,793,217
17	Guangzhou	China	MT	101,521	17	Bremer Ports	Germany	2,712,420
18	Quinhuangdao	China	MT	97,430	18	Gioia Tauro	Italy	2,652,701
19	Ningbo	China	MT	96,601	19	San Juan	U.S.A.	2,333,788
20	Marseilles	France	MT	94,097	20	Yokohama	Japan	2,317,393
21	Osaka	Japan	FT	92,948	21	Manila	Philippines	2,288,599
22	Richards Bay	South Africa	HT	91,519	22	Kobe	Japan	2,265,992
23	Kitakyushu	Japan	FT	87,346	23	Yantian	China	2,139,680
24	Qingdao	China	MT	86,360	24	Qingdao	China	2,120,000
25	Hamburg	Germany	MT	85,863	25	Laem Chabang	Thailand	2,105,262
26	Dalian	China	MT	85,053	26	Algeciras	Spain	2,009,122
27	Kobe	Japan	FT	84,640	27	Keelung	Taiwan	1,954,573
28	Tokyo	Japan	FT	84,257	28	Nagoya	Japan	1,904,663
29	New Orleans	U.S.A.	MT	82,400	29	Oakland	U.S.A.	1,776,922
30	Dampier	Australia	MT	81,446	30	Colombo	Sri Lanka	1,732,855
31	Vancouver	Canada	MT	76,646	31	Tianjin	China	1,708,000
32	Corpus Christi	U.S.A.	MT	75,461	32	Charleston	U.S.A.	1,629,070
33	Beaumont	U.S.A.	MT	75,032	33	Genoa	Italy	1,500,632
34	Newcastle	Australia	MT	73,871	34	Seattle	U.S.A.	1,488,020
35	Tubarão	Brazil	MT	73,182	35	Le Havre	France	1,486,108
36	Tianjin	China	MT	72,980	36	Tacoma	U.S.A.	1,376,379
37	Port Hedland	Australia	MT	72,914	37	Barcelona	Spain	1,363,695
38	Hay Point	Australia	MT	69,379	38	Cristobal	Panama	1,353,727
39	Le Havre	France	MT	67,492	39	Hampton Roads	U.S.A.	1,347,364
40	Port Kelang	Malaysia	FT	65,227	40	Melbourne	Australia	1,327,789

Abbreviations MT=Metric Ton HT= Harbor Ton FT=Freight Ton RT = Revenue Ton
NOTE. The cargo rankings based on tonnage should be interpreted with caution since these measures are not directly comparable and cannot be converted to a single, standardized unit.
Source American Association of Port Authorities

Passengers		Cargo		Movements	
total passengers enplaned and deplaned, passengers in transit counted once.		loaded and unloaded freight and mail (in metric tonnes).		landing and take-off of an aircraft.	
Airport	Total	Airport	Total	Airport	Total
1 ATLANTA, GA (ATL)	6 931 175	MEMPHIS, TN (MEM)	301 984	CHICAGO, IL (ORD)	82 715
2 CHICAGO, IL (ORD)	6 554 489	HONG KONG, CN (HKG)	210 026	ATLANTA, GA (ATL)	78 209
3 LONDON, GB (LHR)	6 130 657	ANCHORAGE, AK (ANC)**	175 008	DALLAS/FT WORTH, TX (DFW)	67 368
4 TOKYO, JP (HND)	6 004 626	TOKYO, JP (NRT)	166 989	LOS ANGELES, CA (LAX)	57 469
5 LOS ANGELES, CA (LAX)	5 658 808	LOS ANGELES, CA (LAX)	148 157	MINNEAPOLIS/ST PAUL, MN (MSP)	47 591
6 DALLAS/FT WORTH AIRPORT, TX (DFW)	4 826 196	NEW YORK, NY (JFK)	143 085	PHOENIX, AZ (PHX)	47 183
7 FRANKFURT, DE (FRA)	4 708 732	INCHEON, KR (ICN)	141 284	PARIS, FR (CDG)	45 834
8 PARIS, FR (CDG)	4 677 340	SINGAPORE, SG (SIN)	140 538	DENVER, CO (DEN)	45 015
9 AMSTERDAM, NL (AMS)	4 149 575	LOUISVILLE, KY (SDF)	138 696	DETROIT, MI (DTW)	44 250
10 LONDON, GB (LGW)	3 553 903	MIAMI, FL (MIA)	136 487	LAS VEGAS, NV (LAS)	43 100
11 DENVER, CO (DEN)	3 532 727	FRANKFURT, DE (FRA)	133 177	CINCINNATI, OH (CVG)	42 691
12 LAS VEGAS, NV (LAS)	3 249 804	CHICAGO, IL (ORD)	112 832	LOS ANGELES, CA (VNY)	41 561
13 MADRID, ES (MAD)	3 242 957	TAIPEI, TW (TPE)	112 685	PHILADELPHIA, PA (PHL)	41 452
14 NEW YORK, NY (JFK)	3 227 036	PARIS, FR (CDG)	110 000	LONDON, GB (LHR)	40 876
15 MINNEAPOLIS/ST PAUL, MN (MSP)	3 221 681	LONDON, GB (LHR)	107 778	FRANKFURT, DE (FRA)	40 668
16 SAN FRANCISCO, CA (SFO)	3 215 654	AMSTERDAM, NL (AMS)	100 129	HOUSTON, TX (IAH)	40 120
17 DETROIT, MI (DTW)	3 154 578	BANGKOK, TH (BKK)	78 912	CHARLOTTE, NC (CLT)	39 708
18 HONG KONG, CN (HKG)	3 150 000	INDIANAPOLIS, IN (IND)	78 720	MIAMI, FL (MIA)	37 958
19 PHOENIX, AZ (PHX)	3 126 697	NEWARK, NJ (EWR)	73 268	AMSTERDAM, NL (AMS)	37 910
20 HOUSTON, TX (IAH)	3 063 539	DUBAI, AE (DXB)	66 104	PITTSBURGH, PA (PIT)	37 647
21 SEATTLE/TACOMA, WA (SEA)	2 958 261	ATLANTA, GA (ATL)	62 298	BOSTON, MA (BOS)	37 573
22 NEWARK, NJ (EWR)	2 895 057	OSAKA, JP (KIX)	60 866	ST LOUIS, MO (STL)	37 304
23 TOKYO, JP (NRT)	2 877 868	BEIJING, CN (PEK)	59 764	ANCHORAGE, AK (ANC)	37 289
24 BEIJING, CN (PEK)	2 797073	OAKLAND, CA (OAK)	58 495	SALT LAKE CITY, UT (SLC)	36 328
25 BANGKOK, TH (BKK)	2 776 687	TOKYO, JP (HND)	57 342	MEMPHIS, TN (MEM)	36 116

26	MIAMI, FL (MIA)	2 759 121	SHANGHAI, CN (PUG)	56 921	NEWARK, NJ (EWR)	36 084
27	TORONTO, OT, CA (YYZ)	2 743 018	DALLAS/FT WORTH, TX (DFW)	55 883	TORONTO, OT, CA (YYZ)	35 348
28	PALMA DE MALLORCA, ES (PMI)	2 640 392	GUANGZHOU, CN (CAN)	53 947	SEATTLE/TACOMA, WA (SEA)	34 735
29	SINGAPORE, SG (SIN)	2 496 030	SAN FRANCISCO, CA (SFO)	48 781	LONG BEACH, CA (LGB)	33 525
30	ROME, IT (FCO)	2 405 232	KUALA LUMPUR, MY (KUL)	46 718	SANTA ANA, CA (SNA)	32 963

[1] *"Port Selection Criteria: An Application of a Transportation Research Framework"*, Paul R. Murphy, James M. Daley, Douglas R. Dalenberg, *The Logistics and Transportation Review*, 1992, vol. 28, no. 3, pp. 237-256.

[2] *"Rotterdam: A Strategic Hub in the Global Trade-Transport Chain"*, T.R. Lakshmana, in Laksmanan et al, *Ibid*, p. 129.

[3] *'Ports widen intermodal reach"*, Philip Damas, *American Shipper*, June 1998, p. 56.

[4] *JOC*, April 7, 1999, p. 1b.

[5] *JOC*, April 8, 1999, p. 1B.

[6] *"Seaports"*, Lourdes Trujillo and Gustavo Nombela, in Estache and De Rus, *ibid*, p. 120.

[7] *JOC*, June 18, 1999, p. 15.

[8] *JOC*, September 2, 1999, p. 1, 5.

[9] *"Report of the Advisory Commission on Conferences in Ocean Shipping"*, US Government, April 1992.

[10] *"Container terminals embrace automation"*, Joseph Bonney, *American Shipper*, August 1998, p. 42.

[11] *JOC*, March 23, 1999, p. 1b.

[12] *"New immigration procedures to speed turnaround at Dalian"*, P.T. Bangsberg, *The Journal of Commerce*, October 15, 1998, p. 12A.

[13] *JOC*, October 7, 1999, p. 3.

[14] *"Strategic Information Linkage"*, Patricia J. Daugherty, in Robeson and Copacino, *Ibid*, p. 760. And *"Competing on Capabilities: the New Rules of Corporate Strategy"*, George Stalk, Philip Evans, and Lawrence E. Schulman, *Harvard Business Review*, March-April 1992, p. 57.

[15] *JOC*, March 5, 1999, p. 3B.

[16] *"Maritime Economics"*, Martin Stopford, 2nd edition, Routledge, London, 1997.

[17] *"Port Reform and Privatization in Conditions of Limited Competition: The Experience in Colombia, Costa Rica and Nicaragua"*, P.E. Kent and A. Hochstein, *Journal of Maritime Policy Management*, volume 25(4), 1998, pp. 313-33.

[18] *"China river-trade terminal opens to accommodate smaller ships"*, P.T. Bangsbert, *The Journal of Commerce*, October 19, 1998, p. 1B.

[19] *"Seaports"*, Lourdes Trujillo and Gustavo Nombela, in Estache and De Rus, *ibid*, p. 118.

[20] *JOC*, March 29, 1999, p. 1b.

[21] *"US views APL as textbook case in Japan port flap"*, Bradley Martin, *The Journal of Commerce*, September 16, 1998, p.1A.

[22] *"Flying Off Course: The Economics of International Airlines"*, Harper Collins, London, 1991.

[23] www.barrowco.com, September 29, 1999.

[24] *"Airports"*, Ofelia Betancor and Roberto Rendeiro, in Estach and de Rus, *ibid*, p. 51.

[24] *"Airports"*, Ofelia Betancor and Roberto Rendeiro, in Estach and de Rus, *ibid*, p. 58.

[25] *"The Air Transport Association Airline Handbook"*, Air Transport Association, 1995.

[26] *"Watch-dogfight"*, *The Economist*, August 22, 1998, p. 16.

[27] *JOC, March 5, 1999, p. 1B.*

[28] *National Association of Free Trade Zones.*

[29] *National Association of Free Trade Zones.*

[30] *JOC, September 21, 1999, p. 1.*

Section III
Import/Export Operations

10. Customs and Regulations

11. Trade Documentation

12. Financing Trade

13. Security

CHAPTER 10
CUSTOMS AND REGULATIONS

One of the most critical parts of an international shipment is the crossing of a border and the regulations that go along with that. This chapter introduces many of the issues that arise from crossing borders. It begins with a general discussion of why borders matter and what governments are trying to do. It helps to understand what they are trying to accomplish with customs regulations in order to fulfill their regulations and stay out of trouble. After laying the foundation of why a Customs agency behaves the way it does, this chapter explains in moderate detail the regulations associated with exporting and importing. This book is not intended as a how-to manual, and this chapter is not intended to supply the reader all of the details on how to clear Customs. What is offered here is an overview to understand the big picture of Customs operations.

CUSTOMS AND REGULATIONS

Customs refers to the government agency and the function of controlling imports and exports to conform to government regulations. This chapter will look at this subject from a variety of viewpoints: the government agencies that are known as 'Customs', the customs function, and the regulations that drive the customs process. This section only discusses the regulations of crossing borders. Regulations that apply to logistics and transportation within a country are dealt with in the other chapters. As with logistics, customs involves both people and cargo, but we will concentrate on the cargo aspects. When the word Customs is capitalized, that means we are referring to the government agency, while the lower case customs refers to the customs function.

US Bureau of Customs and Border Protection
Founded in 1789 by George Washington
Moved from Treasury to Homeland Security after 9.11 changes.
Enforces over 400 laws from 40 agencies.
Funded virtually the entire government for the first 125 years.
Year 2000: $1.4 trillion in imports, and $31 billion duties collected.
Returned $16 for every $1 appropriated by Congress.

The customs function must be understood as the outcome of a political process. While other chapters are best understood in terms of, for example, economics or management, customs should be viewed first as a political issue. The governments that produce the regulations are made up of many different actors, all of whom have their own goals and priorities. The customs regulations are the result of a political process in which the preferences of many people, often conflicting, result in policies that are sometimes confusing or contradictory.

Customs plays a special role for international logistics because the border is a chokepoint. The ports where cargo and people enter and exit a country are the places where a government can control and regulate many activities. For this reason Customs agencies are assigned many tasks, some of which do not seem to have anything to do with trade or travel.

There are two approaches to customs. The US approach is to use the border as the place where laws are enforced. It is believed that once cargo or people enter the country, it would be difficult or impossible to track. Many other places, such as Europe, take a different view. They look at the border as one checkpoint, but much of the same regulatory goals are pursued internally. Europe does not try to enforce at the borders many of the laws that the US enforces. This is because there are several countries within the EU, and one country cannot hope that all other countries would enforce the rules as they would like.

A distinction should be made between customs and immigration. Customs refers to cargo, while immigration refers to people. In the US, there is the US Customs Service that deals with cargo and the Immigration and Naturalization Service (INS) that deals with people. In other countries these tasks may be in the same agency. In reality, the two tasks are overlapping. People stow away in cargo containers. Travelers may be carrying goods that are illegal. Thus we find both agencies, working closely together, in ports.

What Does Customs Do?

There are a few major goals of government regulation that in turn affect customs. While the priorities vary among different countries, the most important goals include national security, revenue collection, competitiveness, managed trade, safety/public order, and protecting national interests.

National security is a broad area that refers mostly to military threats. This includes cargo that has military applications, but also the control of people who pose a military threat. A complicating issue is **dual**

use technology, which refers to products or technology that may have military value, but may also be used for civilian purposes. For example, medical CAT scan technology has been adapted for use in submarine and antisubmarine testing. Note that national security controls apply to the technology as well as the product itself.

Since the 9-11 attacks, there have been changes in the customs process that may or may not be a significant change from past methods. This will be discussed later in the section on security. Customs requires more information for incoming shipments, but the substantive changes since 9-11 are arguable a lot less significant than they appear.[1]

Revenue Collection by Customs was the only source of funds for the US in the first few decades of the country's existence. Taxes collected on trade are often one of the most important sources of revenue for national governments.

Managed trade and competitiveness refer to policies to promote the nation's economy. Free market philosophy believes that a government should not get involved, but all governments do have some rules to control economic activities. Competitiveness is a controversial idea that what benefits other countries results in harm to one's own country. In other words, trade is a zero-sum activity. Other economists strongly disagree.

The two major issues with competitiveness are **dumping** and **foreign subsidies**. Dumping is when cargo is sold in a foreign market at below market prices, normally intended to eliminate the competition. To qualify as dumping, two conditions must be met. First, the cargo is sold below the selling price in the exporter's home market or below the exporter's cost of production. Second, a local company is hurt by the action. If something is being sold, and there are no competitors, and the product is being sold below cost, nobody is being hurt, and thus it is not considered dumping. Foreign subsidies occur when an exporter receives some type of subsidy to improve their competitive position in foreign markets. These subsidies may take any of a wide variety of forms such as outright grants, tax breaks, even the free or cheap use of government facilities or transportation.

The common way governments react to dumping or subsidies are a countervailing duty (CVD) against subsidies or an anti-dumping duty (AD) against dumping. This is essentially a tax to counteract the benefits the offending importer would have gained. The amount of the duty is usually the amount of the subsidy (for CVD) or the difference between what the product is being sold for and how much it should be sold for (AD).

Sometimes an importer has no intention of dumping, and may not even be aware that the purchase was under market value. There have been cases in the US where a shipper, not aware of the market rates, got hit with anti-dumping penalties with little or no warning.[2]

Gray market importing, also known as parallel importing, refers to the practice of importing a product contrary to the wishes of the producer, who normally has their official distributor. Normally, Customs does not get involved with internal company matters, but this is an area where some agencies are beginning to act. High value, branded consumer goods are particularly prone to gray market importing. While some countries have decided not to get involved, others consider this practice illegal. Hong Kong law allows a company to file civil charges for this practice.[3]

Protection of national interests is the catch-all for any goal the nation wishes to pursue. Customs carries out a wide variety of tasks under the orders of the federal government to promote its interests, far above and beyond just national security and competitiveness. This can include the control of wildlife products, pornography, or narcotics. Some of these rules are to enforce safety, such as requiring that all garments be flame-retardant. Others refer to social issues like pornography. Matters of national interest will be discussed further in the section on smuggling.

How Does Customs Operate?

We now shift from what Customs is trying to do to how they do it. This is one agency that has a remarkable variety of tasks that require many talents. On one hand, Customs collects duties, a financial and bureaucratic job. On the other hand, they are controlling smuggling, often dealing with very dangerous people. Customs agencies often look more like military operations, with machine guns and night patrols. Then they also serve to regulate trade, a politically sensitive job involving the competing interests of companies and governments at all levels from local to national, domestic and foreign.

There are many ways that a government can pursue the above goals, but with regards to customs, a few methods are commonly used:

- **Entry restrictions** - certain products may be completely restricted from entry, or only under certain conditions. A total restriction is called an embargo. Conditional entry can include quotas or special documentation required.
- **Rates** - duties vary dramatically. Most are simply a nominal tax, while others are intended to discourage imports of a certain product.

- **Information** - Customs requires that certain information be provided on imports and exports. Government data on economic trends is collected this way. Another reason for submitting information is to control the import/export of some products. Smugglers often get caught when they claim to be shipping something that just does not seem quite right.

The rates charged by Customs are known as tariffs, duties or taxes, all of which mean essentially the same thing in this case. Tariffs, in the case of the US and many other countries, are laws, decided by the federal government. The tariff rates are not just a fee decided by Customs or any other agency. Tariffs are payments that go into the general government fund, just like income tax we all pay. It is not a user fee, such as the fees a carrier pays to use a port. One does not pay Customs for their services.

Quotas are a common way that Customs can control the amount of a product that is imported. This applies to a product, not a specific company, and not just to a particular importer. There are two types of quotas. **Absolute quotas** state a given quantity that may be imported. A **tariff rate quota** states an amount that may be imported at a given tariff rate, and beyond that a different (higher) tariff rate applies. Importers are responsible for keeping track of how much of the quota has been filled so they do not bring cargo into a port only to find that it cannot enter the country. If an importer does accidentally find that the cargo cannot be brought in, there are three choices. The cargo can be A. destroyed, B. returned to the vendor or another country, or C. entered into a bonded warehouse until the next quote year.

Two Types of Quotas:
- Absolute   • Tariff rate

When the quote is filled, the importer can:
- Destroy the cargo
- Return cargo to vendor or another country
- Place it in a bonded warehouse

Visas are one way the exporters' government controls who gets to sell their goods under a quota. A visa is basically a document issued by a government to the businesses in its country to grant them permission to export that product under quota. When a government establishes a quota on imports, it is still up to the exporter's government to decide how to give out visas.

Quota Watch

Current Textile Quotas:

Where the average recent utilization rate indicates full quota utilization within four weeks from August 31, 1998.

Country	Category	Description	Percent Fill
Bangladesh	634	Man-made Fiber Optic Coats, Men's & Boys'	91.23
Bulgaria	433	Wool Suit-type Coats, Men's & Boys'	91.02
Indonesia	314 O	Cotton Poplin & Broadcloth Fabric Other Than Discharge Printed	93.52
Indonesia	350/650	Cotton and Man-made Fiber Dressing Gowns	85.28
Indonesia	625-629 O	Man-made Fiber Filament Fabric Other Than Discharge Printed	91.16
Indonesia	638/639	Man-made Fiber Knit Shirts	88.88
Indonesia	Wool Sub-group	Wool Apparel and Non-Apparel	81.57
Indonesia	435	Wool Coats, Women&s Girls'	84.72
Macedonia	435	Wool Coats, Women's & Girls'	78.42
Malaysia	445/446	Wool Sweaters	70.82
*Mexico	TPL5	Cotton or MMF apparel in Chapters 61 and 62 both cut (or knit to shape) and sewn or assembled in the territory of a NAFTA party from fabric or yarn obtained outside of a NAFTA party	86.77
Mexico	TPL5	Textile & Apparel in Chapters 61, 62 and 63 sewn or assembled in Mexico from fabric cut in the U S from fabric knit or woven outside of the U S or Mexico	99.84
Taiwan..	444	Wool Suits, Women's & Girls'	74.16
Thailand	442	Wool Skirts	71.21
*U.A.E...	334/634	Cotton and Man-made Fiber Other Coats, Men's & Boys;	99.97
*U.A.E.	347/348	Cotton Trousers, Slacks and Shorts	98.60
*On Hold.			

Note: The above list was compiled using U S Customs Service data. Potential adjustments to quotas or charges have not been factored into the analysis.

Fully Utilized Quotas as of August 31, 1998

Country	Category	Description
Kuwait	361	Cotton Sheets
U.A.E.	315	Cotton Fabric, Printcloth
U.A.E.	361	Cotton Sheets

Note: The above list was compiled using U S Customs Service data. Application of adjustments to quotas or charges may allow for reopening of quotas

SOURCE International Development Systems Inc . Washington D C

Source: Journal of Commerce

Sometimes a country controls the amount of a product that can be exported. When both the exporting and the importing country have quota controls, this is known as **dual monitoring**. Why might a country want to limit its exports? The government may have a trade agreement that limits exports as a concession to its trade partners.

In addition to the above methods, **enforcement** by Customs is done in a variety of ways. Its agents routinely board ships, planes, and other vehicles, to inspect the cargo. The documentation, to be discussed in the next chapter, provides many clues to the possibility of violations. Sophisticated agencies are even using computer technicians to search thorough electronic data searching for trends that indicate smuggling. There is always going to be some smuggling, so the goal is not to eliminate it, but simply to minimize it. Cost trade-offs are made in deciding where to search and where to leave alone.

Customs agencies vary in their organization, legal status and operations in different countries. They are virtually always a federal agency, with powers above the local governments. Some countries have similar agencies to control the movement of people and cargo within the country, such as the US Interstate Commerce Commission, but these are very different and not discussed here. The place at which a person or cargo enters the country is the port of entry, which could be physically at the border, such as an overland entry point or port, or anywhere in the country, such as an airport. Note that Customs designates ports of entry; a carrier or port cannot decide on its own to accept foreign traffic. Some countries that restrict international commerce often do so by having few ports of entry, or only placing them in inconvenient locations.

CUSTOMS OPERATIONS

We now look at the Customs function for a shipment. This can get very confusing. To clarify how this all fits together, refer to the box titled "Customs Clearance Process". The next chapter also discusses this process. The emphasis here is on what Customs wants and how they operate. In the next chapter, we show how importers/exporters manage the documentation process throughout the shipment.

Every shipment involves many parties, each of whom has different roles and responsibilities. The following are some of the main roles played. Others will be introduced in this chapter and the next. A person may be playing more than one role. For example, the shipper and the exporter are usually the same, or they may be different parties.

- **Shipper** - Contracts with carriers for movement of goods. This refers to a person's relationship with the carrier.
- **Exporter** - This would probably be the same as the shipper, but it refers to the person's relationship with Customs. This is the person responsible for fulfilling export regulations of that shipment.
- **Importer** - The person who imports the cargo. This may be the consignee, unless the consignee is the bank writing a letter of credit (discussed in Chapter 12).
- **Importer of Record** - This is a legal distinction. See the box below.

Importer of Record

Customs wants one entity that can be held responsible for a shipment. This can get confusing when there is one company producing a product, another buys it, sells it to another that is exporting it, and another company is receiving it on the other side. That is why the Importer of Record is the one party held responsible for fulfilling most of the regulations associated with Customs compliance. This includes:

• Reasonable care	• Record keeping	• Valuation
• Informed compliance	• Classification	• Marking

Export Clearance

Almost every country wants to encourage exports, and have relatively less concern for what is leaving their country than what comes in. For this reason, the rules on exporting tend to be less. However, there are some minimal requirements. Most countries control the export of some goods. The US requires an Export Declaration form (Exdec). This describes what is being exported. They use this form to identify goods that may be restricted from export, but mostly the form is for collecting trade data.

Exporters should be aware of exactly what they are exporting, and where it is going. Governments normally have a website or other source of information on whether there are any restrictions on shipping a given commodity. Importers normally let the CHB handle such questions, but exporters need to either find out for themselves. Freight forwarders may be useful in identifying restrictions, but this is not their job. Whereas the CHB's job is to comply with government regulations, a freight forwarder's job is to arrange for carriage.

Export restrictions may be one of a few types. There may be a complete ban on the product from being exported. Oddly enough, this is quite rare and things one would think would be banned are not. For example, weapons are one of the biggest exports of the US and other countries. The commodity may only be approved for export to certain countries, or to specified importers. This is a soft restriction, in that it is difficult if not impossible to prevent the cargo from being redirected after it leaves the country. Finally, there may be restrictions on the amount, type or quality. For example, some African countries known for their livestock only export males. They do so to control the breed and do not want other countries to establish their own stock.

Some cargo being exported was previously imported. In other words, it came from outside, there was probably some processing, and now it is either being returned or sent on to another destination. In such cases, there may be duties refunded. This will be discussed in the next section, Import Clearance, only because there are other concepts that need to be explained beforehand.

A **carnet** is a document used for cargo not intended for sale, and thus may be shipped without paying customs duties. In other words, the cargo is treated more like luggage. This is common for exhibition goods, or when entertainers travel with their equipment. Carnets do not apply on consumable or perishable goods. If the cargo is left behind, then it is treated as a normal import and duties are paid.

Import Clearance

The import clearance process is far more complex, expensive and difficult. This is because countries are almost always more concerned about what is coming into their country that what is leaving. Some of the differences in import clearance results from the fact that the importer probably has not seen the cargo before and probably did not arrange for the shipping. As for the cargo itself, there is more likelihood of damage or loss toward the end of the shipment than at the beginning.

The port of entry is generally where Customs exercises its controls. This is where importers/exporters must produce the documents that will allow them to move their cargo, and this is where Customs is most likely to inspect or seize the cargo. However, they may also inspect cargo long after it has been imported, far from the port of entry. With exports, their control is naturally gone when the cargo leaves the country.

Customs Clearance Process

A. Where do I clear Customs?
    Port of entry
    In-bond shipments
    Free Trade Zones

B. How do I clear Customs?
    Customs House Brokers
    Harmonized Tariff Schedule
    General Rules of Interpretation

C. Duties are based on three factors.
    Classification of cargo
        General Rules of Interpretation
        Essential Characteristics
    Value
        Classification
        Sale price vs Transfer price
    Country of Origin
        Substantial transformation
        Labeling

D. After Importation
    Informed Compliance
    Duty Drawbacks
        Less value added
    Liquidation
    Customs audits

For the vast majority of transactions, the importer/exporter will only need to file paperwork and pay duties. The duties are a tax, usually on imports and occasionally on exports.[4] In the US, these duties are compiled in a book known as the **Harmonized Tariff Schedule (HTS)**. This has become a massive book, listing how much is charged for imports depending on the product and country of origin.[5] Because of its size and the fact that tariffs change on a daily basis, the 'book' is actually a database. **Duties are based on three factors, the cargo's classification, value, and country of origin.** In all matters, an importer or Customs may seek internal advice or ruling from Customs headquarters, but this is not legally binding.

CHAPTER 61

ARTICLES OF APPAREL AND CLOTHING ACCESSORIES, KNITTED OR CROCHETED 1/

Notes

1. This chapter applies only to made up knitted or crocheted articles.

2. This chapter does not cover:

   (a) Goods of heading 6212;

   (b) Worn clothing or other worn articles of heading 6309;

   (c) Orthopedic appliances, surgical belts, trusses or the like (heading 9021).

3. For the purposes of headings 6103 and 6104:

   (a) The term "suit" means a set of garments composed of two or three pieces made up, in respect of their outer surface, in identical fabric and comprising:

   - one suit coat or jacket the outer shell of which, exclusive of sleeves, consists of four or more panels, designed to cover the upper part of the body, possibly with a tailored waistcoat in addition whose front is made from the same fabric as the outer surface of the other components of the set and whose back is made from the same fabric as the lining of the suit coat or jacket; and

   - one garment designed to cover the lower part of the body and consisting of trousers, breeches or shorts (other than swimwear), a skirt or a divided skirt, having neither braces nor bibs.

   All of the components of a "suit" must be of the same fabric construction, color and composition; they must also be of the same style and of corresponding or compatible size. However, these components may have piping (a strip of fabric sewn into the seam) in a different fabric.

   If several separate components to cover the lower part of the body are presented together (for example, two pairs of trousers or trousers and shorts, or a skirt or divided skirt and trousers), the constituent lower part shall be one pair of trousers or, in the case of women's or girls' suits, the skirt or divided skirt, the other garments being considered separately.

   The term "suit" includes the following sets of garments, whether or not they fulfil all the above conditions:

   - morning dress, comprising a plain jacket (cutaway) with rounded tails hanging well down at the back and striped trousers;

   - evening dress (tailcoat), generally made of black fabric, the jacket of which is relatively short at the front, does not close and has narrow skirts cut in at the hips and hanging down behind;

   - dinner jacket suits, in which the jacket is similar in style to an ordinary jacket (though perhaps revealing more of the shirt front), but has shiny silk or imitation silk lapels.

   (b) The term "ensemble" means a set of garments (other than suits and articles of heading 6107, 6108 or 6109) composed of several pieces made up in identical fabric, put up for retail sale, and comprising:

   - one garment designed to cover the upper part of the body, with the exception of pullovers which may form a second upper garment in the sole context of twin sets, and of waistcoats which may also form a second upper garment; and

   - one or two different garments, designed to cover the lower part of the body and consisting of trousers, bib and brace overalls, breeches, shorts (other than swimwear), a skirt or a divided skirt.

   All of the components of an ensemble must be of the same fabric construction, style, color and composition; they also must be of corresponding or compatible size. The term "ensemble" does not apply to track suits or ski suits of heading 6112.

4. Headings 6105 and 6106 do not cover garments with pockets below the waist, with a ribbed waistband or other means of tightening at the bottom of the garment, or garments having an average of less than 10 stitches per linear centimeter in each direction counted on an area measuring at least 10 centimeters by 10 centimeters. Heading 6105 does not cover sleeveless garments.

1/ See Section XI, Statistical Note 5.

The book/database of tariffs charged by U.S. Customs

Classification

How is the duty amounts determined? The **General Rules of Interpretation** (GRI) are the rules that describe in detail how to read the HTS. For any product being imported, Customs will decide how it is classified by the **Essential Characteristics of the commodity**. One then looks up that commodity in the HTS. There are different duty rates depending on where the product is coming from.

The US HTS was developed over the years from a political process. When a US company feels it is being pressured by foreign competitors, a common response is to seek relief by pressuring the federal government to restrict the imports. The commodity description, with a higher duty rate, is then placed in the HTS that will apply only to that product that if offending the US manufacturer. Over the years the number of such exceptions has grown to the point where the HTS is so large that most users have a computerized version instead of hard copy.

What is meant by 'essential characteristics'? These are those aspects of a product that determine what it is, at least in terms of customs classification. One famous story shows just how the commodity classification process works. A company wanted to import into the US edible panties, essentially a sex toy. Customs had never seen this before and needed to classify it based on the HTS. They had a choice; it could either be called a food, since it was edible, a garment, since it was worn, or a toy, because that was the idea behind it. Depending on what it was classified as, the duty rate would vary dramatically. If it were a garment, it could be subject to quotas designed to prevent foreign garment makers from running US garment makers out of business. As a food product, it would be subject to agricultural and health inspections.

Sometimes a shipment consists of parts that, if assembled, would make a different product. It is not always clear if what are being imported are parts or a product that is left unassembled for shipment. **The Doctrine of the entireties** states that if the entry consists of parts that if assembled would create an object that salable by itself ("commercially different"), then the tariff rate would be for the end product.[6]

Cargo Value

Once the classification of the product is determined, the next step is to determine its **value**, that is, how much is the shipment worth. WTO rules state that the customs value of cargo should be the price actually paid, or payable, when sold for export to the country of importation. When something is imported but it is not being sold, such as transfers within a corporation, there is no sale price and thus **transfer prices** are used. Seventeen percent (17%) of US imports are transfers within multinational companies, according to the Commerce Department.[7] The WTO recognized five methods of identifying the cargo value: transaction value, transaction value of identical merchandise, transaction value of similar merchandise, computer value, deductive value, or other reasonable means. The importer is responsible for declaring the value of the cargo.

Another issue with determining the appraised value is that **accounting systems vary** in different countries. US Customs requires that the price developed be compatible with US accounting methods, but that does not always alleviate the confusion. Some customs authorities use databases of prices, which it has been said leads to arbitrariness and fictitiousness. These databases tend to use the costs that are easiest to obtain, and some data even comes from retail catalogues. These database prices often do not reflect such variables as timing, mode of transport and commercial value of the transaction.

There are many costs that may or may not go into the value of the cargo. The cost of international transportation is not included in the value. Payments to an agent for a given shipment are included. If the importer provided any assistance to the exporter/seller, the value of that assistance is dutiable (this cost is known as an 'assist').

Can cargo be overvalued? Some countries, such as Pakistan, require a minimum price for exports, intended to improve the image of these countries. Apparel and textiles are the most common commodities. There was a case in which a US firm was sued for overvaluing their imports, in which the importer gets a rebate and the exporter meets the national law. The US customs said that trade statistics were skewed as a result of this procedure.[8]

Country of Origin

Robert Reich, who later became the Secretary of Labor, wrote a book called *The Work of Nations*, in which he showed how a so-called "American car" was built from parts from all over the world. He shows how it can be difficult identifying where some products were made when they are the result of a long production process that crosses many borders. It is still important that a country of origin be identified because duty rates and other laws may be different for the products of different nations. For example, when two countries enter into a free trade agreement, they do not want other countries taking advantage of the situation.

Country of Origin
"That country where an article was grown, mined, produced, or manufactured."

The country of origin is "that country where an article was grown, mined, produced, or manufactured".[9] How does one determine where this is if the production process took place in different countries? The country of origin is the last country where the goods underwent a **substantial transformation (ST)**

Matters of ST can become very complicated. Many court cases have resulted from disputes over what constitutes a change in the product. In 1908 a court ruled that cork had not gone through a substantial transformation just because it was dried, treated, and cut into pieces for bottling drinks. The court ruled that a substantial transformation has occurred when the product has a **"new name, character, or use"**, a phrase that is still used today.[10] Recently, emphasis has been on the character more than the other two factors. A common test to see if there has been an ST is to see if there has been an increase in value. GATT in 1994 agreed on country-of-origin rules, in which the country of origin is the last country where the product underwent a change that meant a new tariff classification, known as the **tariff-shift rule**.

In one case, there were jeans shipped to Mexico, baked in an oven, then returned to the US. A legal case was made that there was no substantial transformation, but the case lost. The baking was done to fix the dye, which means the jeans were transformed in a way that increased their value.

Labeling: US law states that a product must be labeled to the ultimate purchaser, so if raw material is imported, undergoes a substantial transformation, then it is an American product.

After Customs Clearance

Clearing Customs may not be the end of an importer's involvement with Customs, and it certainly is not the end of the logistics process. One of the ways that Customs guarantees they will get paid is to require a **bond**. Importers pay for a bond, normally done by paying a company that sells customs bonds. This requires that the importer be reasonably creditworthy, or else no bond dealer would deal with them. Bonds allow the importer to move cargo without delay.

The idea of bonds began in the 1700s because importers wanted to wait as long as possible before paying, when it took days or weeks for cargo to arrive. Now the concern is that Customs may audit the shipment and decide much later that more duties need to be paid. At that point, the cargo could have been sold off, which means they do not have recourse to seize the cargo. One solution may be to pay Customs and wait for their final solution. This would be unworkable because that means holding the cargo until Customs is sure that they have made a final judgment. They do not want to rush into that decision, and they do not want the cargo sitting around. **Liquidation** is when the duties payable are finally decided upon and charged by Customs.

When a product is imported and then re-exported, most Customs agencies allow for a **duty drawback**, a refund of part or all of the duties paid. If there has been some value-added processing done to the product while in the country, the duties paid would be based on the amount of change in the value of the product.

Duty drawback is a two part process. When something is imported, duties are paid. If the product is subsequently exported, the company may be owed a refund. Customs generally will allow a drawback of either a portion or maybe even all the duties originally paid. Each government has a different drawback allowance. Some will give duties back if product was imported and found to be damaged or unwanted. Some will allow products to be tested or repackaged, and received drawback. Some will allow a drawback of duties paid on raw materials used in the manufacture of another product, which is exported. Other governments will not.

There are some special cases where drawbacks may or may not apply. The product must also be in the same condition to qualify for a drawback. **Substitution drawbacks** are when a similar product is re-exported. For example, a company may not know which of its inventory was imported, but it re-exports some of the inventory. Drawbacks also apply to merchandise that does not conform to specifications when it was imported.

What about the reverse, when US goods are exported and then re-imported? US exports that are returned are dutiable as foreign goods except when there has been no ST, or if it was exported for repair/alteration.

In one case, the American space agency NASA imported a valuable lens, installed it into a satellite and launched it into space. They then applied for a duty drawback since it was 'exported'. At first Customs refused, saying that the lens was not exported. A court later found that to be exported, it only needed to leave the country. The law never said that it could not be exported into outer space.

Enforcement and Inspections

Inspections are a routine part of Customs' work. They are usually confirming information from the importer's documents, but they may also be looking for controlled goods. This can include arms and ammunition, dangerous goods, protected wildlife products, narcotics or commercially controlled goods. Customs cannot possibly search every shipment, so they pick and choose based on the likelihood of finding a violation.

Many countries have highly sophisticated systems for deciding which shipments to inspect based on the probability of finding a violation. Companies that are importing for the first time are most suspect. Known shippers are much less likely to be searched. Once a company is caught violating the law, it will be subject to search much more in the future. Importers sometimes accidentally buy from a shipper that also engages in illegal activity. For this reason, US Customs publishes names of companies that are engaged in illegal importing activities. This helps importers avoid doing business with companies that engage in illegal transshipments.

Even a routine inspection can be costly. For example, when US Customs pulls a cargo container aside to be unloaded, inspected and reloaded, the importer pays for this, which can cost several hundred dollars. One controversial method of inspection is to drill a hole into a container to look inside. See the following chapter for more on this.

Pre-shipment inspections are being used, mostly by developing nations, in which the value, quality and quantity of a shipment is determined at the exporter's location, and this information is then transmitted to the importing country's customs authorities.[11] For example, Colombia requires electronic parts inspected by some global company such as Bureau Veritas to prevent poor-quality parts from being imported.[12] The more common reason for pre-shipment inspections is to prevent corrupt customs officials from undervaluing the cargo and taking a bribe.

Many countries have requirements regarding the marking of products, especially garments. For example, US law requires that all foreign-made consumer goods be labeled to identify where it was produced. The labels must be A. conspicuous, B. legible and C. permanent. These labels are not required for US made goods. The reason for this law was to allow consumers to know that the product was foreign-made in the hope that they would prefer US-made products. In another example, Mexico requires that imported consumer products have Spanish-language labels.

Customs agencies around the world are becoming more sophisticated in how they operate. This is necessary not only so they can be more efficient, but also to keep up with the needs of the import/export companies. A company that has a state of the art logistics program does not want to wait for days or weeks while Customs processes paperwork. The future of customs clearance is for a uniform data system in which the customs services of different countries are integrated. International Trade Prototype is a program developed jointly by the US and Britain in which the export data also becomes the import data, eliminating the need to re-enter the data.[13] This means agreeing on how they classify merchandise for duty purposes, and the electronic language used. Another issue is the sensitive and private information of some companies.

In another example of efforts to improve the customs process, the Group of Seven (US, UK, German, France, Italy, Canada and Japan) has been working on simplifying customs rules. Among these countries, it was estimated that 800 pieces of data are used, many of them redundant. They believe that by using common standards, they can reduce that figure to about 140 pieces of data. Japan has strict rules against sharing information with the business community, and has been critical of the project. The European Community may use these standards to develop a European customs standard. Part of the differences between the countries comes from different priorities. There has been a shift from revenue collection and more effort to curb drug smuggling and money laundering.[14]

The World Customs Organization (WCO) is the Brussels-based entity that brings together the Customs agencies in the hopes of coordinating their procedures. The WCO is comprised of 159 countries and 97% of world trade. Michel Danet, Secretary General of the WCO, suggests that the WTO not write standardized processes for customs because each agency needs flexibility.[15]

Accountability of customs agencies has been an ongoing and emotional battle. Importers and exporters often find themselves at the mercy of the authorities that can seize cargo, damage cargo in inspections, or delay shipment. If the shipper complains too loudly, they can become the targets of the inspectors so that 'random' searches are not so random. This happens in almost every country, and there seems to be little difference between developed and undeveloped countries, clean governments or corrupt ones. Sometimes these problems are the result of corruption, but bureaucracy can also be the cause. In all fairness, customs agents are pressured from all directions to catch smuggling, enforce numerous and sometimes contradictory laws, and yet not delay cargo.

While much of the fraudulent behavior is done by the traders, another problem is the role played by some governments. They have it in their best interest to resist other countries' trade laws, and a government is in a very good position to do something about it. China has been reported using Hong Kong to avoid US textile quotas. By one US estimate, China exported $13 billion in textiles in 1993, yet the rest of the world reported importing $23 billion worth of Chinese textiles.[16]

Some countries have programs to allow low-risk cargo to move across borders without delay. This is obviously something done between two countries that already have good relations and work together on trade issues. To qualify as low-risk cargo, there are two basic criteria. First, the company doing the trading must be established, known and not have been caught with any significant infractions. Some types of cargo are typically traded between large, established companies while others are often traded between middle-men and short-lived entities. This affects the risk. Second, the cargo itself should be something that is not commonly involved in illegal operations. For example, car parts can be seen as low-risk because it is unlikely a shipment of car parts is going to be exchanged with something similar but illegal. Pharmaceuticals, on the other hand, can easily be used to transport narcotics.

Intellectual property is a product that has a patent, copyright, trademark or servicemark. This includes such things as music, art, books, and most clothing (particularly designer clothing). Counterfeiting is a common practice and Customs is responsible for controlling the import of such goods. This can be tricky when the product is an almost perfect duplicate. Sometimes governments make intellectual property rights enforcement a high priority, and sometimes they do not bother controlling it at all. The companies that own the labels sometime push governments to enforce the rules more strictly. Companies sometimes work with Customs and teach the agents how to identify counterfeit goods, and identify which shippers are handling such products.

When Customs confiscates counterfeit property, they may ask the owner of the trademark how they want it disposed. For example, one prominent clothes maker had hundreds of counterfeit pants donated to a charity. In another case, a major toy maker had the fake dolls destroyed. If counterfeit products are given away, as the clothes maker did, this may be generous, but there are also concerns of consumer protection. For example, if people are hurt when the counterfeit product ends up being unsafe, you cannot very well sue the importer whose cargo was confiscated.

The standard practice in shipping documents has been to include the commodity. There was a proposal in the US to require invoices to state what trademark is used. For example, the import forms cannot just say '8,000 t-shirts', it would need to say '8,000 Gap shirts'. This would be a powerful tool to combat trafficking of pirated goods, but there is also concern by companies that trade secrets would be revealed to their competitors.

Every day the US Customs Service processes:
1.3 million people
348,000 private vehicles
38,000 trucks and railcars
16,000 containers on 600 ships
2,600 aircraft
Source: U.S. Customs and Business Week

MANAGING THE CUSTOMS OPERATIONS

We now shift our view to that of the importer/exporter that needs to work with Customs and manage the customs functions. In the international logistics process, this is a critical link. Even if an enterprise has developed a high performing supply chain, crossing international borders can disrupt even the best of plans. Luckily, the customs function can be managed proactively to minimize problems.

ACE: Automated Commercial Environment
U.S. Customs is finally beginning a computer system that has taken 10 years and $1.4 billion to develop, an information system that allows all major Customs transactions to happen electronically, including the sharing of appropriate information. Technical difficulties have kept it offline since 1998, but it should be running soon. Given the new demands on Customs and border controls, the need for this system has never been greater.

The first step is to be aware of the rules regarding import/export, and how the Customs agency operates. Working with Customs requires an appreciation for their needs and operating methods. Often these agencies can be bureaucratic, inefficient and even corrupt, but they cannot be avoided. The best solution is to work with them as best as possible.

The second step is to decide where to clear customs. In many cases the importer can choose carriers that enter in different ports. There is also the possibility of either clearing customs at the port or moving **in-bond**. Moving cargo in-bond means that it has not yet cleared Customs. This is done to relieve port congestions, labor for inspections are lower inland than at ports, and to delay paying fees. There is a new concern that this may pose a security problem, but in-bond shipping is very common and it does not look likely to cease.

In the US in the mid 1990's the Customs Service reorganized their operating methods through an act of legislation known as the **Customs Modernization Act**, known as The Mod Act. One of the most important changes was that instead of checking documentation on every shipment, they now require all importers to maintain their own records. If and when Customs inspects these records and there are mistakes, the penalties can be severe. This process of essentially leaving the importers to maintain their own records is known as **informed compliance**.

In almost every country, the laws are constantly changing, which is why the vast majority of shippers use **customs house brokers (CHBs)**. These are intermediaries that handle the customs clearance process. Besides being complicated and ever-changing, customs clearance often requires someone to be at the port of entry in case there are any questions, or the cargo is pulled aside for inspection. This can be expensive for the shipper who is far from the port. About 98% of US shippers use CHBs. The broker usually charges a minimum fee of around $100, and then a certain amount for each shipment cleared through Customs. Customs house brokers provide a variety of services, such as:

- **Computer interface**. Many Customs agencies either offer or require electronic documents. The computer programs for this are not generally available or simple to use for the average shipper.
- **Immediate delivery**. CHBs can clear customs before the shipment arrives, so that it can be taken as soon as it reaches the port.
- **Paperless release**. By using electronic documents, the truck driver can go to the port without any paper documents to pick up the cargo.
- **Door to door service**. CHBs can arrange for local delivery.
- **Classification**. Assist in the classification of the product.

We have been referring to CHBs as individuals, but the companies that offer customs brokerage services are also referred to as CHBs, which can a little confusing. These companies are sometimes associated with freight forwarding companies, transportation companies, or they may be independent. Many specialize in certain industries like computers or garments, where it helps to have a specialized knowledge of that kind of commodity. For example, classifying an obscure computer part can make the difference of higher duty rates. Dual use components with military applications are especially important.

The US Customs Service requires that anyone transacting Customs business on behalf of another party must have a **customs brokers license**. Additionally, to file entries with the government the license holder must obtain a permit. All corporations licensed to be a CHB must have at lease one corporate officer which is licensed, and must have enough licensed employees to prove responsible supervision. Recent court rulings have disturbed the industry because if a company has subsidiaries or related companies (partially or wholly owned), each subsidiary is considered separate. Thus each company would need their own licensed CHB.

Passing the U.S. customs brokers exam has proven to be a major challenge. While there are some years that 50% pass, in other years as little as 2% of testers pass. Ro Leaphart, formerly a logistics executive of The Gap, trains people to pass the exam. She warns her students that "personal stamina and a sincere dedication of time are the two basic ingredients" and that one should study 10-20 hours a week for 12 weeks before the exam.[17] Once the exam is passed, the applicant undergoes a personal background check, because this is a job that gives one an opportunity to work the system. In one case, the daughter of a famous Mafia figure was denied a license.

If an importer fails to keep proper records, there is the possibility of fines or even criminal prosecution. This is one reason why CHBs are used. Even innocent mistakes can be subject to fines because shippers are expected to be able to maintain professional-quality records. Criminal prosecution is designed more for cases where a shipper seems to be trying to cover something up on purpose. While CHBs handle imports, forwarders are used for exports. The forwarder needs to classify the cargo, even though they may not be familiar with it. Sometimes they guess at the proper classification, leading potentially to the shipper being fined.

CUSTOMS POWER OF ATTORNEY

IRS/SS # ____________

Check appropriate box
☐ Individual ☐ Corporation
☐ Partnership ☐ Sole Proprietorship

Is this a Corporate Power of Attorney? ☐ Yes ☐ No If yes, please attach a list of all divisions and/or subsidiaries covered.

KNOW ALL MEN BY THESE PRESENTS: That, ____________
(Full name of person, partnership, or corporation, or sole proprietorship)

a corporation doing business under the laws of the State of ____________ or a ____________
doing business as ____________ residing at ____________
having an office and place of business at ____________, hereby constitutes and appoints each of the following persons ____________
(Give full name of each agent designated)

as a true and lawful agent and attorney of the grantor named above for and in the name, place and stead of said grantor from this date and in all Customs Districts, and in no other name, to make, endorse, sign, declare, or swear to any entry, withdrawal, declaration, certificate, bill of lading, carnet or other document required by law or regulation in connection with the importation, transportation, or exportation of any merchandise shipped or consigned by or to said grantor; to perform any act or condition which may be required by law or regulation in connection with such merchandise; to receive any merchandise deliverable to said grantor.

To make endorsements on bills of lading conferring authority to transfer, title, make entry or collect drawback, and to make, sign, declare, or swear to any statement, supplemental statement, schedule, supplemental schedule, certificate of delivery, certificate of manufacture, certificate of manufacture and delivery, abstract of manufacturing records, declaration of proprietor on drawback entry, declaration of exporter on drawback entry, or any other affidavit or document which may be required by law or regulation for drawback purposes. Regardless of whether such bill of lading, sworn statement, schedule, certificate, abstract, declaration, or other affidavit or document is intended for filing in any customs district.

To sign, seal, and deliver for and as the act of said grantor any bond required by law or regulation in connection with the entry or withdrawal of imported merchandise or merchandise exported with or without benefit of drawback, or in connection with the entry, clearance, lading, unlading or navigation of any vessel or other means of conveyance owned or operated by said grantor, and any and all bonds which may be voluntarily given and accepted under applicable laws and regulations, consignee's and owner's declarations provided for in section 485, Tariff Act of 1930, as amended, or affidavits in connection with the entry of merchandise.

To sign and swear to any document and to perform any act that may be necessary or required by law or regulation in connection with the entering, clearing, lading, unlading, or operation of any vessel or other means of conveyance owned or operated by said grantor;

To authorize other Customs Brokers to act as grantor's agent, to receive, endorse and collect checks issued for Customs duty refunds in grantor's name drawn on the Treasurer of the United States, if the grantor is a nonresident of the United States, to accept service of process on behalf of the grantor;

And generally to transact at the customhouses in any district any and all customs business, including making, signing, and filing of protests under section 514 of the Tariff Act of 1930, in which said grantor is or may be concerned or interested and which may properly be transacted or performed by an agent and attorney, giving to said agent and attorney full power and authority to do anything whatever requisite and necessary to be done in the premises as fully as said grantor could do if present and acting, hereby ratifying and confirming all that the said agent and attorney shall lawfully do by virtue of these presents; the foregoing power of attorney to remain in full force and effect until the ____ day of ____________, 20____, or until notice of revocation in writing is duly given to and received by a District Director of Customs. If the donor of this power of attorney is a partnership, the said power shall in no case have any force or effect after the expiration of 2 years from the date of its execution.

IN WITNESS WHEREOF, the said ____________
has caused these presents to be sealed and signed: (Signature) ____________
(Capacity) ____________ (Date) ____________
WITNESS

(Corporate seal)

Customs Compliance Specialist: A Job Description

Some of the responsibilities of a customs manager or specialist:

Set up and maintain procedures to ensure Customs compliance.

Advise staff on Customs requirements.

Manage relationship between company and Customs.

Develop or revise product classification program for quicker cargo clearance.

Develop use of foreign-trade zone.

Supervise customs brokers to ensure timely and accurate clearance.

Develop duty-drawback program.

Advise company on Customs implications of transactions.

Source: Journal of Commerce

Sometimes companies choose to file Customs entries themselves, using a person whose job title would be "Customs Compliance Specialist".

In many countries, customs is not too complicated, or there may not be a well-developed customs brokerage industry. In the US, the heavy penalties for mistakes have caused some companies to take the brokerage job in-house because they know best what it is they are shipping. The Mod Act also introduced electronic filing, in which the company can submit the forms electronically, another reason that some companies are doing their own customs filings.[18]

A major concern is identifying who is the exporter. Exporters have little control over where their cargo goes after it has left their control, causing some serious problems. For example, some cargo (such as dual use military technology) can be sent to some countries but not others. A buyer has it shipped to one country, but then sends it on to a country where it is forbidden, known as an **illegal diversion**. When a shipment leaves a country, the shipper may be the manufacturer, a reseller, a trading company, or even the NVO. The US has proposed to define an exporter as "an individual, manufacturer, intermediary, or merchant who sells and/or ships goods and services to a foreign importer." One industry leader suggested that a better definition would list the exporter as "the party that benefits most, financially or otherwise, from an export"[19], known as a **Principal Party in Interest** (PPI).

Importers (and to a lesser extent exporters) need to deal with the possibility of **customs delays**. In the case of time-sensitive logistics systems, they should always calculate for the possibility of delays. As Customs agencies become more sophisticated, the computer is doing more of their work. Sometimes computers malfunction, and this may mean that cargo cannot be cleared until the computer is running and thus the documents processed. This is becoming a major issue in some countries. Paperwork delays or computer crashes do not affect ocean transport as much because a ship takes a few weeks to cross an ocean, and lot sizes are much bigger. Trucks and planes, on the other hand, have much smaller lot sizes and cross the border much more rapidly. Hold-ups in the customs would affect them much more drastically. This means that companies with overseas suppliers are less prone and those with continental US suppliers much more prone to this type of problem.

In addition to the legal controls on imports, there are also **non-tariff barriers** (NTBs). These are bureaucratic ways to discourage certain imports. For example, at one time France required that all VCR imports come through a small, out of the way port where they could only clear customs at very limited hours. Japan required 'weighing fees', in which incoming cargo containers were weighed for no particular reason. Shippers can get around such restrictions, but the whole point is that it makes trade not cost effective.

Cargo inspections are not just done by customs agencies. Many shippers and carriers are doing their own inspections, or hiring security companies to do it. In other words, they would rather catch a problem and deal with it internally that let the government officials catch them and see the company's good name associated with a drug bust. Some companies are having their cargo x-rayed in case someone tried to place contraband in the shipment. This may be done by an employee or someone not associated with the company.

A nightmare for any trader is the possibility of a **customs audit**. Much like a tax audit, this is when the customs agency asks a company to either answer some questions, provide supporting documentation, or a full inspection of the records and interviews with the personnel. In those countries that have a proper legal system, it is usually recommended to maintain an open and honest approach. Often lawyers or accountants are brought in to represent or advise the client during the audit. There is some debate as to whether a lawyer or an accountant is appropriate. Lawyers note that an audit is checking that the company is acting according to the law, yet accountants note that the adherence to the law is essentially a matter of good accounting.

Prior disclosure is when an importer notifies customs of a mistake without being asked. The key word here is 'mistake'. If an importer violates the rules with the intention of making a prior disclosure, this is clearly an abuse of the system. There have been cases of prior disclosure where the court believed that the defendant intended to defraud the government from the beginning.[20] This can be seen as a sort of plea-bargain where customs does not need to pursue a lengthy investigation and the shipper is willing to pay something for the mistake. It is very practical and wise for a government to have some sort of prior disclosure process because traders are humans, and mistakes occur. By treating every incident as if there were some criminal intent would be unfair to some shippers and disrupt the customs process. However, in those countries where the legal system is not well developed and officials are less than honest, the trader must make hard decisions about what information to release and how to work with the officials.

Companies need to anticipate customs audits. The US Customs recently reduced the number of information items it looks at from 220 to 110, though it tightened up the standards by only allowing for one mistake. For example, the following are the most common mistakes:

- An importer's listing of the product's value off by more than .5%.
- An importer's payment of a dumping duty is off by more than .5%.
- More than 5% of products are misclassified.

U.S. Customs Service
Opinion on Elephant Jock Straps

HQ 086388
April 25, 1990
CLA-2 CO:R:C:G 086388 CRS
CATEGORY: Classification
TARIFF NO.: 6114.30.3060(EN)

Ms. Jean Maguire
Area Director of Customs
U.S. Customs Service
New York Seaport
6 World Trade Center
New York, NY 10048-0945

RE: Reconsideration of NYRLs 830911, 831612 and 835595; Elephant Jock Strap, as an article of fancy dress, is of a class of merchandise separate and distinct from toys and is excluded from Chapter 95 by virtue of Note 1(e), Chapter 95.

Dear Ms. Maguire:
This is in reply to your memorandum dated January 22, 1990, file 830911, in which you requested reconsideration of New York Ruling Letter (NYRL) 830911 dated September 16, 1988, and NYRL 835595 dated November 24, 1989.

FACTS:
The merchandise at issue in the above-referenced rulings consisted of various articles of novelty apparel. NYRL 835595 involved the classification of an article of knitted fabric of oversized proportions designed to cover the male genital area. The article, marketed under the name

"Superstud," was classified in subheading 9503.90-6000(EN), Harmonized Tariff Schedule of the United States Annotated (HTSUSA).

NYRL 831612 concerned the classification of a nylon knit article with an exaggerated frontal section described as "super man shorts." NYRL 830911 concerned, in part, a jock strap of knit and woven man-made fabric, the pouch of which was adorned with the likeness of a king with crown. Although the rulings were issued prior to the entry into force of the Harmonized System, both articles were classified, on an advisory basis, in subheading 9503.90-6000.(EN)

Pursuant to your request for reconsideration of the above rulings, you have submitted a sample of an "elephant jock strap," a 100 percent knit nylon groin pouch with a seven inch "trunk." There are large "elephant ears" on either side of the pouch, and attached to the front or "face" of the article are "eyes" and a tuft of "hair." An elastic strap or G-string girds the article around the wearer's loins.

The elephant jock strap is unrelated to the NYRLs at issue but is representative of the articles which were the subject of those rulings. The jock strap was imported from Taiwan through the port of Los Angeles and was classified in subheading 9503.90-6000(EN), HTSUSA.

ISSUE:

Whether the article in question is an article designed to amuse such that it is classifiable as a toy of Heading 9503, HTSUSA; alternatively, whether it is classifiable as fancy dress of Heading 9505, HTSUSA, as underwear of Heading 6107, HTSUSA, or as other knitted garments of Heading 6114, HTSUSA.

LAW AND ANALYSIS:

Heading 9503, HTSUSA, covers other toys; reduced-size ("scale") models and similar recreational models, working or not; puzzles of all kinds; and accessories. The official interpretation of the Harmonized System at the international level (four and six digits) is embodied in the Explanatory Notes. The General Explanatory Note to Chapter 95, HTSUSA, provides that the "Chapter covers toys of all kinds whether designed for the amusement of children or adults."

The fact that an article can be worn does not prevent it from being considered a toy. In United States v. Topps Chewing Gum, Inc., 440 F.2d 1384 (1973, it was held that buttons, designed for wearing and on which were written humorous sayings, were dutiable as toys in item 737.90, Tariff

Schedules of the United States Annotated. In Topps, Judge Lane stated that, "[i]f the purpose of an object is to give the same kind of enjoyment as playthings, its purpose is amusement, whether the object is to be manually manipulated, used in a game, or as here, worn." 440 F.2d 1384, 1385 (1971). While the article in question doubtless has considerable amusement value, we do not believe that the fun or amusement that it stimulates "is essentially the same kind of frivolous enjoyment [that children] would derive from objects which we commonly think of as toys." Id at 1385. Rather, it is Customs' view that the article in question contemplates an entirely different form of entertainment and is therefore a class of merchandise separate and distinct from toys.

The NYRLs discussed above were issued during the transition from the Tariff Schedules of the United States Annotated (TSUSA) to the HTSUSA. Under the TSUSA, there was no provision for "entertainment articles" as distinct from toys. Now, however, Heading 9505), HTSUSA, covers festive, carnival and other entertainment articles, including fancy dress articles of textiles. Mary Brooks Picken's, The Fashion Dictionary, 134 (3rd ed. 1973), defines fancy dress as a "costume representing a nation, class, calling etc., as worn to a costume ball or masquerade party." However, Note 1(e), Chapter 95, HTSUSA, excludes fancy dress of textiles of Chapter 61 and 62 from the coverage of Chapter 95. The elephant jock strap is a form of costume to the extent that it represents a fashion of dress appropriate to a particular occasion. Consequently, we are faced with the question of whether this particular item of fancy dress is properly classifiable in either Chapter 61 or 62. If so, it is excluded from classification anywhere in Chapter 95, either as a toy of heading 9503 or as an article of fancy dress of heading 9505.

Heading 6212, HTSUSA, covers various foundation and body-support garments, whether or not knitted or crocheted, including jock straps. The article in question, although loosely referred to as a jock strap, is not designed to provide support and is therefore not classifiable in heading 6212.

Heading 6107, HTSUSA, covers men's or boys' underwear. The difficulty in classifying the elephant jock strap as underwear of heading 6107 is that the article is undoubtedly designed for display and thus is not an undergarment within the common meaning of that word. Moreover, Customs' Guidelines for the Reporting of Imported Products in Various Textile and Apparel Categories, 53 FR 52563, 52570, state that the term "underwear" refers to "garments which are worn under other garments and are not exposed to view when the wearer is conventionally dressed for appearance in public, indoors or out-of doors." Here, the article's sole purpose is display.

Heading 6114, HTSUSA, covers other knitted garments which are not included more specifically in other headings of Chapter 61. According to the Explanatory Notes, which are the official interpretation of the Harmonized System at the international level, (four and six digits) these other garments include, inter alia, such diverse articles as aprons, boiler suits, smocks and other protective clothing, academic robes, jockey's silks, ballet skirts and leotards. Although a novelty article such as an elephant jock strap is not ejusdem generis with those garments listed above, the enumerated articles have no common or unifying theme. Since heading 6114 is a residual category, and since there are no more specific or appropriate headings in the HTSUSA, Customs is of the opinion that articles of novelty wearing apparel such as the jock strap in question are classifiable therein.

HOLDING:

The elephant jock strap is classifiable in subheading 6114.30.3060(EN), HTSUSA, under the provision for other garments knitted or crocheted, of man-made fibers, other, other, men's or boys', and are dutiable at 16.1 percent ad valorem. The textile category is 659.

Pursuant to section 177.9, Customs Regulations (19 CFR 177.9, NYRLs 830911, 831612 and 835595 are modified in conformity with the foregoing.

The designated textile and apparel category may be subdivided into parts. If so, visa and quota requirements applicable to the subject merchandise may be affected. Since part categories are the result of international bilateral agreements which are subject to frequent renegotiations and changes, to obtain the most current information available, we suggest that you check, close to the time of shipment, the Status Report on Current Import Quotas (Restraint Levels), an internal issuance of the U.S. Customs Service, which is available for inspection at your local Customs office.

Due to the changeable nature of the statistical annotation (the ninth and tenth digits of the classification) and the restraint (quota/visa) categories, you should contact your local Customs office prior to importation of this merchandise to determine the current status of any import restraints or requirements.

Sincerely,

Jerry Laderberg, Acting Director

Commercial Rulings Division

[1] "Ship Shape", San Francisco Chronicle, October 17, 2001, p. B1.

[2] "Customs Update", The Journal of Commerce, September 23, 1998, p. 12C.

[3] Albaum et al, Ibid, p. 252.

[4] The U.S. Constitution forbids any tax on exports, but other countries do tax exports.

[5] Duties apply to the country of origin, which may be different from where the cargo is coming from at that time.

[6] Schaffer et al, Ibid, p. 444.

[7] JOC, April 9, 1999, p. 1a.

[8] JOC, October 15, 1999, p. 1, 4.

[9] Schaffer et al, Ibid, p. 450.

[10] Schaffer et al, Ibid,, p. 451.

[11] "EU, US critical of pre-export inspections", John Zarocostas, The Journal of Commerce, November 6, 1998, p. 8A.

[12] JOC, March 29, 1999, p. 3a.

[13] "Customs uniformity urged", Bill Mongelluzzo, The Journal of Commerce, September 23, 1998, p. 1B.

[14] "Nations try to simplify customs rules", American Shipper, September 1998, p. 58.

[15] "Security vs. Trade", R.G. Edmonson, JoC Weekley, November 26-December 2, 2001, p. 34.

[16] "China Using Hong Kong to Evade US Trade Limits", San Francisco Chronicle, April 1, 1997, p. A10.

[17] "CHB Exam Prep", Ro Leaphart, Leaphart and Associates, Alameda, CA, 2001, p. viii. See also www.leaphart.com.

[18] "Do-it-yourself customs brokerage", Chris Gillis, American Shipper, August 1998, p. 56.

[19] JOC, June 18, 1999, p. 1A.

[20] JOC, September 27, 1999, p. 1, 8.

CHAPTER 11
TRADE DOCUMENTATION

This chapter provides an overview of the documentation process, which is helpful to understand why government requires certain documents and what they do with the information. However, this chapter is not meant as a guidebook for clearing customs or handling documents. After describing in general terms the role of documents in an international shipment, the chapter goes into moderate detail on the main ones, particularly the Bill of Lading and Waybills. The final section discusses insurance, why it is needed and what are the shipper's options for protecting ones self. [1]

THE ROLE OF DOCUMENTATION

Documentation plays a special role with logistics at the international level as compared to the national level. While there are documents used for a shipment within a country, there are much more documents used in crossing borders, and their enforcement tends to be stricter. Do not confuse documentation with regulation. Documentation is simply one way shippers comply with regulations.

What is the purpose of documentation? There are a wide variety of documents, some required by the government, others are the organization's internal policy. Some are required by business partners, customers, banks and others. 'Documentation' is not just a piece of paper, but the process that leads to their creation and use. The document is a management process as much as a result. The basic principle of trade and documentation are **control** and **liability**. As we saw in the discussion of selling terms and Incoterms, whoever has control of the cargo should be liable for it. The documents indicate who is in control or liable at any point in the shipment. Among the main purposes of documentation are:

- **Fulfill regulations**, This is the most common reason.
- **Manage risk**. Documents are often associated with insurance or bank policies that limit a party's risk in an international shipment. The government regulations just mentioned also serve to identify and limit a party's risk.
- **Common understanding**. There are things that do not necessarily need to be written down, but international transactions are complex and work with different languages and cultures. Documents serve as a common agreement.

- **Record keeping**. The term 'documentation' implies record keeping, maintaining official records of an organization's activities. Some record keeping is legally required, but much is the organization's choice.

There are **four types of documents**. Transportation documents are used for the physical movement of the cargo. Banking documents are use for the financing of the shipment. Commercial documents are for the purchase of the goods. Government documents fulfill government regulations.[2] Another way to look at it is that there are two main roles of documentation, business-related and law-related. In other words, documents serve the needs of a business and the government.

Four Types of Documents
- Transportation   - Banking   - Commercial   - Government

A **documentary sale** refers to the sale of goods in which ownership on the paper documents is more important than the physical possession of the goods. When you buy something in the store and walk out the door with it, there is no confusion that you now possess it. In a documentary sale, you have bought something that may be on the other side of the world, and the only thing to show that you are now the owner is the documents.

Information technology is fundamentally changing what is considered a "document". There are paper documents and electronic documents, or there may be one document in both formats at the same time. This means a policy is needed to determine which is to be the official document. An important question is, which is better, paper or electronic? There are no clear answers here.

Paper documents do not conflict with different computer systems. It is easier to see which is the official copy. They can be physically destroyed, but do not get corrupted or deleted like a computer glitch. **Electronic documents** can be transported anywhere in the computer world in seconds, an infinite number of copies can be made, also in seconds. Yet it is not always clear which is the official record, and they can be harmed by computer glitches and, becoming more common, viruses. They also require any user to have both the document and the program to read or alter it.

It may seem that documents can solve a lot of problems, which is true, but they come at a **cost**. Documents and the documentation process is expensive. While the costs are much lower now given computers, it is not the creation of the document that costs money. It is the management systems needed to collect and process the information. If this is information that an organization would need anyway, there is little extra expense. However, there are often times when information needs to be collected just for the document. This can be described in one word, bureaucracy. This problem does not just occur with the government. Private companies are also struggling to control documentation.

The different modes of transportation affect the documents used, but also the process of documentation. An ocean passage takes days, sometimes weeks. This gives the parties involved time to work on documents. Air cargo, by contrast, moves extremely fast and the documents need to keep up with the shipment. These will have some interesting consequences to be discussed later.

Some forms of economic integration, like free trade agreements, affect documentation. The goal of free trade agreements is to eliminate barriers to trade, and one of the easier ways is to limit the amount of documents needed. A common market would mean that all governments within the agreement are as one market. That means the only documentation that should be required would be those documents that are national in nature. In other words, a common market should not have any documentation needed just to cross the border.

Documentation is commonly used as a **non-tariff barrier** (NTB). It is used to discourage imports and increase their costs, sometime to the point where trade is no longer profitable. The WTO has been active in eliminating documentation, and tries to identify when documentation rules are an NTB.

The biggest example of a common market is Europe. Documentation has been largely eliminated and a Single Administrative Document has been introduced. One problem though is the many languages in use. There are fewer documents needed, but those that do exist often need to be translated into many languages. In North America, Canada complained to the US that their beef should be labeled "imported".[3] This was a result of conflicting rules, NAFTA on the one hand and domestic labeling requirements on the other.

One of the roles of documentation was to control risk, and this includes cargo **security**. Requiring proper documentation can control theft and the trafficking of stolen goods. On the other hand, thieves have discovered that they simply need to learn the game. False documents are a large industry. The insurance industry traditionally referred to losses as "hull" or "cargo", in which it was either a ship that was damaged/lost or the cargo. However, that description is not longer accurate for the current environment. If a ship sinks, both the hull and the cargo are lost. Mueller and Adler refer instead to documentary frauds, frauds in connection with charters, scuttling, and cargo theft.[4]

In response, the documentation process usually includes some verification process. Documents are often cross-referenced, so the smuggler would need to have a false document but also get into the official database. Electronic documents seem to be safer than paper documents, but still not fool-proof.

Confidentiality is a concern in all aspects of documentation, when communicating with the government and other companies. Some information is a matter of public record, some may be semi-private, while others can be very sensitive. If the public were able to gain access to the shipping records of a shipper, companies could collect valuable intelligence on each other. This then creates an incentive for legitimate shippers to lie on their documents. Documents circulating among private companies also raise concerns about privacy. One major difference in the private sector is that, while public agencies usually have an obligation to protect privileged information, private actors do not have a general obligation. Private actors are usually restricted only by non-disclosure agreements, and often these are not in use. Chapter Sixteen, Information Systems, elaborates on this issue.

THE LEGAL STATUS OF DOCUMENTS

Not all documents are government controlled. However, in this section we look at only those documents that are official, to see how their status is determined, and how countries agree on what documents to use. Some documents are used only within a company or within the private sector, and not subject to government control. The problem arises when there is a dispute, and the legal status of those documents are called into question. If the document is not government controlled then the organizations involved can agree how it should work.

There is a special challenge in the **legal status of trade law**. Each law has its own legal system, which was discussed in Chapter Five. What if there is more than one legal system involved and the laws are not the same? Virtually all legal systems have rules pertaining to conflict of laws. Many contracts state that in case of a dispute which legal system will be used to resolve the dispute.

What makes things even more confusing is that each country has multiple offices that have their own regulatory systems, and thus create their own document needs. For example, the US has federal, state, county and city laws, as well as many special districts and jurisdictions. Even within the federal level, there are over 40 federal agencies that have jurisdiction over imports and exports. It becomes vital that the Customs or some other agency take control to prevent rampant increases in paperwork.

At the international level, the **U.N. Convention on Contracts for the International Sale of Goods** (CISG) sets the standards for most of the documents we are concerned with. The UN simply sets the standards, but individual governments need to write it into their own law. It would be extremely unlikely that any country hoping to do business internationally would disregard these standards. In the US, the **Uniform Commercial Code** (UCC) has been since 1951 the basic law governing sales and contracts of sale. However, the UCC is being replaced by the CISG, which became effective in 1988.

The **U.N. Commission on International Trade Law** (UNCITRAL) is the primary international organization that seeks uniformity in trade law, including documentation. For example, when the ICC established new guidelines for Bills of Lading (to be described later in this chapter), the UNCITRAL adopted them, thus insuring they would be the global standard.

Over the past 20 years there has been growth "disuniformity" as the US follows a somewhat different law. The US codified the Hague Rules with the Carriage of Goods by Sea Act (COGSA), but these are different in some ways from Hamburg/Hague-Visby rules (to be discussed soon). This has created uncertainty and confusion on legal matters of lost or damaged cargo. The UNCITRAL has been trying to revise the rules to create something acceptable to the US and the rest of the world.

What are the issues that are being debated? The "nautical fault" defense has been used by carriers, in which navigational errors may be used by a carrier to avoid liability. This may be eliminated. There is also the issue of "performing parties". Some want only the carrier that issued the BL to be liable, but others want any carrier that handles the cargo to be liable. Forum selection is always a major concern in international law. This refers to where a case is heard. If there is more than one jurisdiction, then it is possible to move the case to the country that would be most advantageous. There seems to be little agreement on how to decide where a case is to be hard given that the current legal conflict.[5]

For the rest of this chapter, any rules mentioned are based on the CISG unless otherwise stated. Treaty law takes precedence over local law. However, recall that each country decides how it is going to enact treaties. Note the contradiction. There is a treaty stating that it takes precedence over local law, but it is the government that decides if that treaty is going to be enacted into local law.

The CISG applies when there is a commercial sale of goods between two parties in different countries. The place of business of the buyer and seller must be in a country that has ratified the CISG. The place of business requirement can mean that two companies in the same country can possibly be subject to the CISG if the contract is performed outside of that country. The convention does not cover many transactions, including goods bought at auction, ships and aircraft, electricity, goods associated with assembly contract, and others.

The U.N. Commission on International Trade Law
www.uncitral.org

CONTRACTS

The first step in an international transaction (also known as the performance cycle) is the creation of a contract. While this may seem straightforward, it is not always clear when there is an agreement. As Dick Locke notes, it can be hard "getting to No", in which you know that there is not an agreement.

Legally, parties must have **four parts to a legally binding contract**. First, there is an **offer and acceptance**. Silence does not mean that an offer was accepted. Second, the deal must involve something of value, known as **consideration**. Third, parties must have **legal capacity**. They must be competent to engage in the contract, and if they are acting as agents for someone else, have their principal's approval. Finally, a contract may **not be for illegal purposes**.[6]

Four Parts of a Contract
• Offer and Acceptance • Consideration • Legal Capacity • Not for Illegal Purposes

The **mirror image rule** states that what is offered must exactly be what is accepted. Any changes may be considered a counteroffer. This becomes problematic in the real world where the forms used by many companies vary, even when the intent is the same. This is knows as the "battle of the forms". The CISG accepts different forms as the same agreement if it is "sufficiently definite and indicates the intention of the offeror to be bound" (Article 14). If the parties do perform, then it is assumed that a contract was made, and the last form used takes precedence. Among the most common problems in the "battle of the forms" is when some issues are mentioned in one form but not mentioned at all in another.

The CISG does not require that a contract be in writing. Countries vary in their requirements. For example, the US requires a written contract for any sale over $500. A common requirement in many countries is that even if there is nothing in writing, a witness is required to prove the existence of the contract.

Breach and Relief from Contract

Sometimes a contract cannot be fulfilled, and sometimes one side refuses to fulfill their obligations. Probably the single most important aspect of a contract is that it is an obligation, so no party may simply walk away from it. There are remedies for breach of contract, in which other side is refusing to fulfill there obligations. There are also conditions in which you may walk away from it.

Under what conditions can you claim that the other side has committed a breach of contract? It must be a **fundamental breach**; you cannot claim a breach just because of a little problem. Buyers and sellers have the **right of avoidance**. A buyer can simply not accept a shipment if there is a breach, but this may only be done after the seller has been notified. The seller has the **right to remedy**, which is to fix the problem, and the right to additional time to perform.

There are three general ways a party may be **excused from fulfilling a contract.**[7] They are excused if it is physically or legally impossible to fulfill, if the underlying purpose of the contract no longer exists, or if there has been a change in circumstances that makes it commercially or financially impossible to comply.

Force majeure clauses are a way to claim that something remarkable has happened to prevent a party from fulfilling a contract. The contract often states what is considered an event so remarkable as to allow one to breach the contract. Courts are reluctant to allow companies to drop their obligations, so many contracts include a force majeure clause. These typically include such things as wars, blockades, fire, acts of governments, inability to obtain an export license, acts of God, acts of public enemies, failure of transportation, quarantines, strikes, and so.

THE DOCUMENTATION FLOW

This section provides an outline of an example shipment and the documentation flow that accompanies it. This example assumes that the cargo is not under a Letter of Credit (LC), which will be discussed in the next chapter. This is only a brief outline since it is difficult to describe the flow and at the same time describe the documents in sufficient detail. The reader may want to read this section, read the next section on the major documents, then return to this.

1. Order Processing and Shipment
2. Export Clearance
3. Shipment
4. Import Clearance
5. Delivery

Major Documents

Purchase Order A contract between buyer and seller, which includes the agreed terms of sale.

Bill of Lading The most important document in the entire process, this is a document of title, a receipt of goods, and a contract of carriage.

Waybill Closely related to the BL, a waybill does not convey title. It used to be mostly for air cargo, but ocean carriers are using it more.

Invoice A document listing the cargo and its important characteristics.

Certificate of Origin A certificate, typically provided by the government of the exporter, certifying the nationality of the cargo.

Sanitary Certificate Commonly used in food shipments to certify that it is hygienic. Also known as a phyto-sanitary certificate.

Carnet Permission for sample goods to enter without paying duties as regular cargo.

Export Declaration The form submitted to the government by the exporter providing basic information on the cargo.

The outline just provided was meant as a brief overview. We now follow the process again, explaining the details. The emphasis here is on the documents. Details of the customs clearance was discussed in the previous chapter. This is a three step process. Recall that there are three channels, communications, cargo and documentation. The seller is at this time delivering the cargo to the carrier. The documents are prepared and delivered to the carrier, and the exporter.

1. Order Processing and Shipment

The first step in logistics is when a PO is made. This is a communication of some kind that the buyer sends to make a purchase. For example, when one called a sales catalog to make a purchase, that is a PO. In the business realm, there are more formal methods. Between companies that have an ongoing relationship, there are often electronic PO systems. The PO would include the selling terms, or at least propose one. This is also the beginning of the performance cycle. A PO can be a contract, or it can be one communication between parties that have an ongoing contract.

The selling terms should be determined at this point, assigning who is responsible for arranging the transportation and customs clearance. Even if it is not part of the agreement, this is where each person is making their own logistical arrangements. In other words, both parties now know that a shipment is about to be made.

In this section it is assumed that the shipment is not being financed. Letters of credit and the documentation process is discussed in the next chapter.

2. Export Clearance

The exporter may use a freight forwarder, who in turn books the cargo with a carrier. The forwarder is normally experienced with all the documentation and is the best source for current information. However, the forwarder is not like a CHB in that their job is not to help the exporter fulfill regulations.

When the exporter contracts with a carrier, the carrier creates the Bill of Lading. This is the single most important document in the entire trade process, and is discussed in detail in the following section. The BL, which is normally electronic with most developed countries, is transmitted electronically to the importer.

The Export Declaration (Exdec) is used in the US (with similar documents in other countries) to tell the government what is being exported. Its primary purposes are to keep statistics on trade volumes, and to control the export of some commodities. The US Exdec states "These commodities licensed by the United States for ultimate destination (country name). Diversion contrary to U.S. law prohibited". The reasons why some commodities are controlled is discusses in Chapter Ten.

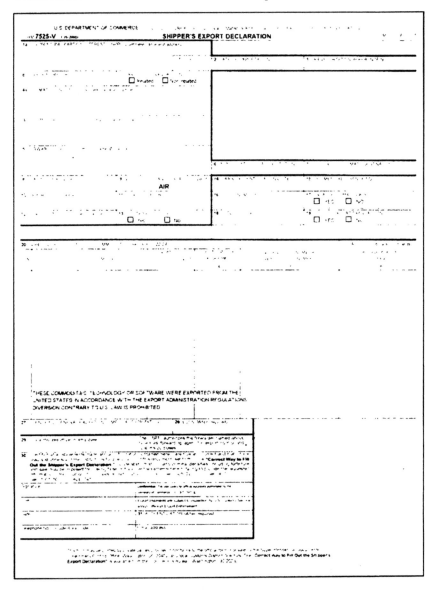

An invoice is a document stating the buyer, seller, and details of the goods being exchanged. It is normally issued by the seller. A **commercial invoice** is used for the buyer and seller. A consular invoice is essentially the same thing but it has all the specific pieces of information needed for customs. Most customs agencies specify exactly what is to be included on the invoice.

The **Certificate of Origin** states the origin of the cargo. This is important for countries that assess duties in part based on the cargo origin, and because goods from some countries are controlled.

Many countries require **sanitary certificates** (also known as hygiene or phyto-sanitary certificates) for some food and live animals to prove they are free of diseases.

A **carnet** is used when sample goods are being imported, to avoid paying duties. Sample goods may in some cases enter a country free of duties as long as they are being re-exported. A carnet is permission for cargo to be imported as sample goods.

3. Shipment

While the cargo is in the process of moving to its destination, ownership is transferred at some point agreed upon in the selling terms. Meanwhile, the documents are being transmitted to the Seller.

Bailment is a legal term for when the owner of property (the bailor) transfers possession to someone (the bailee). Carriers are responsible for delivering the cargo in the same condition as was received (known as strict liability), and are liable for loss or damage even if it was not their fault. Exceptions to this rule are as follows:[8]

- Acts of God
- Acts of public enemy or terrorist
- Acts of government intervention or court order
- Acts of the shipper, such as improper packaging
- Inherent characteristics of the cargo, such as perishability
- Carriers own disclaimers

4. Import Clearance

The documents should have arrived before the cargo arrives, except in the case of air cargo. The Customs Clearance process is where documentation is most important. However, given the current US program of informed compliance, mistakes that occur at the time of clearance may not appear until much later.

The Buyer may, when presented with documents, may refuse them for things that may seem a technicality, such as shipment date different from that which was agreed, inadequate insurance coverage, etc.

Inspections normally occur at the time cargo arrives in a port. The Certificate of Inspection, Certificate of Weight, or Certificate of Analysis (for chemical industry) may be required by the importer. For example, every major garment retailer in the US requires an inspection of apparel coming from Hong Kong.[9]

Sometimes the cargo does not arrive. It may have been lost or stolen in transit. The importer has many protections in this case, which are discussed in the next chapter. A carrier may only deliver cargo to the holder of the original copy of the Bill of Lading. **Carrier's Misdelivery** is when the carrier gave the cargo to the wrong party, in which case the carrier is liable. If a buyer sues the exporter for nondelivery or breach of contract, he is asking for damages equal to the price of the difference of the contract price and the fair market value.

5. Delivery

When the importer is notified that the cargo has arrived, he can normally send a truck to the port to pick up the cargo. There seems to be a variety of practices, which are more or less protective of the cargo. Many ports do not require any special documents when the truck shows up to pick up the cargo. This may be changing given increased security. Some ports are familiar with the local truck drivers, but large, modern ports need more controls. After delivery, the importer in the U.S. is still required to keep all records according to the standard of informed compliance.

Carrier's Lien
Carriers have a lien on the cargo to insure they get paid. If the shipping fees are not paid, carriers normally sell the cargo at auction to get paid

BILLS OF LADING AND WAYBILLS

The **Bill of Lading** (BL) and **Waybill** are the single most important documents for international trade. We will discuss the BL first, and the waybill later. The BL is created by the carrier, but this is not always a simple matter. If intermodalism or other multi-modal transport is used, the company that organized the overall shipment would issue the BL. Forwarders and NVOCCs sometimes issue their own BLs. This means the party issuing the BL acts as if they are the carrier, but other companies are actually performing the carriage. Why is this important? There is heated debate about liability. Some countries hold all parties to a shipment liable, whereas others only hold the issuer of the BL liable. The BL serves three key roles:

1. **A contract of carriage**. It is signed by the shipper and thus becomes a contract between the shipper and carrier for the voyage.
2. **Documentary evidence of title**. When a shipment is spending large amounts of time in between buyers and sellers, it may not be clear who owns the cargo at any one time. Whoever is stated on the BL owns the cargo. Think of this as the pink slip to a car (for the US).
3. **Receipt of goods**. This is the receipt given by the carrier to the shipper to show that they received the goods.

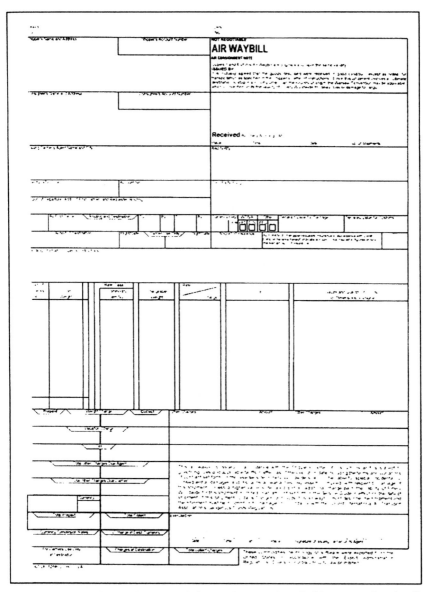

As the 'basic document' for an international shipment, the details can be very important. Interpretation of the legal meaning and changes in practice has major global impacts. For example, the shift toward electronic transmission of documents have raised the issue of which is the primary document, the paper or electronic BL. Fraud and smuggling often entail misuse of a BL.

One key part of the BL is the **cargo description**. An inherent conflict exists in the description of the cargo. The BL must state the true nature of the cargo. The carrier is responsible for writing this down, yet they are not the experts on the cargo. Only the exporter is. Disagreements often arise when the exporter says one thing, and the carrier says another. If the BL needs to be 'clean' in order for the shipper to sell it on documentary credit conditions, then the differences will have serious consequences.

What is meant by 'clean'? When the cargo is delivered for shipment, the BL should at that point state exactly what is the cargo and its condition. If the cargo is in good condition and the BL shows no changes, it is clean. During the voyage, there may be damage, or their may be changes to the BL. The BL then becomes 'dirty', which means that it is not exactly correct for some reason. This can be very troublesome.

Electronic Bills of Lading

One of the major changes in trade in the past several years has been the dramatic increase in electronic documents. The problem in the past has not been the technical aspects of transmitting data. The concern was creating a large enough pool of companies willing and able to handle electronic forms, and for the laws in each country to accept them. To make an electronic document 'real', governments needed to agree on what constituted a legitimate document, signature, etc.

Legal standards for BLs are provided in ICC's **Uniform Customs and Practice for Documentary Credits**, known as UCP500. It is currently published with a supplement for electronic BLs, known as eUCP. In 1990 the Comité Maritime International adopted *CMI Rules for Electronic Bills of Lading*. Rules relating to the eUCP are independent of any technology. Electronic transmission may only be used if all parties approve. If the transmission is corrupted, the receiver may request that it be re-presented.

The carriers are the crucial player in the process of handling eBLs. The CMI rules calls for a "Private Key". This key was any technically feasible manner of insuring the authenticity of an electronic transmission. The carrier provides to the Holder (the party in possession of the BL) the key, which in turn could transmit it via the carrier to another party. The key is unique to each party, which means that whenever it is transferred, a new key is created. If a party does not want to participate in electronic transmission, a paper BL must be provided on demand.

There is an alternative system run by an organization called BOLERO. Instead of using electronic keys, messages are routed from seller to buyer through a Trusted Third Party (TTP). Authenticity of the messages is confirmed by electronic signatures. BOLERO's job is to confirm the identity of all parties involved and that the messages have not been changed.

An electronic signature has been defined by the ICC as "a data process attached to or logically associated with an electronic record and executed or adopted by a person in order to identify that person and to indicate that person's authentication of the electronic record."[10] How does a bank determine if a document is original? If it is presented as original, then it is assumed to be so. The bank is under no obligation to confirm if a document presented as original is in fact so.

Types of Bills of Lading
Clean BL No notations by the carrier of problems, otherwise it is foul
Onboard BL Signed by the carrier that it is loaded
Received-for-shipment BL Means only that it was received by the carrier
Straight BL Nonnegotiable, cargo delivers only to the consignee.
Forwarder's BL Created by a forwarder, who is acting as an intermediary.

Waybills

A **waybill** is similar to a BL in that it acts as a cargo receipt and contract of carriage. The critical difference is that it does not convey title. It is mostly used for air cargo that moves so fast that a document conveying title is not practical or needed. However, ocean shippers are finding that waybills make the documentation process much easier.

Since it does not convey title, there is less concern transmitting it. Importers can pick up their cargo more easily. While it may seem that there are less protections, traders have found that the security provided by a BL is not as important as previously believed. Now that there are eBLs, it is less clear what the original is, and even why it should matter. If someone were to try to steal cargo by creating false documents, a false BL is just as easy to create as a false waybill.

There are no negotiable air waybills. A negotiable waybill would mean that the cargo would need to wait for the bank to approve release of the cargo, yet airports do not have space for traffic waiting for bank releases. The 1929 Warsaw Convention governs the use of air waybills, though a recent international agreement is changing that to allow for paperless waybills. For the past 28 years, a piece of paper was required to accompany every shipment.

INSURANCE AND LIABILITY

"Imagine the incredible intricacies of a case involving any given product sold in a country A to a buyer in country B, who makes credit arrangements at a bank in country C, with the product loaded on board a ship registered in country D, and with a master from country E. The ship sinks off the coast of country F, while her "sunken" cargo appears in country G. And this listing of nationalities does not even account for the dozen or two dozen countries from which the sailors hail who might have to be summoned as potential witnesses...The question is, where is the venue? Which country can dispose of the case?"[11]

The whole point of terms of sale was to determine liability. It must be clear who owns the cargo at any point in time in case there is loss or damage. If there were never any loss or damage, the entire point about ownership would not matter. Historically, cargo would be in the possession and control of a ship, and thus the ship's captain, for months at a time. It was all but impossible for the carrier to prove that the losses were due to natural disasters, negligence or theft, so the carrier was held strictly liable. In effect the carrier was also the insurer.

As the shipping industry became more economically powerful they included disclaimers in the BLs that all but eliminated any liability. This resulted in a period when shipping was extremely unsure. In 1892 the US government passed the Harter Act, still in effect, that limited how much a carrier can disclaim any liability. The **Hague Rules** of 1924 also define the limits of carrier liability. COGSA codified the Hague Rules into US law in 1936.

Liability for losses was based on gold. The US used the gold standard until 1974, when it went off the gold standard. At that time the liability was arbitrarily set at $9.07 per pound of cargo, and a maximum of $500 per "package". By current standards this is extremely low, which is why courts have been quite liberal in interpreting a 'package' as the smallest unit in a shipment. One shipper claimed that a container with 4,400 suites, each of which was on its own hanger in a plastic cover, as 4,400 packages, for a maximum liability of $2.2 million.[12]

A new set of liability rules were established in a UN convention in 1978, known at the **Hamburg Rules**. They assign much more liability on the part of the carrier. The convention was intended to protect the interests of developing nations that do not have large shipping fleets. As of 1994, 22 countries have ratified the rules, making it effective in those countries, but most developed nations, including the US have not approved of these rules and do not appear likely to do so.

However, the **Visby Amendments** have been applied to the 1924 Hague Rules by some countries, notably the UK, Canada, Belgium and Germany, raising the per package liability to $1,000 per package, and the US looks likely to approve these. It also makes carriers liable when they are "reckless" in the operation and navigation of the ship. The liability amount is based not on gold or US dollars but on the "special drawing rights" of the IMF, a fluctuating unit of currency based on the economic strength of IMF members.

Even a well-written contract and clear identification of who owns the cargo does not eliminate disputes over matters of liability. Problems occur in identifying who caused the loss or damage, and to what extend the different parities (shipper, carrier and consignee) could have affected the situation. According to COGSA, any BL that lessens the carrier's liability is automatically void. Crew and stevedores cannot be held liable, under the Himalayan clauses, named after a famous court case in the US, but in other countries (United Kingdom and Canada) they can.

There is some debate over whether a carrier can be held liable for what the shipper states in their BL. The carrier has the option of weighing and opening the container to confirm the shipper's statements, and courts have found carriers liable in such cases.[13]

A **material deviation** from the terms of the BL makes the carrier exempt from any limitations on its liability. For example, an Israeli ship dumped a load of clocks on a dock in order to return to its country to join the war effort. Since the load was dropped off in a port that was not part of its route, this was a material deviation. The clocks were dumped on the port and became rain-soaked, so the carrier was liable. On deck stowage has also been considered a material deviation, but containerized cargo on a modern ship is considered reasonable.[14]

The choice of venue refers to the issue of which jurisdiction a case is decided. Contracts normally have a clause stating where any disputes will be decided, but often this clause is missing or the contracts contradict each other. Foreign arbitration is extremely expensive, so some companies would settle early just to avoid litigation. Small shippers, especially those trading with developing areas, are at risk of this tactic.

Claused But Not Dirty[15]

Certain conditions can be placed on cargo that does not make the B/L dirty. In other words, some types of cargo can have 'problems' that are not really problems, and thus do not mean there is any true problem with the cargo. For example, the following is considered 'normal' and does not make the B/L dirty:

Commodity	Clauses
Oil raw material	'Old dented drums'
Dairy Products	'Soiling of undamaged cases caused by old broken cases (eggs)'
Poultry	'Some cartons dented'
Cotton	'Dirty bales'
Sawn wood	'Shipped wet'
Iron and steel	'Slightly rusty'

---

[1] *The author is strongly indebted to two sources for the information in this section, Schaffer, Early and Agusti's "International Business Law and Its Environment", and Bagley's "Managers and the Legal Environment".*

[2] *Albaum et al, Ibid, p. 500.*

[3] *"Canadians have a beef with US "label" bill", Courtney Tower, The Journal of Commerce, August 17, 1998, p. 3A.*

[4] *Mueller, Ibid.*

[5] *"Attack on 'disuniformity", JoC Week, April 15-21, 2002, p. 14.*

[6] *"Managers and the Legal Environment: Strategies for the 21" Century", Constance E. Bagley, West Publishing, St. Paul, MN, 1999, p. 215.*

[7] *Schaffer et al, Ibid, p. 181.*

[8] *Schaffer et al, Ibid, p. 226.*

[9] *Schaffer et al, Ibid, p. 204.*

[10] *"UCP 500 + eUCP", ICC, ICC Publishing, Paris, 1994, p. 57.*

[11] *Mueller, Ibid.*

[12] *Schaffer et al, Ibid,, p. 239.*

[13] *Schaffer et al, Ibid,, p. 237.*

[14] *Schaffer et al, Ibid,, p. 241.*

[15] *"B/Ls and back letters: An intrinsic conflict", Gerard Verhaar, American Shipper, July 1998, p. 24. à*

CHAPTER 12

TRADE FINANCE

In this chapter, we look at the aspects of finance that are unique to logistics and trade. Of particular importance is the financing of a foreign purchase. The first section is an overview of international trade finance. We discuss the fundamental reasons for financing transactions and how companies deal with risk. The next section is a step by step description of a transaction using a letter of credit. In the previous chapter we went step by step through the documentation process of an international shipment. We now go through the same process, only in this case there is financing involved. Because the transaction is complicated, there is first a brief and general description of the transaction, followed by a discussion of the details. The last sections involve some special areas of finance, including insurance, currency issues and security.

THE ENVIRONMENT OF INTERNATIONAL TRADE FINANCE

Many purchases are financed. This is much simpler when the purchase is done within a country than in an international transaction. A domestic purchase involves only one legal jurisdiction, which makes debt collection easier. When multiple countries are involved, things get more complicated and the risk increases. In this case, risk includes:

- **Seller's risk-** non-payment
- **Buyer's risk-** not receiving goods or goods not in proper condition.

Risk is much greater in international transactions for a number of reasons. First, multiple legal jurisdictions make it more difficult to understand one's right and options. Second, collecting on a debt in another country is much more expensive. Third, the long distance also complicates matters because debts that may be cost effective to collect nearby may not be so when there are long and expensive plane rides involved. Overdue accounts may be collected in one of three ways. Either the seller collects, a lawyer collects, or a collection agency is used. All of these options can become expensive to the point where any profit margin is lost.

On one extreme, a buyer pays in advance and accepts all the risk of not receiving the goods, a partial shipment, damaged goods or some other problem. This is known as **cash in advance**. On the other extreme, the buyer pays on delivery or even later, known as **cash on delivery**. In this case, the seller faces all the risk of non-payment. Either of these two extreme cases may be used when the buyer and seller are related, such as subsidiaries of the same company, or if they have an ongoing relationship and trust each other. Cash in advance is often used in the following cases:

- Buyer is in an area of instability
- Buyer has bad credit
- Exchange rate controls
- Goods made to order

Financing international trade is a special market, very different from financing a house or a factory or anything else. When you finance a house, the bank is making a long term commitment, and can take the house if the loan is not paid. When financing a factory or some other business transaction, the bank can look at the creditworthiness of the company and the prospect of success for the transaction. In financing international trade, you can see there are some fundamental differences:

- It is **short term**. The exporter typically needs money to buy materials and make the product, and cannot wait to be paid on delivery. Or an importer needs the loan to buy something, which will later be resold or used by the importer. These actions do not take years and years like a house mortgage.
- In case the load is not paid, **seizing the cargo** is generally not as easy as foreclosing on a house. The shipment may be in a foreign country, missing or it may not have any resale value for the lender.
- **Multiple jurisdictions** exist. Sometimes it is only the two countries of the origin and destination, but it can also be other countries, such as the countries the cargo passes through enroute. This could become an issue if there were a problem with the shipment enroute.

Credit is granted when a party, usually a bank but not always, either lends money or accepts risk. What do we mean by accepting risk? This is when the creditor is not lending or paying money, but offering to make a loan in certain circumstances, such as if the buyer fails to pay. The main types of credit include:

- **Consignment** Goods are shipped to a company for resale. This is usually done between related parties.
- **Open Account** Buyer has an account for a certain monetary amount of goods. This is used when there are ongoing shipments.

- **Letter of Credit (LC)** Banks write a LC guaranteeing payment.
- **Seller's Credit** The seller accepts the risk.

The LC is by far the most common method of financing trade, and will be discussed in detail. An LC may be defined as "a letter addressed to the seller, written and signed by a bank acting on behalf of the buyer. In the letter, the bank promises it will honor drafts drawn on itself if the seller conforms to the specific conditions set forth in the LC".[1]

Rules relating to LCs are published by the ICC (which also publishes the Incoterms). The book, *UCP: Uniform Customs and Practice for Documentary Credits*, contains the rules used by banks and merchants worldwide. The rules set out the legal format of letters of credit, rules by which banks handle LCs, and defines the rights and responsibilities of the parties. This book is not law in any nation, yet there is general agreement and thus credit documents refer to the UCP, just as selling documents make use of Incoterms. Laws in some places, such as New York, specifically state that if the UCP is used, the UCC (the normal source of US law) is *not* applicable. In other words, New York has decided that the UCP takes precedence over some established laws.

Governments often provide credit to promote exports. The US has a variety of programs, such as the Export-Import Bank, the Foreign Credit Insurance Agency, and the Commodity Credit Corporation. These agencies have different missions, such as providing insurance or providing credit. The World Bank provides a wide variety of financial and technical assistance to developing countries to promote economic development. Sometimes the assistance a government provides to its own companies is considered a subsidy, which raises concerns of free trade. Recall the discussion in Chapter Ten on dumping and countervailing duties. There are many programs to assist exporters that are considered 'legitimate', but some are considered a subsidy.

What about the interests of the bank? So far we have been discussing trade finance from the perspective of an exporter. Banks are in the business of providing capital or credit, and trade finance is one of many markets. There are thousands of banks in the world, and the US alone has about 800. Yet only a small percentage of these banks are involved in trade finance. That is because there are some special demands, and potential risks involved. Financing an international shipment is not the same as financing a local enterprise. The bank needs to understand the nature of the risk. This means understanding the general business conditions of the countries involved in the trade, the different legal systems, and, hopefully, the companies involved in the transaction.

Despite what some believe, banking is not a matter of simple financial calculations. The **Bagehot Problem** refers to the banker's need to understand the customer's moral character, which is a highly subjective process, in order to make a very quantitative decision on how much money to loan. For example, a Swiss trading company had two partners, a Swiss national and someone who was from a country associated with instability.[2] The Swiss individual would go to the bank when they needed financing because they got better conditions than when the other partner went to the bank, even though they both had identical qualifications and were presenting the same business plan.

Advantages of a Letter of Credit
Only need to check credibility of bank, not importer.
Banks generally immune from political stability or exchange rate controls.
Less uncertainty because requirements for payment stipulated on LC.
Pre-shipment risk: if order is cancelled during manufacture, bank is still liable.
Ensures that someone will buy the product once shipped.
Guards importer to terms of sale and condition of goods.
Bank has responsibility for oversight.
Because LC is good as cash in advance, importer can get good prices.
Some exporters will only sell on LC.
No cash tied up.
If prepayment is required, it is better for the bank to hold the money than exporter.

THE LETTER OF CREDIT TRANSACTION PROCESS

We have already discussed an international transaction and the documentation flow. There are some changes when the cargo is being financed. The following is a brief description of the process, with an emphasis on those aspects that change due to financing. In order not to repeat what was already described in Chapters Ten and Eleven, it may be assumed that anything not mentioned is the same as if the cargo was not being financed. In this chapter, we use the terms Buyer/Seller, which mean essentially the same as Consignee/Shipper or Importer/Exporter. The Seller is also known as a Beneficiary, because they are being guaranteed payment.

Overview of LC Transaction

Buyer sends a purchase order.

Buyer applies to an Issuing Bank for an LC

LC delivered by the Issuing Bank to Notification Bank

The Notification bank notifies the seller that an LC was issued.

Seller ships goods and gives documents to Notification Banks

Notification bank delivers documents to Issuing Bank. Issuing Bank accepts documents, which include a Draft.

Issuing Bank pays Seller (or Accepts the Draft), and Buyer pays Issuing Bank.

Seller receives Acceptance or payment.

The L/C Transaction Process

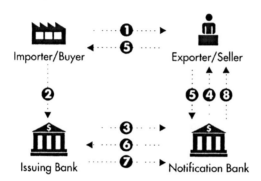

When a purchase is made, the Buyer goes to a bank, the **Issuing Bank**, to request an LC. Note the comments on applying for credit. The Bank may want to hold the Buyer's money immediately, or for the Buyer to have the money in an account with the Bank. This would be a case where the Buyer was very risky, and the seller wanted assurance to be paid. If the Buyer was more credit-worthy (in the opinion of the Bank, not necessarily of the Seller), the LC may require credit at some time in the future, such as when the cargo arrives.

The LC states the terms that must be fulfilled for the Seller to get the money. It may be that as soon as the Seller ships the cargo and presents the Bank with the documents proving that the cargo was shipped, he gets paid. Or the LC may state that after the cargo has arrived and the Buyer accepts it, the Seller will get paid. There are many variations on the LC terms. Note that if the LC requires the Buyer to accept the cargo first, the Seller is not at risk of the Buyer taking the cargo without payment, but there is still the risk of the Buyer refusing to accept the cargo. This will be discussed further.

Once the LC is approved, the Issuing Bank transmits it to the Seller's bank, known as the **Notification Bank**. Banks only talk to other banks for a good reason. If the buyer's bank was communicating directly with the seller, there may be a conflict of interest. Also, banks know how to communicate with each other easily. Therefore, the Issuing Bank tells the Notification Bank that an LC has been issued, who in turn tells the Seller.

The Seller then ships the cargo. When the cargo is delivered to the Carrier, the Carrier creates a **BL** or **Waybill**. In this case, we will assume it is a BL to make the explanation simpler. The Notification bank receives the BL from the Carrier, a **Draft** from the Seller, and other documents such as Insurance Certificates, Invoices and so on. The Draft is a letter that basically says "Pay Us". For example, when you receive your bills each month, those are Drafts.

The documents are sent by the Notification Bank to the Issuing Bank. The BL normally states the Issuing Bank as the consignee, which means the cargo is owned by that bank. The Issuing Bank can then hold onto it until the Buyer pays.

At this point, the Issuing Bank has the documents, and the cargo is still enroute. In the case of ocean cargo, this may take days or weeks. In the case of air cargo, the process takes as little as a few hours for the cargo to arrive.

When the Issuing Bank received the documents, the Draft states whether it is to be paid immediately or at some time in the future. In the case of immediate payment, if the papers are in order, the Issuing Bank will pay at that time. The money is returned via the Notification Bank to the Seller.

If the Draft calls for payment at some time in the future, the Issuing Bank will stamp the Draft, making it an **Acceptance**. An Acceptance is a Draft that has been received and acknowledged by the Issuing Bank, and thus the Issuing Bank will pay that amount at the specified time in the future. The Acceptance is returned via the Notification Bank to the Buyer. The Buyer may hold onto the Acceptance, or sell it at a discounted price. It is thus a monetary instrument that can be traded on the open market.

When the cargo arrives at the final destination, clears Customs, and arrives at the final destination, the Buyer picks up the cargo and either inspects it at the port or warehouse, or takes it to his own location. He inspects the cargo and may refuse it if the terms of the contract are not meant. The problems associated with refusing cargo are a major issue. A buyer who changed his mind may use any excuse to refuse it, or pretend to refuse it to renegotiate a lower price. The LC states how much time the Buyer has to decide whether or not to accept the cargo, normally not more than 15 days. If the LC called for immediate payment, then the Seller has already been paid and refusing the cargo puts the Buyer at risk of not getting a refund.

LETTERS OF CREDIT

The LC is essentially a letter from a bank stating that it will pay for the cargo when presented with documents from the Seller. The bank is then accepting the risk that the Buyer will not pay. From the Seller's perspective, it is only a letter promising payment. Why is it that a bank can write a letter, and businesses will respect that, yet a letter from the buyer would not be given the same treatment? Simply put, banks are special. In most countries, banks are carefully regulated and their existence depends on being reputable. Still, there are many countries where even the banks do not have the stability or trust to write LCs that are respected.

> **LC:** "A conditional undertaking by a bank, issued in accordance with the instructions of the account party, addressed to or in favor of the beneficiary. The bank promises to pay, accept, or negotiate the beneficiary's draft up to a certain sum of money, in the stated currency, within the prescribed time limit, upon the presentation of stipulated documents."[3]

The LC is credit, which means that the buyer needs to be creditworthy. For a first-time applicant, this can be difficult, especially if it is not a well-established business. Sometimes the bank requires that a certain amount of money be deposited with them. Some commodities more easily lend themselves to credit, such as something that could be repossessed by the bank. Other commodities like perishables entail higher risk. The LC has some variations:

- **Documentary LC** A draft submitted with documents such as invoice. Non-documentary LCs are only for noncommercial transactions.
- **Irrevocable LC.** All parties must agree upon any changes. Revocable LCs are only good because it shows a bank thought the importer creditworthy.
- **Confirmed LC.** Both issuing bank and seller's bank agree to honor draft if needed.
- **Revolving LC.** When it is automatically approved for ongoing shipments.
- **Back-to-back LC.** When the issuing of one LC is contingent on the buyer's customer also getting an LC approved for buying the product.

There are often changes to the transaction, such as the buyer sells the cargo while it is still in transit, or the terms change. Agreements between buyer/seller have no legal significance. They all must be through the issuing bank as an amendment. If there is a problem in the cargo, that is between the Buyer and Seller.[4] The Issuing Bank may not negotiate with or deal with the Seller in any manner. The Advising bank has a limited role. It only passes on the information, and its only responsibility is to take reasonable care in confirming the LC is authentic. It does this by checking the signature on the LC with the signature on file with the bank.

The **Rule of Strict Compliance** states that all documents must comply with the description on the LC, and any that do not are called **discrepant**. It has been estimated that over half of all LC transaction involve discrepant documents[5] Typographical errors are usually acceptable, but only if it is clearly a simple clerical error. In some cases typos are used as an excuse for refusing a shipment. If the issuing bank accepts a document that is discrepant, it may not be able to collect from the buyer. Cargo descriptions on other documents may be described in more general terms that are not contradictory.

Most discrepancies are resolved by the buyer signing a waiver. However, if the buyer wants to back out of the deal, he can easily find a discrepancy as an excuse. Once the bank has announced that it will not honor the draft, the buyer can then offer to waive the discrepancy, but only for a generous discount off the sale price. One observer familiar with this process said that any banker who cannot find a discrepancy is not "worth her/his salt".[6]

The shipment must conform exactly to the LC. For this reason, some sellers wait to see the LC before packing the goods or arranging transportation. There are many details, for example, there is the exact seller. Subcontractors or agents are now allowed to be listed as the seller. The cargo must be of the exact type, size, packages in a certain way, transported in a certain manner, etc.

Some countries have banks that are not creditworthy, either because the banks are not professional or the political/economic situation is risky. A seller may then ask to have a **Confirmed LC**, in which the LC is confirmed by another bank in the seller's own country. China at one time forbade Confirmed LCs, believing that it was an insult to the country.

There are Irrevocable and Revocable LCs. An Irrevocable LC is one in which the bank cannot decline from its obligations. There is not much protection with a Revocable LC so it is rarely used and then normally between subsidiaries of a company.

Drafts

The Draft is a letter saying, "pay this". Is is referred to outside the US, particularly in the UK, as a Bill of Exchange. After the product is sent out, the draft is sent to initiate the payment process. It is only used if there is a credit arrangement since it would not be relevant if prepayment is made. It enables the exporter's bank to act as collection agent. US Law says Drafts must be:

- In writing
- Signed by the issuer
- An unconditional order to pay
- A certain sum of money
- Payable on demand or at a certain time
- Payable to the to order of the bearer

There are different types of drafts. The **sight draft** means payment is due immediately. A **time draft** is payable at a certain time in the future. A **Clean Draft** has no documents with it, and is mostly to press for payment. A **Documentary Draft** includes documents, such as Bill of lading, Commercial invoice, Consular invoice and so on.

Sight draft/documents against payment, SD/DP means that the consignee gets the documents when they pay the draft. **Sight draft/documents against acceptance, SD/DA,** meant that they get the documents just for signing the acceptance.

CURRENCY ISSUES

Different currencies make an international transaction more complicated, and must be carefully managed. Currency rates are determined by the supply and demand for currency relative to each other, the details of which are beyond the scope of this book. Instead, we will only consider how to deal with these fluctuations.

Exchanging currencies can be expensive since it usually involves a transaction charge. One way companies avoid this is by paying in their home currency. That just transfers the problem to the seller. Another alternative is to have a bank account in the local currency. Customer service is affected when it leads to confusion and risk. In one example, United Parcel Service offered a global rate sheet that quoted everything in US dollars to eliminate risk from currency fluctuations. Previously they worked with 140 rate charts.[7]

Trading in some currencies is government controlled, which usually means that exchange rates are fixed. This does not mean the currency's value is fixed, since free markets determine the value. There may also be controls on the amount of a currency that may be exported.

Currency risk refers to the possibility that the exchange rate may change during the time of the transaction. This would benefit one party and harm the other. There are financial tools designed to control this risk. Forward contracts allow one to buy a certain amount of currency at a certain price at a certain time. In other words, the exchange rate is fixed. Forward options give one the option of buying the currency at the given exchange rate. In either case, these tools cost money. Another solution is to have the sales contract in one's home currency, and thus the other party will face the currency risk.

Trying to anticipate future exchange rates is a lot like forecasting the stock market; do so at your own risk. Also note that when you buy financial tools to counteract currency risk, the price is based primarily on what the market anticipates of the currency fluctuations.

There are some currencies that are more commonly used in global trade. It is more likely that they will be acceptable for payment. The US dollar is by a large margin the most widely used currency in the world. There are more dollars in circulation outside of the US than in the US. In some countries, notably Russia, Panama and Cuba, the US dollar has been or still is the 'unofficial' currency, universally used but not officially recognized. Europe has just introduced the Euro, a currency whose value is determined by a combination of the national economies that are party to the agreement. It is not clear what effect the Euro will have on global commerce, but it will very likely be a competitor to the US dollar as the world's leading currency. In some regions, other currencies are predominant. The German Deutschmark was common in Eastern Europe prior to the Euro. In Asia the Japanese Yen is in common use.

Exchange rates can have operational effects on international logistics and trade. During high inflation, the rules of inventory management change drastically. If a currency is not going to hold its value, many choose to stockpile goods instead. Management decisions are based on monetary values, and those need to be stated in a common currency. Exchange rate shifts should be seen as a symptom and not a cause of changes in the global economy. Still, the volume and patterns of trade are affected by changes in relative wealth.

Risk is an often misunderstood concept. It is not a yes/no question of whether something is risky, but how much risk is involved. There are an almost infinite variety of risks and everyone faces some kind of risk at any time, anywhere. The question is what type of risk are we talking about and what is the level of risk? Statistically, we call this an **expected average**, the average amount of loss over the long term. There may be an extremely high risk of small loss, or an extremely low risk of high loss. Logistics entails many risks, including the general business risks that any enterprise faces. Carriers face many risks, including damage to their assets (the vehicles) and the operators. In this case, we confine ourselves to the risks associated with an international shipment, particularly:

- Non-payment of goods
- Non-delivery of goods
- Theft
- Damage of goods from any cause

Risks are affected by the actions of all parties involved, including the shipper, carrier, and consignee, and many others. Some risks can be controlled by the shipper in things like packaging or mode of transport. In other areas there is little choice. In one major Pacific storm in 1998, APL alone reported 270 containers lost and 550 damaged.[8] One can reduce risk by not shipping during storm season, but this only serves to show how reducing risk comes with a cost. The auto shipping industry has gone through a major change from past years when one-third of the cars arriving in the US from Japan needed repairs. This was the result of a cooperative agreement between manufacturers, carriers and the longshoremen. Handling cars have become a fine art. The longshoremen are trained not to touch anything on the car unless necessary and they back the car in to make for easier exit.[9]

All carriers have a required minimal liability. COGSA limits carrier's liability to $500 per package. However, this led to a dispute over what a 'package' is. Courts have been interpreting that as the smallest possible unit. For example, if one ships a 20-foot container with three pallets, each of 100 boxes, then this one shipment would consist of 300 packages.[10] Many shipments are much more valuable than the carrier's liability, which is why the shippers must manage the risk themselves.

There are **two basic ways to manage risk**: accept the risk or buy insurance. One may accept the fact that over a given time period there will be accidents and include money in the budget to pay for those losses. The alternative is to buy insurance. This does not eliminate the risk, as some think. The insured pays the insurance company the expected average plus overhead costs. If one did not pay for insurance, over the long term they would pay less, because they are paying for long term average losses but not the insurance company's overhead.

Why then would anyone pay for insurance? The long term expected average is only an average. There is that chance that a major loss would be more than the company could pay for. That would result in bankruptcy. Insurance protects against that. Another reason for insurance is that it allows a company to keep to their budget. They know what their insurance premiums are, plus a small amount extra for the deductible. Without insurance, they may not have any losses and gain, or they may have large losses and suffer. Companies usually prefer stability, which also tends to keep stock prices higher.

If cargo is covered in an LC, insurance is almost always required, just as a bank that offers you a mortgage insists that you keep insurance on a house. **Peril** is the insurance term for anything that may cause loss or damage. The term **"average"** refers to the extent of insurance coverage. There is **particular average**, which only covers a particular shipment, and **general average**, which also covers the entire voyage and the ship. The York-Antwerp Rules cover the rules of general average. They are not covered in any treaty nor have they been enacted into national law, but have traditionally been used in contracts of carriage and BLs.

Why general average? This is becoming less common, but traditionally every shipper with cargo on a given ship shared the risk. Whatever losses were incurred, including the loss of the entire ship, was shared by everyone. This is less common now. Policies are often free of particular average (FPA), which means insurance companies do not pay for partial losses.

The average can vary widely depending on what kind of risk one wants covered. There are standard packages, plus some special risks that can be added on. Special amendment for strikes, riots and civic commotion (SRCC) are very expensive or impossible to get when they are needed most.

Insurance covers the shipment as long as it is in **continuous ordinary transportation** (COT). If the cargo stops enroute for some reason, such as to be repacked or processed, it is not being transported, that is, not in COT. Customs-seized cargo is not in COT.

There may be a conflict of interest. The buyer is not liable for anything that happens to the cargo before she takes ownership, and the shipper is not responsible for anything after she relinquishes ownership. This handoff means that damage may not be discovered until later, or the shipper may not take appropriate precautions for that part of the trip that they are not responsible for. Insurance does not give one the right to be reckless with packaging and other safeguards. The shipper is obligated to take precautions as if no insurance were in effect.

When damage occurs, the next step is to file a claim. **Surveyors** are independent consultants who inspect the damaged cargo and file a report. This is then used to assess how much will be paid. Insurance companies can pay the claim promptly, or they can delay payment for a long time. There are mixed feelings about how well carriers and insurance companies handle claims, and how important it is in the first place. One study showed that "claims handling" was last of ten selection criteria for airports by air cargo managers and airport managers.[11]

Normally when an insurance claim is made, the insurance company hires an **adjuster** who checks out the damage and reports back to the insurer. Based on the adjuster's findings, the insurer then decides how much to pay the shipper for the losses. In this case, the forwarder can act as the adjuster, simplifying the process. When a forwarder sells insurance, they can also act as an adjuster.

Freight forwarders and **NVOCCs** are not held liable for damaged cargo, though the carrier is, according to most international legal norms. Still, forwarders have a strong interest in making sure that their customers have adequate insurance. For this reason, at least one insurance company has created an internet-based database available to forwarders to show them what shippers have in terms of insurance coverage. The back of the BL states that domestic air carriers can limit their liability to 50 cents per pound of damaged or lost freight, international air carriers to $9.07 per pound, and ocean carriers to $500 for the entire container.

The insurance industry can play a major role in maintaining standards of safety in the transportation industry. If a carrier cannot get insurance, it will almost certainly go out of business in a short period of time. Shippers would not want to use an uninsured carrier, and banks would not lend money. The IMO recently produced what has been described as "the most far-reaching and dramatic safety initiative ever" to strengthen the rules to prevent ship collisions and groundings. The key to gaining

code was for the insurance industry to require ocean
his certification or face the loss of their insurance.
ance industry replied that they were not enthusiastic
were already enforcing high standards.[12]

e industry could cover just about anything, though
_ is by far the most common policy. There is a trend to more
extensive coverage, including coverage of risks that businesses assumed were
part of general business risk. Open cargo policies cover all shipments by a
shipper of certain types of cargo for certain destinations over specified
routes. In such cases the shipper has the authority to print out insurance
certificates to send along with the BL.

An 'all-risk policy' can cover everything from disrupted supply
lines to market volatility. Just about anything that may threaten expected
earnings, like weather-related losses or the cost of raw materials, can be
covered. Only if the company's earnings are more than 5% less than
expected does the insurer pays. These policies are typically 'all-risk' which
means that it covers anything not specifically excluded. Such policies are
just beginning to be offered in the late 1990's, so they can be expected to
change and evolve. Another company offers evacuation insurance so
earnings will be protected if a company office needs to be evacuated, even if
there turns out to be no damage.[13]

War and political risk insurance has become an important topic.
Recall that premiums are based on an expected average of historical loss.
However, war and political risks are so vague and unpredictable that
insurance is haphazard and extremely expensive. The cost of the attacks of
9-11 was over $50 billion.[14] It would be statistically invalid to calculate the
expected average loss when you have one day loss of this magnitude, along
with long periods of no loss. In cases where there seems to be a good reason
to expect an attack, insurance companies simply refuse to offer a policy. As
of Spring 2002, as an example, general war-risk coverage was 2.75 to 5 cents
(US$) per $100 coverage on the London market, but can be over 25 cents
when moving cargo in troubled areas.[15] Aviation insurers increased their
rates after 9-11 1,500 percent for hull and passenger liability coverage.[16]

Unlike other purchases, it does not always make much difference
where one buys the insurance. Insurance is not a commodity that can be bought
cheaper elsewhere. There may be slightly lower administrative costs, but that is
relatively very small compared to the expected-value part of the premiums. A
good insurance company needs to be in a country where the insurance industry

is well regulated (not heavily regulated, but well regulated). Like banks, the industry needs to have some controls to prevent companies from making short-term profits and then go bankrupt when there are large claims. Imagine a new insurance company offering very low premiums, keeping a large share and then, when there are claims, going bankrupt.

Some countries do not have well developed insurance industries. An insurance company from such a country should be considered carefully. Ideally it will be affiliated with an insurer from a more developed country. China does an enormous amount of trade, yet they are still in the process of building institutions for a capitalist economy. Insurance companies are among the institutions that it needs to develop. In 1999 the government looked at the insurance industry and found that 71,000 firms were selling insurance as a side-line to some other business. This means that their insurance business may not be of the proper standard, and 12,000 of them were shut down.[17]

SUPPORTING A FLOATING CITY[18]

Royal Caribbean Cruises, based in Miami, is faced with some unique logistics challenges. How does one support a floating city that is constantly moving all around the world? And how do they maintain their reputation for having the finest foods, in all its freshness, supplied from all over the world? "Recently 1,500 pounds of lamb didn't make it to a vessel on time, so a chartered plane flew the lamb to the ship in Mexico" said Jim Walton, director of materials and logistics.

Royal Caribbean Cruises Ltd owns 17 vessels, supported by Walton's 22-person staff in an office that operates 24 hours a day, 7 days a week. Traditionally, cruise ships had 90 days worth of frozen and dried foods on board. Newer ships are emphasizing fresh food, more passenger space and less storage space. Now the standard is 14 days worth of provisions on board.

Planning is the key. Despite the difficult nature of supplying the ship, there is enough time to prepare. The ship's itineraries are planned two years in advance. Delivery dates are set well in advance for the diverse commodities. This includes chilled produce, frozen food, dry goods, bonded marine items, gift shop and hotel items.

Resupply is made at many of the ports of call around the world. Containers are shipped to foreign destinations to meet up with the ship. These containers are scheduled to arrive five to seven days ahead of time to insure they arrive when the ship arrives. Supplies are positioned so that an intense unloading and loading operation can be done without the ship being delayed. License plates on every pallet means that each pallet loaded can be logged in and checked for accuracy.

Up to forty pallets can be loaded in an hour, for a total of 200 in five to six hours. Meanwhile 2,000 passengers are boarding on the gangways. Loading is labor intensive, requiring about 16 people on board and another 20 outside. Reverse logistics is also part of the job. Immigration officials board to inspect what is coming off. This can include hazardous waste, engine parts, etc.

Shipments come in but must be broken down for different ships. For each ship there are different sectors, such as beverage, hotel, gift shop, marine, entertainment, cruise programs, aquatics and hotel services. These goods are loaded through different doors of the ship. The ship includes a poultry room, a fish room, a wine and liquor room, and a fresh vegetable room. This is due to US public health rules regarding cross contamination.

Royal Caribbean's partners are another key to successful logistics. About 400 vendors are used, including such major third-party logistics providers as Ryder. Distributors or carriers are responsible for failure to get the cargo there on time. If the supplies miss the ship, the vendor is responsible for chartering a plane to make up.

---

[1] *"Multinational Financial Management"*, Alan C. Shapiro, Allyn and Bacon, Boston, 4th edition, 1982.

[2] The author does not want to contribute to the stigma by identifying this country.

[3] Schaffer et al, Ibid, p. 263.

[4] Schaffer et al, Ibid,, p. 266

[5] Schaffer et al, Ibid,, p. 274.

[6] Schaffer et al, Ibid,, p. 275.

[7] JOC, June 15, 1999, p. 8.

[8] *"Container industry braces for lawsuit"*, Tom Baldwin, The Journal of Commerce, November 12, 1998, p. 1A.

[9] JOC, April 8, 1999, p. 1A.

[10] *"Closing a Cogsa Loophole"*, Peter M. Tirschwell, JoC Weekly, November 12-18, 2001, p. 30.

[11] *"Improving International Trade Efficiency: Airport and Air Cargo Concerns"*, Paul R. Murphy, Douglas R. Dalenberg, James M. Daley, Transportation Journal, Winter 1989, vol. 28, no. 2, pp. 27-35.

[12] JOC, September 7, 1999, p. 16.

[13] JOC, September 23, 1999, p. 11.

[14] *"Tough Choices"*, Special Report on Risk Management, JoC Week, June 24-30, 2002, p. 19.

[15] *"Payback Time"*, Special Report on Risk Management, JoC Week, June 24-30, 2002, p. 20.

[16] *"Looking for Answers"*, Special Report on Risk Management, JoC Week, June 24-30, 2002, p. 23.

[17] JOC, September 20, 1999, p. 13.

[18] Based on *"Logistics of a floating city"*, Robert Mottley, American Shipper, December 1998, pp. 24-29.

CHAPTER 13
SECURITY

Security has become increasingly important. Even legitimate businesspeople must be aware of the regulations. This is because every shipper must comply with the same laws, and because one does not want to become accidentally involved in illegal activity. This chapter begins with a discussion of corruption in the customs agencies. The amount of corruption varies dramatically among countries. Since international shipments involve multiple countries, one is likely to deal with corruption somewhere. The next section introduces dangerous goods, which include a wide variety of cargo that many one would not think of as hazardous. Worldwide concern for safety of society and the environment makes this an increasingly important topic. The next section is on security, which refers to cargo theft but also to rules designed to deter terrorism. The last section on smuggling describes how smugglers operate and how to protect oneself from getting involved.

CORRUPTION

One of the main concerns with government agencies, and particularly customs, is corruption. While many agencies are highly professional, there are also many where corruption is likely, or even the standard operating procedure. This becomes a major problem for shippers and society. Corruption is the misuse of entrusted power for private benefit.[1]

Where is one likely to encounter corruption? It usually involves the government because they are the ones with a public duty. Private business may be corrupt in that it is dishonest, but that is a general issue beyond the scope of this book. However, there has been a trend toward privatizing many functions that have traditionally been government jobs. We are concerned particularly with the Customs agencies. This is where the public, and particularly the importer, encounters the government, yet is most vulnerable. The government has far more power in this situation than an individual shipper, and thus there is the chance of abuse.

Corruption is the misuse of entrusted power for private benefit.

The problem is partly the direct cost of the bribes. What is more serious is the lack of certainty and abuse of the legitimate governmental services. The lack of corruption is often referred to as **transparency**, in which the rules and procedures are open for all to see, and there is a clear way of conduction business that all must follow. The primary organization that fights corruption and researches this issue is the German-based Transparency International.

It must be emphasized that corruption in the industrial world is the exception, and not the rule. The reason the wealthy parts of the world are that way is specifically because of a culture of doing business in a transparent and legal manner.[2] Even the most efficient forms of corruption are still far less efficient than legitimate forms of business.

A couple misconceptions of corruption should be addressed. First, corruption does not always imply lawlessness, but that the laws and tradition are out of alignment. Samuel Huntington, a prominent political scientist, notes that all societies have rules, some written and others not. Corruption is typically a symptom that the old method of business, which involved personal contacts and paying fees to move one's cargo along, conflicts with new rules that fit the modern conception of how business and government agencies are supposed to operate.[3]

Second, corruption may be bad for society, but it can be very good for an individual. It is often said from those accustomed to working in places where a bribe can get thing done, how frustrating it is to deal with bureaucracy and not be able to pay one's way out of the situation. The worst situation is when one is dealing with the old-style system of personal contacts, but they cannot accept bribes and they will not perform their duties. This happens when anti-corruption rules are enacted in a highly corrupt government to prevent the lowest level officials from getting bribes.

In international logistics one deals with multiple countries, so there is a good chance one must deal with bribery. Even the best customs agencies encounter some level of corruption. There is another reason that the international trader will be more likely to encounter corruption. As a foreigner, one is in much less of a position to defend one's rights.

Bribery is voluntary, as contrasted to extortion, which is forced. This means that bribery requires two willing parties. The importer does not need to pay a bribe, but does so expecting to get something in return. There are two fundamentally different types of payments. **Grease payments** are intended to make someone do what they are supposed to do. The purpose is only to speed up the process, insure the official does a quality job, and so on. A **bribe** is when a payment is made for someone to do something illegal. For example, getting a shipment through customs quickly would mean a grease payment, but getting a certain commodity into a country that is normally forbidden is a bribe.

The US Foreign Corrupt Practices Act (FCPA) took on this issue directly. If one country's companies are forbidden from making such payments, it would only mean that other companies from other countries have a big competitive advantage. Still, business ethics is important. Therefore, the FCPA says that grease payments are allowed, but bribes are not. Obviously there is a gray area between which is which.

Companies and individuals can choose from **five general policies in dealing with corrupt government agencies**. The first is to never participate. This may make it difficult or impossible to operate in some areas. At the very least it gives less honest competitors an advantage. However, some would even disagree that honesty puts one at a disadvantage. Some businesspeople have found that if you give the impression that you are willing to pay bribes, officials create situations where a bribe is necessary. But if you have a reputation for never paying bribes, and that you will protest such behavior, officials show you more respect.

The second approach is to only pay grease payments, but not bribes. This is meant to minimize the negative social effects of corruption. Third, one can actively pay whatever bribes or grease payments necessary to be competitive. The fourth policy is where you not only pay whatever bribes are necessary, but that you actively create opportunities for bribes. In other words, instead of defensive bribes, you are practicing offensive bribery.

The final policy to based on the saying, "when in Rome, do as the Romans". This is to follow whatever policy is considered normal in that area. If bribes are routine, then you use bribes. If not, then you do not offer them. This strategy could fit any one of the first four policies, but the difference is that you adjust your policy depending on the area. The problem with this strategy is that it is rarely clear what is 'acceptable', and one can easily justify whichever of the four policies one wants to follow.

Policies on Bribery
- Never
- Grease payments only
- Bribes as necessary
- Bribes whenever possible
- Match local standards

Corruption operates along the same rules as any other economic activity, and, from the government's perspective, there is an optimal level of bribe taking. To provide a theoretical example of ideal types, imagine a dictator that wants to take as much as possible from traders. He will not take too much, or else he discourages business and his overall income from bribes is less. Therefore, there is an optimal level of bribery. We see in reality many areas of the world where highly corrupt agencies manage the situation so they do not take too much.

One problem in some places is that there is no central leadership. When nobody has an overall interest in maximizing income, every corrupt official tries to take as much as possible, knowing that if they do not take it, someone else will. This is when corruption is most inefficient. In a supply chain, officials are only concerned with the commerce that moves through their domain. If every official in a long supply chain is trying to take a bribe, the overall result is to make the business unprofitable. It is very unlikely that shipments moving through many areas are going to have the protection of one regime that will encourage an efficient level of corruption. In other words, in international trade we tend to see the least efficient and most destructive forms of corruption.

DANGEROUS GOODS[4]

Increased concern for security and environmental protection has led to strict controls on the transportation of dangerous goods (DG). What is a DG? According to the US Department of Transportation (DOT), it is anything "capable of posing an unreasonable risk when transported in commerce to health, to safety, and to property." This is further divided into three groups, based on its threat to health, flammability, and reactivity. Furthermore, there is a 5-point scale. Zero is minimal hazards and five is "severe hazard".

More than three billion tons of hazardous materials are transported by air, rail, vessel or highway, annually in the U.S, more than 800,000 shipments every day.[5] There is now a large industry of logistics and transport professionals specializing in this area. Shippers need to be aware of what they are shipping and the regulation that apply. In most industries logisticians do not need to worry about much of the technical aspects of their cargo, but this is not the case with DG.

Regulations of DG come from a variety of sources. We already saw how US law tends to act independently of UN or other international law. In this case, the US transportation law on DG is the *Hazardous Material Regulations*, known as the HMR (the book containing these regulations is 49 CFR Parts 100-185). The HMR applies to all transport of DG within the US and shipments leaving the country. In Canada, there is another set of laws, known as the *Transportation of Dangerous Goods* (TDG). Both of these are quite compatible with international law. The ICAO has established transportation standards, published in a book called *Recommendations on the Transport of Dangerous Goods*, updated every two years.

The two major modes of international transportation, air and ocean, also have their own specialized sources of DG rules. ICAO has *Technical Instructions*, and IATA also publishes additional operating instructions. Ocean cargo is governed by the UN's *International Maritime Dangerous Goods* (IMDG) Code adopted in 1965. Environmental concerns were added to IMDG in 1985 when a section was added to prevent marine pollution. In 2002 the format of IMDG's Code was drastically reformatted, though the content is generally the same. While most of these regulations are compatible, the following refers to the HMR unless otherwise specified.

The mode of transport makes a difference. Air transport has the most stringent rules regarding the amount and packaging. If anything goes wrong with DG on an aircraft, there is not the option of quickly putting to the side of the road. ICAO's rules include a distinction between those goods that may be put on a passenger aircraft, those that may be put on a cargo but not passenger plane, and those that may not be put on an aircraft at all. HMR requires that certain DG be loaded in such a manner that it is accessible to the crew during flight operations.

Most of the regulations on transporting DG are simply matters of common sense. Looking at the following list, we see that much of this is not at all controversial. The following are the main areas in the regulations:

- Identification
- Classification
- Marking
- Labeling
- Packaging

- Documentation
- Emergency response information
- Training
- Transportation

If there is an overall principle of these regulations, it is **identification and classification**. Most of these regulations are simply the details of how this identification is to take place. DG must be properly identified, with its **proper shipping name** and, if required, the technical name. The trade name (the brand name) is not used. For example, Flammable Liquid, N.O.S. (Not Otherwise Specified) is a shipping name, but it also requires a technical name.

Containers and some vehicles with DG must be **placarded** on the exterior based on class and amount. These placards identify the hazardous class of the DG in the unit. The basic purpose of these placards is to assist emergency responders in the event of an accident involving the container. The packages within the container or shipping unit must also be marked and labeled if required. They are similar in appearance to placards but are smaller for use on various types of packaging. Whereas other commodities may be concealed for security reasons, this is not an option for DG. The

DOT states that text on placards is optional. Since many people do not speak English, the emphasis is on the symbol and not the text to explain what is in the shipment.

Packaging is different for DG. Many types require performance-based specification packaging (also known as UN tested packaging, but there are some differences in the two terms). UN regulations specify the strength and characteristics required. An overpack refers to packaging that is meant to consolidate smaller shipments that could be shipped individually. The overpack must conform to the labeling rules as well as the packages inside (note this would not apply to non-DG shipments). One of the more interesting aspects of the IMDG Code is that labels should be readable after three months immersion in the sea.

Training is required of anyone "handling" DG. In the US, the term "handling DG" covers everyone from the person preparing the shipment to the dockworker actually loading the container. They need to take a certified initial training course and recurrent training periodically (three years in most cases, but for air cargo it is every year). This may be done through an in-house program or through a third party training course.

Documentation is another major area of the regulations. Most DG shipments include a signed certificate from the shipper stating that the goods are properly packed according to the regulations. Vehicles carrying any DG must have a manifest of all the DG onboard, including the type of commodity and other details. When the same container is carried by truck in the US the trucker must have a signed copy of the Dangerous Goods Declaration and a signed packing certificate. These documents can be combined into one form as long as all required information is listed and the DG Declaration and CPC statement are both signed on the form. In the US, many DG shipments require a 24 hour emergency telephone contact on the shipper's declaration. There are companies of hazardous cargo experts who act as the contact on behalf of the shipper, such as Chemtrec.

Although shippers are ultimately responsible for preparing DG shipments, **carriers have special responsibilities** when it comes to accepting a DG shipment. With non-DG goods they are not responsible for damage resulting from inadequate packaging. That is not quite the same with DG. The carrier is required to refuse any shipment if there is reason to believe that DG is being improperly transported and the shipment has not been offered according to the appropriate regulations.

The regulations vary depending on the type and class of DG, and the amount in a given shipment. For example, explosives of certain categories may only be carried in certain quantities. Many of the goods have different rules depending on whether it is a small quantity, a limited quantity, or a bulk shipment.

A good example is that some types of Class 1 explosives are not allowed to load vessels that are carrying other specific types of DG when the net explosive weight exceeds 10 kilograms. If the net explosive content is less than 10 kilograms, the container is allowed to load the vessel with these specific commodities that otherwise would not be allowed. Other examples relate to the size and type of packaging. Most DG commodities are restricted in the type of packaging that can be used. These restrictions are generally based on the type of DG, the net weight of the DG cargo, and the specific construction of the packaging.

The US has been using the term 'hazardous materials', internationally the term used is 'dangerous goods'. Americans are slowly converting. For example, the Hazardous Materials Advisory Council, and industry association, has renamed itself the Dangerous Goods Advisory Council. http://www.hmac.org/

Document	Mode/Destination
ICAO Technical Instructions	Shipments by Air
IMDG Code	Shipments by Water
TDG	Shipments from Canada
HMR CFR 49	Shipments from US
UN Recommendations	All modes

Dangerous Goods Classification

Class	Description
1	Explosives
2.1	Flammable gases
2.2	Non-flammable gas
2.3	Toxic gas
3	Flammable liquid
4.1	Flammable solids, self-reactive substances, solid desensitized explosives.
4.2	Substances liable to spontaneous combustion
4.3	Substances which, in contact with water, emit flammable gases
5.1	Oxidizing substances
5.2	Organic peroxides
6.1	Toxic substances
6.2	Infection substances
7	Radioactive material
8	Corrosives
9	Miscellaneous Dangerous Goods

SECURITY

Security in this case refers to threats of attack to both cargo and the people involved in the logistics process. This usually means theft, but vandalism and other forms of attack also occur, such as politically motivated attacks. Issues of security in international business can become quite broad, so discussion will be limited to matter of cargo security and incidental threats to personnel managing the cargo. The importance of security can be seen in the variety of threats that are presented in the supply chain:

- Shrinkage disrupts the supply chain.
- Often unplanned and last-minute.
- Puts personnel at risk.
- Reflects an inability to control one's operations.
- A company can be accused of collusion.
- A client's cargo is often irreplaceable.
- Individuals and institutions are often legally required to take steps to control crime.

Cargo theft is one cause of loss where a lot can be done to control it, yet it has been called the transportation industry's 'dirty little secret'. According to the FBI, the cost of cargo theft in the US is estimated to be around $12 billion a year, second only to health care fraud ($25 billion).[6] Two things put cargo at its greatest risk, **dwell time and changes in control**.

Two Greatest Threats to Cargo
- Dwell Time   • Changes in Control

Dwell time is when the cargo is not physically moving. It is extremely difficult to steal cargo while it is moving. Besides the obvious difficulties of getting into a truck while it is moving, the carriers have limited personnel on a ship or plane, and it would be hard for anyone to get away with cargo stolen while enroute. That means anything that keeps the cargo moving also prevents theft.

Changes in control refer to those points in a shipment where cargo is being passed from one party to another. This may be at a port, where different modes of transport connect. It may also be when drivers are switched. Every transaction is a risk point. The receiving party should check the cargo, but often this does not happen. Often it is impossible to check the cargo when cargo is handed off. The cargo may be taken at a port by someone posing as a carrier or the consignee.

Moving cargo through customs quickly has another important effect, it promotes cargo security. One of the most vulnerable points in a shipment is when the cargo is in the port area, or under the often-weak control of customs. The customs agencies are notorious for not taking responsibility for loss or damage of cargo under their control.

It is important to note that theft cannot be totally eliminated except in a few special cases. What we are talking about is controlling theft, which means to minimize it and its effects as much as possible. Generally speaking, that means that one should spend resources that reduce the loss until the cost of the control exceeds the savings. Just as with any other business venture, one invests money only until that point where the returns are equal to or less than the costs.

The great majority of theft involves insiders, especially with containerized cargo. This is self-evident since the thief would need to know which container to take, or take any container with a very low chance of getting anything with resale value. Some law enforcement officials believe that a majority of the cargo thieves are ex-narcotics smugglers who see the increased returns and lighter enforcement from stealing cargo. Cargo theft is an inside job 50% to 85% of the cases[7] and often involves the logistics department. These are the people who know which containers have the valuable merchandise.

Some companies accept the situation and do little to find stolen cargo. Some companies do not want the bad publicity. Insurance coverage makes it easy to become blasé about it. Others are actively trying to find stolen cargo and prosecute the offenders. Some shipping companies and shippers use private police like TransPro Group in Whitehouse, NJ to find stolen cargo, bypassing insurance companies.[8] Insurers are becoming more involved in theft prevention, investigation and prosecution. One company switched from surveyors to investigators, mostly ex-police, and recovered in one year $1 million cargo stolen from containers. Alan Spear, a security expert, notes that "surveyors usually give you the 'yup, it's gone' incident report".[9]

When theft becomes too costly, insurers will not offer insurance, and that effectively means that business will stop. Shifts in trading practices can be seen in many places in the world. There are regions in the world that traders of high-theft items avoid or stay out of. Even in the US, there is talk of moving shipments of small electronics and other high-value goods away from Miami.

A **security plan** is something that has not been commonly done by companies, who instead contract out security and that contractor worry about it. Yet companies are finding that it is important for them to control the overall security, or at least understand what their needs are. This goes back to the saying that you can delegate authority but you cannot delegate responsibility.

The **human factor** in a security plan is one of the least understood. Carriers spend millions of dollars on high-tech cameras and X-rays machines, yet on nights and weekends they leave the facility in the hands of security guards that are poorly paid and occasionally untrustworthy. It is not clear how much theft is committed by security guards themselves, but it is significant. There are no national standards in the US and most other countries for security guards.

Security companies should screen their employees carefully before hiring them and take certain steps to maintain security. Fast and frequent job changes are a warning sign. Personal contacts between the guard and the employees of the company they are to protect are a key issue. Shorter shifts are often used to keep the guards alert.

Security is not just a matter of theft; terrorism and the possibility of unintentionally getting involved is also a problem. Many national laws forbid companies from doing business with people suspected of terrorism, arms dealers, and so forth. This means a trader needs to be careful to identify any business partners or customers. One way to do this is by software that tracks which individuals/companies are on black-lists in various countries. A list of banned companies and individuals may be found at www.bxa.doc.gov.

Food is the number one item to steal because it is difficult to track and easy to get rid of. Second most popular targets are electronics, expensive clothing, computers and parts, perfumes and cosmetics. Major cargo theft zones are the New York/New Jersey corridor, southern California, southern Florida, with Memphis and Chicago on the rise. Many of the thieves get 90% of the value of the cargo because the buyer does not know it is stolen. Companies that have identified one of their drivers as the link in several thefts hesitate to publicize the fact because of the negative publicity.[10]

Over 200,000 stolen cars are shipped from US ports annually, according to US Customs and the National Insurance Crime Bureau. Those cars which are shipped overseas in containers are worth an average of $20,000, while those driven across the border are only worth $8,900. The most popular destinations for these cars are Russia and China.

Transporting stolen cars can be difficult when it comes to getting documentation. Instead, many of the cars are stripped as parts, which are all but impossible to identify. About 90% of the cars smuggled from the US into Mexico are sent on to Central America or shipped overseas. There are even cars being stolen from Mexico and brought into the US. These are either cars stolen off the manufacturers lot and/or used in drug smuggling.[11]

Computer chips are now competing with cocaine as some of the most valuable yet concealable of cargo. Spear notes "it's perfectly possible now to put $500,000 of product in a 24-inch by 24-inch cardboard box. That's one reason why eight different categories of gangs are operating now in Chicago in cargo theft". He believes that half of all major heists are planned in logistics departments. Few are ever harmed in cargo heists. The average bank robber gets $5,000 and goes to prison for 25 years. Most container thieves get $25,000 and go to jail for five months.[12]

Since the attacks of September 11, there has been a 'paradigm shift' in the way security in general, and especially transportation, is managed. This applies both in the US and elsewhere. The major regulator change in the US has been the creation of the Transportation Security Administration (TSA). It is scheduled to have 67,000 personnel, making it the largest new federal agency since WWII. Much of the agency's work is at passenger airports, but they also deal with cargo. One of the more controversial proposals is that cargo manifests will go to the TSA instead of Customs.

First, the priority has been on accurately identifying who are the parties involved in any shipment. Whereas before it was possible to buy or sell, to import or export with people that one did not know hardly at all, that is no longer advisable. One should have some form of confirmation about the identity of trading and business partners. Most governments have a list of entities with which it is illegal to business.

The **known shipper** is a concept designed to identify legitimate shippers. It simply means that that entity has imported or exported before. The TSA maintains a database of known shippers. New shippers face obstacles. For example, forwarders and carriers are required to physically visit the premises of shippers, unless they are already on the list of known shippers. Customs are also much more suspect of new shippers and are likely to inspect their cargo.

The other major change is that shippers need to be more careful to protect their cargo from being used as a **naïve host**. This is when smugglers or others put their illegal shipments inside legal shipments without the legitimate shipper knowing about it. If the illegal cargo is caught, a legitimate shipper may be held liable. In such cases, it is difficult to find who the real smuggler is. This will be discussed further in the next section.

As dawn breaks behind the mountains of Oman's Musandem peninsula, a peculiar scene unfolds in the tiny port of Khasab. Round the headland and into the harbour comes a procession of 12-foot dinghies, powered by outboard motors and steered by Iranian peasants. After two nerve-wracking hours spent dodging the Iranian coast guard on their way across the Strait of Hormuz, the smugglers have reached their destination. As they approach the shore, an unearthly thudding and yelping drifts across the water: each boat is packed with 20-30 seasick sheep or goats destined for the dining tables of rich Gulf Arabs.[13]

Sheep smuggling

Smuggling is defined as "conveying goods or persons, without permission, across the borders of a country or other political entity".[14] This means that only international transfers are smuggling. Earlier we discussed corruption and security issues. Corruption is the relationship between the shipper and a government official. Smuggling refers to the relationship between a smuggler and the government. Legitimate shippers still need to be concerned because many of the customs controls that all must comply with are designed to catch smugglers. Also, smugglers are now using legitimate shippers to hide their illegal goods.

Trafficking, unlike smuggling, is the buying/selling and transport of something, yet the term has become synonymous with smuggling. The term 'trafficking' has become synonymous with smuggling because that is the most common term used for the illegal movement of goods within a country. Also, sometimes the goods may be legal (as in the example of sheep), but the transport is illegal. We are not concerned in this chapter with cases where the goods are illegal (as in narcotics) but there is no transportation involved.

Legitimate and innocent shippers need to be concerned about smuggling as much as anyone else. Understanding how this works helps explain why the rules are as they are. Customs regulations are designed to accommodate legitimate trade while stopping smuggling, but this involves tradeoffs. What may seem like unreasonable bureaucracy to an honest shipper is designed to control the dishonest ones. Second, and more serious, is the possibility of being used as a naïve host. This is when illegal goods are placed in legitimate shipments without the shipper knowing about it. If the illegal cargo is caught, the honest shipper may be blamed, and cannot reveal the smuggler's identity because they do not know.

The movement may be secret, but that does not necessarily mean it is illegal. It may be illegal to one country but not another. It may be a trade-secret but legal. There are even cases where governments support the secrecy of a shipment. For example, the shipment of high-value goods may be done secretly as part of its security.

It is quite possible to get away with smuggling. The control of smuggling is based on a cost and benefit tradeoff. All smuggling could only be stopped by spending massive amounts of money and effort, and all but stopping trade. Therefore the rules are designed to prevent smuggling as much as possible without stopping trade. What smart smugglers are doing is take advantage of this, by finding ways to move the goods that are simply too expensive for the Customs agency to pursue.

Smuggling is found mostly in the following conditions:

- Large differences in tariffs between countries.
- Large differences in wealth.
- Low to medium political controls.
- Borders that are difficult to control.

Differences in tariffs between countries create the opportunity for arbitrage. This is the same as other forms of arbitrage, in which one buys cheap in one country and sells at a higher price somewhere else. The larger the difference in price, the greater is the incentive to overcome anti-smuggling efforts. Differences in wealth means you have people poor enough to want to engage in smuggling and wealthy people creating the incentive. Low to medium political controls on smuggling provides the opportunity; high levels of control make it difficult. Finally, borders such as that between the US and the Caribbean are porous and difficult to patrol.

Smuggled products are usually high value/low volume in order to move faster and easier. Other controlled products would be candidates for smuggling but are too bulky, heavy or otherwise difficult to handle in a cost-effective manner. However, anything can be smuggled. The products that are smuggled include but are not limited to the following major categories:

- Highly taxed goods
- Narcotics
- Weapons
- Garbage
- Hazardous waste
- Stolen goods

Garbage and hazardous waste would seem to violate the statement just made that smuggled goods tend to be high value and low volume. Yet there are cases where garbage is being taken into countries with lax controls for disposal, in order to avoid high cost disposal in the country of origin. This will be discussed in Chapter Seventeen in the section of Trafficking of Hazardous Waste. In such cases, it is not clear that this is even smuggling so much as large scale corruption in the receiving country.

Transportation is a key part of smuggling. The movement is virtually always in one direction, though borders can have different commodities flowing in both directions. In other words, legitimate cargo sometimes goes into a country for processing and is then returned, or at least shipped on to another place. In smuggling, on the other hand, one does not want to be moving the goods any more than necessary. Small, non-commercial forms of transport are common, such as small passenger planes or cars, though commercial planes, trucks and even trains may be used. In fact, commercial vehicles are sometime preferred because they can be harder to search, in some border crossings, are too numerous for the inspectors to keep track of. The value of one load of narcotics can be so great that the plane or vehicle used is abandoned. There are three general ways that contraband crosses borders:

- **Open run**. Sometimes in full sight of the authorities, smugglers move too fast to be caught. On the Mexican/US border, sports cars show up at the border crossing, and speed off before the police can follow.

- **Secret crossing**. In order to cross the border without being detected, isolated points of entry are used, with small planes/vehicles, typically at night.
- **Concealed cargo**. The illegal cargo is hidden inside the carrier. This is the most common method of moving large loads. Concealment can also simply mean that the documentation is false, but the officials are not expected to visually inspect the cargo.

K. Hawkeye Gross describes in *Drug Smuggling*, his memoirs of sixteen years of smuggling, how the transport of marijuana from the Caribbean and Latin America can be done using planes or boats.[15] The smugglers only take care of getting the cargo from the source into the U.S, and do not get involved with end users. Thus the smuggler is a logistics professional, not be confused with drug dealers. Planes and boats are only used once, and it would be unusual if they are not seized or damaged to make another trip. This part of the industry is unique in that the shipper is also the carrier. The person who gets the marijuana from growers, known as the Source, is responsible for arranging an airstrip, bringing the cargo, and bribing anyone that needs to be paid off. This Source is the single most important person for the smuggler.

For transporting cash within the US, Gross highly recommends the US Postal Service, which he commends for being extremely reliable with packages. Buses and trains are better than cars because they are not subject to being stopped or searched like a car driven by someone with a smuggler's profile.

It cannot be assumed that both countries at a border have the same priorities. As the story of Oman and Iran shows, what may be smuggling to one government may be legal to the other. Canada is concerned about guns entering from the US, while the US is concerned about drugs entering from Canada.

Enforcement against smuggling is one of the largest industries in the world. The US spends an estimated $40 billion a year, including local and federal forces, to fight smugglers. Customs agencies struggle to keep pace with the smugglers, and it is often considered a losing battle. Each year over 16,000 marine containers enter the US, so catching contraband is a lot like looking for the proverbial needle in a haystack. The smugglers have far greater resources than governments. While it may seem odd that even a major government can be outwitted by civilians acting on their own, consider for a moment the budget of most Customs agencies, and then take away from that budget all the other tasks that must be performed besides chasing smugglers.

Now consider the vast amounts of money earned from smuggling, and the far greater number of people engaged in the smuggling. Between 1996 and 1998, US enforcement agencies seized about 1 million pounds of narcotics, and of that 60% was seized by Customs.[16] Yet the amount that was not caught was far greater. Smuggling was a main attraction for Russian mafia, who had billions of dollars at their disposal.[17] In another case, a billion dollar smuggling ring was discovered in Xiamen, China involving about $9.5 billion of goods. Over 700 investigators were used, presided over by a senior Communist Party official.[18]

The key to stopping this trade, at least for the US, has been information. This is a battle that is hardly ever fought with guns. Even Hawkeye Gross advised against even carrying a weapon. Law enforcement agents spend much of their resources on things like accountants and computer technicians. In Miami-Dade County, the police have been using a database system to check every vehicle that is shipped through the port against a database of stolen cars. This simple method was credited with recovering 873 stolen vehicles in a 12-month period. This was a 32% increase over the previous year.[19]

The authorities that deal with smuggling are faced with a unique task. Customs officials sometimes do this, but their training and primary task is administrative, identifying cargo and assessing duties. Local police are also not accustomed to the task, and the military are definitely not appropriate. Given the unique combination of skills necessary, many countries have created anti-smuggling police. The US uses customs agents but there are special units within the agency that have police/military training. China created a special force but that was because so many of their top officials where involved in the smuggling. The government estimated that tens of billions of dollars worth of cargo was being smuggled into the country.[20] This indicates that their definition of smuggling is quite broad given the controlled economy, and the involvement of officials.

Ideally, Customs and legitimate shippers work together. Some customs agencies have a good relationship with businesses, minimizing interference and discussing with industry leaders how they can work together, while in some countries there is an antagonistic relationship. Customs looks at all shippers as suspect, and in return shippers are fearful and uncooperative to customs. "Importing goods into this country is not a right, it's a privilege", notes one air freight executive (referring to the US).[21] Others would disagree, saying that one does in fact have a right to conduct business free from government interference.

Border officials sometimes drill holes in truck trailers and cargo containers looking for false walls. This damages the container itself, and leads to water damage when water gets into the holes. Shippers are generally reluctant to file for claims, which they could do, because the cost of filing a claim is too much bother. Other methods such as x-rays are being developed to avoid drilling. In Laredo, where up to 6,000 trucks cross the border every day, there are very few loading docks where cargo can be unloaded and inspected.[22]

Free trade regimes have played a role in allowing for increased smuggling by reducing the controls of cargo crossing the border. After NAFTA's implementation, the volume of cargo across the Mexican/US border increased dramatically, and so has the incidence of smuggling. The most common mode of transport is intermodal containers on trucks, which are difficult to search.

X-Ray Machines
These machines are not just used by airports for the passengers walking on the planes. Customs officials are now using them to check cargo, and shippers themselves are scanning their own shipments in case someone inserted contraband without their knowledge. The new generation of machines that scan a 12,000-pound main-deck air cargo pallet in 30 seconds[23], and cost between $3 and 4 million each.[24] One of the new models uses a frequency at the higher end of the color spectrum to get a clearer picture in dense materials. One concern is that because the scanning process can be done so quickly, that means these expensive machines are sitting unused for most of the day.

Documentation plays a key role in smuggling. The normal practice is to use forged documents, or legitimate cargo and documents but a small amount of contraband as well. One way to prevent forgery of documents is to make them electronic, so they can be stored in a database. The US apparel controls require a visa for importing apparel. Forged visas and quota documents cause a lot of trouble for legitimate traders, which are why countries are eager to be included in an electronic system called Electronic Visa Information Service (ELVIS).[25]

Transshipment is a common business practice that plays a special role in smuggling. One common method of identifying contraband is to look at shipments from problem areas. Smugglers use a transshipment point, somewhere where security is lax enough to get through but has a good enough reputation so as not to attract attention. Also useful are those ports which have large volumes of traffic so one shipment would not gain as much notice.

Another benefit of transshipment is that even if the cargo is seized, it is much harder to identify the origin, and hence the smugglers. Every time a container is shipped through a port, the time and effort required tracking it increases, and law-enforcement agencies make cost-benefit decisions about which cases to investigate.

It should not be necessary to say this, but I am going to say it anyway. This information is not intended to help or encourage anyone to smuggle. The information provided here was collected from easily obtained public sources. It is intended for informational purposes only. For those foolish enough to think this information will afford you a way to get started, remember that knowledgeable smugglers and their law enforcement counterparts are operating at a level of sophistication far beyond what is described here.

-The Author

Armenia's Gateway to the World: The Geopolitical Significance of the Port of Poti

By Jeffrey Engels, US Agency for International Development

During Soviet times, the small land-locked country of Armenia traded almost exclusively with other USSR Republics, but in 1991 when the Soviet Union collapsed so too did Armenia's export infrastructure and opportunities. Today, due to the transit and trade blockade imposed by Turkey and Azerbaijan after the Nagorno-Karabakh war, Armenia is completely dependent on the Black Sea port of Poti, Georgia. It remains the only viable road-rail-maritime channel through which the Republic of Armenia can export goods to the Commonwealth of Independent States (CIS), Europe, and beyond.

The port of Poti, Georgia, is located on the eastern side of the Black Sea and is the principal maritime gateway for the transshipment of a wide range of cargoes to and from Central Asia to Europe and the CIS. The port was officially founded in 1858, but its history dates back thousands of years: in classical times it was known as the port of Phazisi, where Jason and the Argonauts landed to steal the Golden Fleece. With its strategic geographic location, it grew from a small trading center to a vital link in the chain of the ancient Silk Road trade route.

Poti was a favorite caravan stop for traffic from the East--Iran, India, and the North Caucasus. The Greeks settled there and cross-traded with the shores of the Ionian and Aegean Seas, shipping Georgian handicrafts, wood, leather products, wine, and silk. By the 7th century, Poti was closely linked with Constantinople (Istanbul), the capital of Byzantium, and played a strong role in the economic development of the region that lasted hundreds of years. By the mid-nineteenth century a formal facility was designed and constructed, and by the 20th century this modern port was active with international commercial traffic.

Today, Poti has the current capacity of annually moving over 7 million tons of cargo. It has both road and rail links and from its 14 berths moves railcars, bulk grain shipments, containers, general cargo, perishable goods, and even passengers. A few berths can handle ships holding 25,000 tons and are equipped with portal cranes for efficient loading. The port also has warehouses available for storage of goods such as cotton, saltpeter, *etc*. Given its prime Black Sea location, the port of Poti is as important today as any time in history, and no more so than for Georgia's neighbor, the Republic of Armenia.

Armenia is credited as one of the cradles of civilization: at one time the country stretched from the Mediterranean in the West to the Caspian Sea in the East, and encompassed present day Turkey and Iran: spanning 120 provinces, many races, cultures, languages, and commercial interests peacefully coincided to bring rapid growth and prosperity to an Armenian Empire. Then came incorporation in the Roman Empire, the Turkish Empire, and later still into a Soviet Federated Republic, along with Georgia and Azerbaijan. Today, covering an areas of 29,800 square kilometers (Slightly larger than the state of Maryland), completely independent since 1991 and the first ex-Soviet Republic to privatize its land, Armenia exports its agricultural products.

From early on Armenians have been cultivating their land. According to Biblical legend, after the Flood Noah disembarked from his Ark atop Mount Ararat and descended to the surrounding Armenian plateau to grow grapes for wine. Antiquated pictograms and 9th century B.C. Urartian inscriptions confirm early agriculture throughout Armenian valleys. Greek historians Herodotus, Xenopon, and Strabon all wrote about farming in the region, and agriculture remains the core of the Armenian economy.

Since the collapse of the USSR, Armenia has implemented one of the most comprehensive land reforms in the CIS. Large Soviet-era agro-industrial complexes have been replaced by small-scale farming, and 321,000 small private farms and over 200 processing enterprises have been created. A new liberalized economy has allowed agriculture to contribute to about one-third to GDP and accounts for 42% of employment. Agricultural producers and processors are confident in capturing export markets with their excess capacity, but the country is land-locked. Armenia's ability to trade is entirely contingent upon its relations with surrounding nations and the capabilities and limitations of their transportation systems.

Concurrent with political independence in 1991 came fighting between Armenia and Azerbaijan over the disputed region of Nagorno-Karabakh. The region, largely an Armenian enclave, had been incorporated in the territory of Azerbaijan under Russian ruler Joseph Stalin (1929-1953). Before a cease-fire was brokered in Moscow in May of 1994, 150,000 Azeris were expelled from Armenia and more than 500,000 Armenians fled Azerbaijan. Although a cold peace has held ever since, border closures and a trade and transport embargo have been imposed on Armenia from two of its largest encircling neighbors, Azerbaijan to the east, and Turkey to the west and southwest. Only through Georgia does Armenia have a vital export channel through which commercial goods can flow freely to CIS and European trading partners. Airfreight is too expensive to move most goods, so road and rail links that access the port of Poti are essential for international trade.

Under Soviet Rule, Armenia traded almost exclusively with the Union's other Republics. A foreign trade organization (FTO) controlled each product group, and exports by each Armenian enterprise were determined by the State Planning Committee (commonly known by its Russian acronym, Gosplan) in Moscow. Businesses had no control over the size or destination of shipments of their products. In 1989 the FTO monopoly was removed, and Armenia signed bilateral trade protocols with former Soviet Republics and operates under these today.

One of Armenia's key export commodities is cognac, derived from native Armenian Areni grapes that provide its remarkable flavor. In Soviet times Armenian-produced Ararat *konyak* was hard to come by unless you were a member of the elite. When British Prime Minister Winston Churchill attended the Yalta Conference in 1945, Joseph Stalin served it and Churchill was so taken with the spirit it became his favorite brand. The industry has survived war, revolution, and even Mikhail Gorbachev's perestroika-era anti-alcohol campaign. Today, the Yerevan Brandy Company is owned and operated by France's Pernod Ricard group. More than a dozen types of cognac are produced and the company has enough spirit to last at least 20 years, including significant quantities aging in the factory's cellars since the 1930's, 1950's and 1960's. By far the Russian market remains the most lucrative for Armenian cognac and represents 70% of its total sales.

To reach Moscow consumers, a combination of road and maritime transport is required. Wooden crate packed bottles are loaded in canvas-tarped trucks and sent north toward the Georgian border. Each bottle is sealed with an excise stamp, provided by the buyer, and the driver carries all the necessary export documentation. Given the value of the cargo and the high risk of hijacking or pilfering, a single driver accompanies the load the entire route, and on Georgian soil the truck is escorted through the country by police and customs officials, typically consisting of four police officers and three customs officials traveling in two cars, one in front and one in back.

Under this protection, the convoy connects with Route M27 directly to Poti. Once a week a Swiss-owned common carrier calls Poti, loads the trucks, and sails to Novorossiysk, the largest seaport of the Russian Federation on the Black Sea. Here Russian customs thoroughly checks all documents and cargo. Once cleared, the trucks proceed to a bonded warehouse on the outskirts of the capital for off loading. Pending unforeseen transit or customs delays, cargo usually arrives in about 10 days.

The fruit and vegetable canning industry is another strong force in the food and agriculture sector. Since independence, Armenian agro-processors have reorganized their enterprises and retooled their factories. All canneries are now in private hands and over the last few years these companies have succeeded in capturing the local market and have expanded for export capability. Armenian sweet jams, juices, tomato paste, and vegetable marinades are all current or potential export items. Since a shipment of these products is not as valuable as cognac, local processors use rail and maritime transport when cargo is bound for Kiev, Ukraine.

The Armenian Railroads (ARM) did not exist before the collapse of the Soviet Union, but instead have been created as a part of the Soviet Union Railways (SZD), which crossed Armenian land. The Armenian Railway is in the same isolated position as the highway network due to the Nagorno-Karabakh dispute, but rail links provide a transportation corridor for deliveries to and from the port of Poti. With close cooperation with the Georgian Railway (GR), rail links, adjacent to canneries' loading docks, can carry either mixed or straight shipments of product loaded in 50' box cars directly to the port. These cars are designed to hold up to 60 tons and/or 5000 cubic feet of cargo. Each is equipped with weather-resistant sliding doors. Since the Caucasus railways were all built to the same gauge and standard, rail transport to Poti is relatively easy.

Once portside, the boxcars are then loaded on a steamship vessel bound for the deep-water berths of Illichivsk, Ukraine. Upon arrival there, all taxes and customs duties can be paid, or postponed for the final inland destination. At Illichivsk, a few kilometers southwest of Odessa, the boxcars resume transit on the Ukrainian rail network that takes them to the capital city of Kiev.

The Georgian port of Poti is presently the exclusive operational outlet a land-locked Armenia has to export to its trading partners. Changes in the border situation would drastically alter trade opportunities, attract foreign investments, reduce unemployment, allow for planning within the economy, and bring stability to the entire Caucasus. However, until a mutually beneficial political solution is found to settle differences with Azerbaijan over the territory of Nagorno-Karabakh, the trade embargo will continue. And as long as it continues, Poti will be critical to the economical survival of the Armenian economy and the only logistical solution the country has for international trade.

Jeffrey E Engels is Marketing Manager of the USDA's Marketing Assistance Project (MAP) in Yerevan, Armenia. MAP assists farmers and agribusinesses in production, marketing, and exporting food and related products to increase incomes, create jobs, and raise the standard of living for Armenians working in the agro-processing sector.

[1] *Transparency International.*

[2] *"Culture Matters: How Values Shape Human Progress", Lawrence E. Harrison and Samuel P. Huntington, editors, Basic Books, New York, 2000.*

[3] *"Political Order in Changing Societies", Samuel P. Huntington, Yale University Press, New Haven, CT, 1968.*

[4] *This section was written with the assistance of Jeffrey Beason of OOCL and an anonymous reviewer*

[5] *"Hazardous Materials Transportation Training Modules", US Department of Transportation, 2000 edition*

[6] *JOC, June 18, 1999, p. 8.*

[7] *Ibid.*

[8] *"Shippers turn to private cops to find stolen goods and bypass their insurers", Chris Barnett, The Journal of Commerce, September 2, 1998, p. 1A.*

[9] *American Shipper, June 1998, p. 24.*

[10] *"An engine of prevention", Angela Calise Dauer, JOC, October 27, 1998, p. 9A.*

[11] *"200,00 stolen vehicles shipped from US ports", Lisa Shenkle, JOC, October 22, 1998, p. 7A*

[12] *American Shipper, June 1998, p. 24.*

[13] *"A goat, a rug, special price?", The Economist, April 3, 1999, pp. 38-9.*

[14] *"Dictionary of International Trade", Edward G. Hinkelman, World Trade Press, San Rafael, CA 1998, p 178.*

[15] *"Drug Smuggling: The Forbidden Book", K. Hawkeye Gross, Pladin Press, Boulder, CO, 1992.*

[16] *"Customs bullish on drug war", American Shipper, March 1998, p. 58.*

[17] *"Kremlin Capitalism: Privatizing the Russian Economy", Joseph R. Blasi, Maya Kroumova, and Douglas Kruse, Cornell University Press, Ithaca, NY, 1997, p. 119.*

[18] *"Billion-Dollar Smuggling Ring Tests China's Will to Stop Corruption", Charles Hutzler, San Francisco Chronicle, January 21, 2000.*

[19] *Shenkle, Ibid p. 7A.*

[20] *"China Forms Police Force on Smuggling", San Francisco Chronicle, July 17, 1999.*

[21] *"Customs bullish on drug war", American Shipper, March 1998, p. 58.*

[22] *"A boring nuisance for importers at border", Kevin G. Hall, The Journal of Commerce, September 25, 1998, p. 1A.*

[23] *JOC, September 2, 1999, p. 3.*

[24] *"Customs bullish on drug war", American Shipper, March 1998, p. 58.*

[25] *JOC, June 15, 1999, p. 3.*

SECTION IV
LOGISTICS MANAGEMENT

Chapter 14
Intermediaries and Alliances

In the beginning of the book, we described shippers, carriers and consignees, the three main parties of a shipment. Now we elaborate on the many other parties to an international transaction. Intermediaries refer to the many companies or individuals that facilitate trade. Some of them work for the shippers, some for the carriers, and some for the consignees. Sometimes these arrangements become more elaborate, such as alliances. Working with intermediaries and alliances can be rewarding, but there are also some things to be careful about.

THE ROLE OF INTERMEDIARIES

Competitive pressure has made many companies return to their core competencies and outsource functions that others can do better. Many logistics companies have appeared that provide integrated services to shippers. There is a wide range of services being offered, such as finding the distribution system. The basic economic rule is that a company should outsource when the benefits of another company doing the same job are greater than the cost of supervising that provider. There is a cost of supervising any person or office within a company, but this cost is relatively small compared to supervising the performance of those outside the company. Therefore the outside service providers should be able to do the job well enough to cover the cost of supervising them.

Intermediaries are the 'middle men' or the 'go between' of the logistics world, companies that offer services to assist in the logistics process. They provide a broad and ever growing variety of services between the shipper and the consignee. They are becoming more common as companies try to gain full advantage of logistics, and because there has been a strong tendency to contract out parts of the work.

Third party logistics, commonly known as **3PL**, is a general term referring to an outside company providing the service of managing logistics. Not all intermediaries are 3PL, which will become more obvious as we describe the different types of intermediaries later in this chapter. 3PL has become a huge industry in itself. Cass/Prologis estimates that 3PLs managed $50 billion worth of expenditures in 2000. Note that this does not mean companies spent that much money on 3PLs, but the companies delegated that much of their operations.[1] The 3PL industry is growing at 24% a year.[2] Seventy five percent (75%) of US manufacturers are either using or considering the use of 3PL service providers according to a survey by Exel Logistics and Ernst & Young LLP.[3] It is estimated that 10% of US companies completely outsource their logistics activities, and 40% of companies outsource their European logistics services.[4]

The 3PL industry has grown to the point where there are now **4PL** logistics providers. These are logistics companies that service other logistics companies. One of the best examples of this is where alliances and partnerships have caused some carriers to take out their ships or planes from a trade lane, and use their partner's vessels.

There are three basic issues in dealing with 3PL providers, **control, compensation,** and **expertise**. Control refers to the ability to make the contractor (we may refer to 3PL providers as contractors in this chapter) behave as if they were within the company. It should never be assumed that a contractor is going to behave exactly as the principal (the company hiring the contractor) would wish. Compensation refers to the amount and manner in which the contractor is paid. Some may be paid on commission, while others are paid an amount regardless of performance. Ideally all contractors would be paid on performance, but when this is difficult, costly or impossible to measure, other forms of payment would be required.

Expertise means that the contractor has knowledge that the principal does not have, which is a major reason for hiring the contractor. However, the two parties need to agree on the transfer of information. Is the contactor expected to divulge all information, or may they keep it private? For example, sales agents often want to keep their contacts private or else they would be cut out of the process.

Many 3PL providers are asset-based. For example, imagine an ocean carrier that acted as a consultant on how to set up the most efficient ocean transportation. This compromises their independence, which has been an issue in the 3PL industry.

The use of a 3PL provider does not mean the company out-sources all of its logistics. It can outsource only part of the process, and it can even have different 3PL providers doing different functions. 3PL providers usually come out of (that is, the company is created from) five sources of integrated logistics services:[5]

- **Carrier-based**. Carriers such as ocean shipping companies or air cargo carriers need to understand the underlying needs of their customers in order to make a good sales effort. This means they are in a good position to help their customers though 3PL services to actually conduct those logistics activities.
- **Warehouse based**. Similar to carrier based, only these are warehousing companies. Warehouses are much more sophisticated that just a large storage building, as will be discussed in Chapter Fourteen.
- **Forwarder/broker based**. Freight forwarders are two of the most common intermediaries, to be discussed later in this chapter.
- **Information based**. These are information providers or information technology companies.
- **Customer based**. Some companies have developed very sophisticated logistics abilities, which they then offer to other companies. Such a 3PL provider would probably not work for a competitor to the home company.

There has been remarkable growth of 3PL providers since the early 1990's, and should continue in the same fashion into the new millennium. New 3PL providers sometimes even outsource their own work. This is a dangerous tactic given that start-ups are usually low on money and staff.

A number of issues are raised in considering intermediaries, and can be viewed in two different ways:

1. They are fundamentally logistics companies, unlike other companies, such as a manufacturer, that does some other job but has logistics as one of its tasks. In other words, logistics intermediaries are like carriers in that they play a central role in logistics.
2. They also need to be considered for how they interact with other companies in the conduct of logistics.

Intermediaries are companies in their own right that need to make decisions about where they conduct business. One of the most important decisions is which countries to become involved with. There is a strong incentive to be present wherever their clients are located. However, just because, for example, a manufacturer can enter a highly risky country and do business does not mean that intermediary could also do the same. One common strategy is for an intermediary to follow a large company into a developing nation. For example, Siemens' strong move into East Germany was followed by many supporting companies, and Johnson & Johnson's entry into Latin American markets encourages others to do the same.[6]

Fritz Companies' Moscow office

It is difficult to say exactly how many intermediaries are operating worldwide, or even in any given country, because it is not clear what constitutes a 'logistics intermediary'. Some of these are companies that offer a business service that is used by the logistics industry. Others are subsidiaries of a company that is not an intermediary. According to Richard Armstrong, there are about 1,000 companies in the US providing 3PL services, of which 56 provide two thirds of the gross revenue. The 3PL industry is earning about $34 billion a year in the US and growing at 20% a year.[7]

While it is not clear who is the largest such company in the world, one company deserves mention. In the People's Republic of China, Sinotrans was the one official forwarder up until 1997. Since then about 1,500 licensed forwarders appeared, while another 500 operations are working without a license. Still, competition is limited and Sinotrans holds 60% of the forwarding market with their 64,000 employees, 57 branches and 508 sub-branches.[8]

Virtually every type of company with a logistics need can or does use the services of intermediaries. Small companies may use them because they do not have the resources in-house. Yet even the largest companies use them because they can do some things more efficiently than could be done in-house. Companies that are based in one region but trade worldwide are more likely to outsource more. This makes more sense for them than to re-engineer their operations as a global company would.[9]

TYPES OF INTERMEDIARIES

The following are the major types of intermediaries. This is far from a conclusive list, especially since changing needs result in new companIies and new services are offered.

Freight Forwarders

Freight forwarders, more commonly called 'forwarders', are the travel agents of the freight world. When you have cargo that needs to be moved anywhere, they consider the origin and destination, and its special handling characteristics, and arrange for the transportation. They find the most efficient, cost-effective mode of transport and routing.

Another way of looking at forwarders is the difference between wholesale and retail. Carriers sell wholesale space, since it is being bought directly from the provider. Forwarders act as retailers. Note that forwarders only act as agents, not resellers. NVOCCs, to be discussed later in this section, are resellers.

Airfreight forwarders tend to operate as a special group because of the special demands of the air cargo industry. Forwarders are mostly for international shipments, since these are more complex. Still, there are **domestic forwarders,** who only handle domestic shipments. This also presents difficulties. Domestic cargo, at least in the US, tends to be more time-sensitive because of competition from integrated carriers and truckers. There is no need for customs work, and many customers want same vendor for domestic and international shipments.

No certification is required of forwarders, but the US National Customs Brokers and Forwarders Association offers a program leading to the designation of 'Certified Ocean Forwarder'. This entails a lengthy exam covering all aspects of the industry, including documentation, Incoterms, regulations and hazardous cargo. A minimum of three years experience is also required. Shippers are also allowed to take this test, and receive the title of 'Certified Ocean Shipper'.[10] The Federal Aviation Administration certifies air freight forwarders. Although this is not legally required, the airlines generally will not work with a forwarder that is not certified.

When business is bad for the carriers, this does not necessarily mean that intermediaries will also suffer, particularly the freight forwarders. There seems to be a trend in which during bad times, the forwarders can gain better deals from the carriers. Ironically, the worse off the carriers, the better for the forwarders. This cannot be generalized across all intermediaries, however. If the level of trade is down, then everyone in the industry will suffer from lower demand. But when rates are being lowered due to excessive capacity, which happens to many of the carriers, then the issue is how the fixed pie is divided.

POWER OF ATTORNEY
EXPORTED (U.S. PRINCIPAL PARTY IN INTEREST)/FORWARDING AGENT

Know all men by these presents, That   the
                              Name of U S Principal Party in Interest   (USPPI)
(USPPI) organized and doing business under the laws of the State or Country of
and having an office and place of business at   hereby authorizes
                              Address of USPPI

           the (Forwarding Agent) of
Forwarding Agent   Address of Forwarding Agent

to act for and on its behalf as a true and lawful agent and attorney of the U S Principal Party in Interest for and in the name place and stead of the U S Principal Party in Interest from this date in the United States either in writing electronically or by other authorized means to

Act as Forwarding Agent for Export Control Census Reporting and Customs purposes Make endorse or sign any Shipper's Export Declaration or other documents or to perform any act which may be required by law or regulation in connection with the exportation or transportation of any merchandise shipped or consigned by or to the U S Principal Party in Interest and to receive or ship any merchandise on behalf of the U S Principal Party in Interest

The U S Principal Party in Interest hereby certifies that all statements and information contained in the documentation provided to the Forwarding Agent relating to exportation are true and correct Furthermore the U S Principal Party in Interest understands that civil and criminal penalties may be imposed for making false or fraudulent statements or for the violation of any United States laws or regulations on exportation

This power of attorney is to remain in full force and effect until revocation in writing is duly given by the U S Principal Party in Interest and received by the Forwarding Agent

IN WITNESS WHEREOF   caused these
                   Full Name of USPP (USPPI Company)
presents to be sealed and signed

Witness   Signature
                                   Capacity
                                   Date

Customs House Brokers

See Chapter Ten for a discussion of customs house brokers.

Non-Vessel Operators and Non-Vessel Operating Common Carriers

Non-vessel operating common carriers (NVOCC) buy space from carriers and resell them. NVOCCs essentially act as if they were a carrier, but they do not own or control any of the ships, planes etc. They issue their own bill of lading. Legally, in the US and many other countries, they have similar obligations of a common carrier, such as filing rates with the government and practicing non-discrimination.

A similar type of company to the NVOCCs, NVOs are carriers but not a common carrier. They are not the same as forwarders and they make a point of keeping that distinction. Forwarders arrange for the transportation, while NVOs consolidate small shippers (see Consolidators below), doing the physical work, and also act as a re-seller of ship space. Sometimes the term NVO is used as an abbreviation of NVOCC.

NVOs in the US appeared in the 1970s to do consolidation work that was too costly for the ocean carriers using unionized labor. In response, the International Longshoreman's Association (ILA) fought for exclusive right to stuff or strip (that means to load or unload) any container within 50 miles of a port. The case went all the way to the Supreme Court before the union lost.[11]

Some NVOs are associated with forwarders, while others are neutral. Neutral NVOs can offer prices 5-10 percent lower than forwarder-based ones, but they cannot offer the door-to-door services of a forwarder.

NVOCCs and NVOs sometime face different regulations from carriers. For example, under the OSRA, they cannot enter into service contracts with shippers as carriers can. This puts them at a disadvantage.

Consolidators

Consolidators, a specific activity often done under NVOs, take small shipments and consolidate them into larger shipments. This is usually for intermodal cargo, to fill an intermodal container. Other types of shipments that are not charged by the container generally do not need this service. Consolidators, if they are not associated with an NVO or NVOCC, would use a forwarder or carrier to arrange for the transportation, as if they (the consolidator) were the shipper.

It is very common to mistake a forwarder for a consolidator. In fact, one of the best selling textbooks used for college marketing courses in the US refers to forwarders as those entities that consolidate small shipments into big shipments. This is not correct. Some forwarders may offer consolidation services, and the tasks involved may be similar, but there is a difference.

Shippers Associations

Shipping Associations are organizations, usually not-for-profit, in which shippers join together to get better negotiating leverage against carriers. Often small companies find it difficult to compete with larger companies for overseas markets because they have higher transportation costs. This way, even smaller shippers can get lower rates as if they were a large shipper. Members of these associations are usually small shippers, though some can still be quite large.

These associations are becoming much more common in the US with the deregulation of the ocean shipping industry. Traditionally, ocean liners needed to file their rates and they were thus public knowledge. If a large shipper got a low rate, the small shippers could insist on receiving the same or similar rates. With deregulation, shippers and carriers can sign confidential service contracts. Small shippers now do not know what rates their large competitor is getting. Therefore they are finding it necessary to come together in shipping associations to insure they are getting the lowest possible rates.

Export Management Companies and Export Trading Companies

Export Management Companies (EMC) and Export Trading Companies (ETC) assist companies in marketing their product in other countries. This is a valuable service for companies that may be good at manufacturing their product, and marketing it domestically, but do not have the expertise in international marketing and logistics.

The primary difference is that the EMC acts more like an advisor to the exporter, but rarely takes ownership of the cargo. The ETC is actively buying, selling, marketing and transporting the products by itself.

EMCs and ETCs generally have a long term relationship with their suppliers/customers. For an EMC, it would be a customer. For an ETC, that same 'customer' would be considered a supplier. Forwarders often work on a single shipment for a customer, but that would be unusual for an EMC/ETC. If you think about how EMCs and their customers work together, that makes perfect sense. Except in the case of a product that is extremely expensive or unique, the EMC/ETC and producer would want to work together over a period of time to make the relationship worthwhile.

Professional Services

Lawyers and **accountants** are offering an increasing variety of services to the logistics and international trade community. There has been an increased specialization in these fields so that there are lawyers and accountants, and even law firms and accounting firms, which work just on international trade.

These services are not cheap, but they can make a critical difference for the company. Lawyers are not just used when there is a problem, but can advise on how to avoid problems in the first place. Accounting firms have created departments that advise clients on top-level strategies that improve supply chain operations and reduce taxes. Transfer pricing and international taxes are key areas where the shipper would probably do well to get expert advice.

Integrated logistics implies a top-level strategic approach to business planning, and thus logistics planning. In response, accounting firms (and to a much lesser extent law firms) are creating international trade services that offer planning in a wide variety of areas that affect money. This includes network design and supply chain management designed to take advantage of international tax policies and related regulations.

Customs audits often require the support of professional services. Lawyers and accountants have been debating who is more appropriate for the task. Accountants tend to be more useful in the ongoing process of maintaining customs documentation, and that is why, they would suggest, the audit is also part of their job. Lawyers would suggest that an audit is essentially a legal process, and thus a legal professional is needed. One important distinction between the two is that there is a special lawyer-client privilege to keep company information secret from the government. This is something that accountants do not have.

MANAGING ALLIANCES AND SUBCONTRACTORS

In logistics, at least as much as any other industry, there has been a strong trend toward alliances in order to promote efficiency. Sometimes this involves intermediaries, sometimes not.

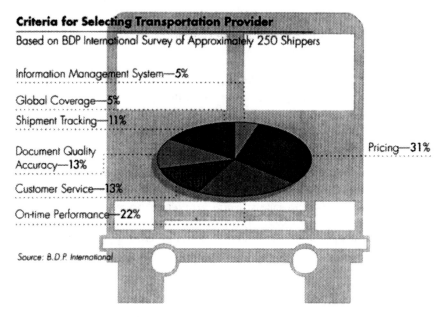

Criteria for Selecting Transportation Provider

Based on BDP International Survey of Approximately 250 Shippers

Information Management System—5%

Global Coverage—5%

Shipment Tracking—11%

Document Quality Accuracy—13%

Customer Service—13%

On-time Performance—22%

Pricing—31%

Source: B.D.P. International

There are some **barriers to implementation of a logistical alliance or partnership** that must be overcome. Many business functions are affected, since logistics is fundamentally about coordinating the activities of many different company departments. Supply chain leadership is affected. If the company was working toward leadership of its supply chain, how will things change when another company is coming in to take over operations? There is a common fear of being locked into the relationship. Alliances are inherently committing, which is a genuine concern when it comes to planning for the future. Finally, there is the question of benchmarks. Different companies judge success differently. The alliance members need to agree on how they are going to assess the performance of the partnership.

One of the major concerns with alliances is the management of information. In the inset on the Art of Alliances, Kanter notes that information should flow freely between partners. That recommendation is quite controversial. Protecting one's information is one of the most important things any company can do. This will be discussed in Chapter Sixteen, but it should be noted at this point that providing information is a calculated risk.

The Art Of Alliances[12]

Rosabeth Moss Kanter's research suggests that effective alliances between companies are not entirely rational and business-like. Much like a family relationship, companies need to be flexible, involved and personal. She offers the following criteria of a good match:

Individual Excellence. Each partner has something to offer, and their motivation for entering into the arrangement is to pursue opportunity instead of escape a problem.

Importance. The alliance is important for each of the partners.

Interdependence. The partners need each other, which means that each offers something that the other needs but does not have.

Investment. Partners are willing to invest in the alliance.

Information. Communications are open and both sides are honest and generous in providing information.

Integration. The partners have many connections and shared operational procedures at different levels.

Institutionalization. The alliance is given formal status, with clear objectives and procedures.

Integrity. Trust is an intangible but vital element of an alliance, so the partners do not do anything to violate that trust.

While it helps to see what it takes to make an alliance successful, it also helps to see what were the causes of an **alliance gone bad**. According to one estimate, one out of three users of 3PL services cancels at least one contract. There is even talk of an industry providing mediation and arbitration services for outsourcing contracts.[13] One expert discussed the possibility that a 3PL provider gains a client, but that the partnership ends in failure. He suggested that 'a (logistics) provider who doesn't get enough information (from the potential client) up front should turn down the business'. He also suggested that the provider have a few people asking questions when they are negotiating with the potential client and looking for potential trouble areas in the new relationship. Salespeople from the 3PL provider should be compensated not just for acquiring a new client but for retaining the client. That means the salesperson should stay involved even after the contact begins.[14]

TIACA notes that inter-professional communications is an issue that needs to be addressed to promote overall efficiency. Each industry, and professional fields within an industry, has very different training, skills, issues and priorities. Professionals trying to solve the same problem often work at cross purposes. They suggest that "inter-professional relationships, especially with regard to traditional freight services, need to be clearly redefined in order to establish who does what for whom and for how much."[15] Note that communicating across the profession of air cargo is one of TIACA roles.

In one famous case, OfficeMax terminated the account of Ryder Integrated Logistics and sued for $21.4 million. Ryder countersued for $75 million. Ryder ranked first in three of the last four years in a survey by Mercer, so clearly this was a large and well established 3PL provider. Among the accusations, Ryder claimed that OfficeMax denied them important information. OfficeMax then said that if they did not get the information needed, they should have refused the business. To make the situation more complicated, the company providing OfficeMax before Ryder also had a lawsuit stating that Ryder has stolen proprietary data.[16]

Contracting out the logistics work, just like contracting out any part of a company's operations, carry some risks. 3PL providers can be seen as a threat to in-house logistics departments. When a 3PL provider is used, it does not mean that the entire logistics department is eliminated. There certainly needs to be someone within the company to coordinate the work even if almost all the work is outsourced. Still, there are many cases where parts of the logistics function is outsourced, and the remaining in-house employees are wondering if their jobs are going to be eliminated later. Even the logistics executive's job is not safe. One consultant advises that the logistics executive should bring the possibility of outsourcing to management's attention before an outsider (a potential 3PL provider) does.[17]

Criteria for the Selection of 3PL Providers[18]

- Financial Stability
- Management Depth and Strength
- Strategic Direction
- Operations
- Quality Initiatives
- Chemistry and Compatibility
- Business Experience
- Reputation with other Clients
- Physical Facilities and Equipment
- Information Technology
- Growth Potential
- Cost

Most Commonly Outsourced Services:[19]

- Outbound transportation
- Warehousing
- Freight Consolidation/Distribution
- Selected Manufacturing Activities
- Product Returns and Repair
- Information Technology
- Inventory Management
- Customer Service

- Freight Bill Auditing/Payment
- Inbound Transportation
- Cross-Docking
- Product Marking, Labeling, Packaging
- Traffic Management/Fleet Operations
- Product Assembly/Installation
- Order Fulfillment
- Order Entry/Order Processing

---

[1] "Managing Logistics in a Perfect Storm, 12ᵗʰ Annual State of Logistics Report", Robert V. Delaney and Rosalyn Wilson, Cass Information Systems and Prologis, Washington D.C, June 4, 2001.

[2] Ibid.

[3] "Use of 3ʳᵈ-party logistic rising", The Journal of Commerce, October 13, 1998, p. 16A.

[4] "Outsourcing: A Megatrend in the Making?", Frank J. Quinn, Traffic Management, February 1990, p. 7.

[5] Bowersox and Closs, Ibid,, p. 112.

[6] "Global logistics market trebles", American Shipper, October 1998, p. 40.

[7] "We, The People, Demand Logistics Productivity", 9ᵗʰ Annual State of Logistics Report, Robert V. Delaney, 1998.

[8] "Chinese forwarding: From old regime to new", Philip Damas, American Shipper, March 1998, p. 36.

[9] "Global logistics market trebles", American Shipper, October 1998, p. 40.

[10] "Education of a Forwarder", Chris Gillis, American Shipper, November 1998, p. 40.

[11] "Don't bet against NOVs' future", Chris Gillis, American Shipper, May 1998, p. 33.

[12] "Collaborative Advantage: The Art of Alliances", Rosabeth Moss Kanter, Harvard Business Review, Vol. 72, no. 4, July-August 1994, p. 100.

[13] "Messy Divorce", Robert Mottley, American Shipper, March 1998, p. 26.

[14] "Third-party logistics failures", Robert Mottley, American Shipper, September 1998, p. 44.

[15] "Global Voice of Air Logistics: Manifesto", the International Air Cargo Association, p. 23.

[16] "Messy Divorce", Robert Mottley, American Shipper, March 1998, p. 22.

[17] David Anderson, as quoted by Mitchell E. MacDonald, in Frank J. Quinn, "Outsourcing: A Megatrend in the Making?", Traffic Management, February 1990, p. 7.

[18] "Logistics Outsourcing: A Management Guide", Clifford F. Lynch, Council of Logistics Management, Oak Brook, Il, 2000, pp. 58-60.

[19] "Third Party Logistics Services: Views from the Customers", Glenn A. Dalhart, C. John Langley, Jr, Brian F. Newton, University of Tennessee, 1999.

CHAPTER 15
INVENTORY MANAGEMENT

It has been said that logistics is the "management of inventory in motion or at rest."[1] This chapter discusses the multiple roles of inventory, some of which are well known and others that are quite subtle. Guidelines for the management of inventory are reviewed, although given the international focus of this book, this section is brief. In the final sections we discuss economic aspects of inventory management, including some of the more common algebraic equations used for generating inventory statistics.

INVENTORY TYPES AND CHARACTERISTICS

The people in Dushambe, the capital of Tajikistan, appreciate inventory management. When this country was part of the Soviet Union, the author was traveling in the area, and visited the city's main department store. In the shoe department, which was dark for lack of light bulbs, a thousand pairs of shoes filled the shelves. This department was as well stocked as one could hope for. Only there was a problem. Every one of those shoes was XXXX-large heavy black leather work boots. There was only one type of shoe, in one size, which almost nobody could wear.

While inventory management is an important part of logistics, it can be hard to get excited about this field. It does not offer the same glamour as planes or ships. Many think that inventory management is only about boxes sitting in a warehouse. This is because we take for granted that products are available in stores when we want them.

This was not always the case. Up until only a few decades ago, the major challenge of inventory management was simply to get the products into the store. Now customer expectations have increased, and not only do they want the store shelves full, but they expect a wider and wider assortment of goods. Whereas the average store was holding 15,000 SKUs in the 1970s, today a supermarket may have 40,000 SKUs, and the large department stores may have 100,000 SKUs. Inventory investment, the value of all inventory, was over $1.485 trillion in the U.S. in 2000.[2] Because of the large amount of money involved, managers cannot avoid the importance of good inventory management.

Ironically, while the goal of retailers is to offer a wider assortment of goods, the overall goal of inventory management is to eliminate inventory to reduce **carrying costs**. What has been the net effect? Throughout the 1990s, it seemed that overall inventory levels (in the US at least) were going down. Lower inventory levels were considered the basic measure of good logistics. Yet Ginter and LaLonde found that inventory has gone up in some industries, which they believe is the result of demand for more variety of goods.[3] There are some very different issues when one looks at international logistics.

According to Bowersox and Closs, "the objective (of inventory management) is to achieve the desired customer service with the minimum inventory commitment, consistent with the lowest total cost".[4] To put it another way, customers want inventory so that they can find what they want when they want it. Meanwhile the store does not want to keep too much inventory because it costs money. Accommodating these contradictory needs is the essence of inventory management. Inventory consists of the following parts:

- Raw material
- Components
- Work-in-process
- Finished products

Average inventory is the average amount that is held over a long period of time. This includes two parts, cycle inventory and safety stock. **Cycle inventory** is the inventory that is consumed and then replenished. **Safety stock** is the cushion to prevent a stock-out. A **stock-out** (also known as an outage) is when inventory runs out and an order is either not filled or must be backordered (ordered from the supplier). An **inventory turn** is when the entire inventory is consumed and then replenished. It can be viewed as the inventory equivalent of a performance cycle, which was already introduced. A high turnover rate means that the cargo does not sit in inventory for very long before it is consumed, and there needs to be more frequent replenishment.

Why is it necessary to hold inventory? Although costly, it plays a vital role in the supply chain by allowing for geographic specialization, decoupling, the balancing of supply and demand, and to buffer uncertainties:[5]

- **Geographic specialization**. The different operating units of a company have their own needs in terms of inputs. Inventory allows them to collect their own set of inputs. Geographic specialization also refers to the fact that different locations specialize in the production of different products.

Inventory allows for the dispersion of production to wherever it is most efficient.

- **Decoupling**. Decoupling refers to the separation of the different operating units so they do not need to work at the same pace as each other. Imagine two factories in which one provides the raw material for the other. Decoupling means that these two factories do not need to coordinate their operations. This way, each working unit can operate at its most efficient rate and stockpile the excess if needed.

- **Balancing supply and demand**. Consumption and production are not always balanced, and inventory helps fix this. The best example is seasonal foods. There are certain seasons for the production of, for example, oranges, and some are stored for use throughout the year.

- **Buffer uncertainties**. Buffer stock or safety stock prevents outages resulting from variations in short term demand.

Reorder Quantity

A. Order Quantity of 200 Units

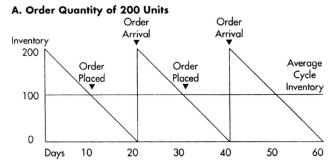

B. Order Quantity of 100 Units

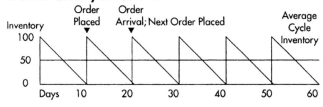

Inventory needs change depending on the **type of industry**. Businesses operate in a competitive environment, which means they need to insure that their inventory capabilities are comparable to their competitors. The three main industry sectors based on inventory needs are manufacturing, wholesale, and retail. The **manufacturing** sector may hold the most types of inventory, from raw materials to finished products. The **wholesale** sector tends toward finished products, but there are also wholesalers of parts or raw materials. Wholesalers are challenged by seasonality of some products and large increase of products on the market (i.e. SKUs). The **retail** sector ranges from small stores to modern retailers that may have 100,000 SKU's at any time.

International logistics affects inventory management is some important ways. There is a longer performance cycle if one is sourcing from a foreign supplier. As a direct result of longer performance cycles, in-transit inventory will be greater because of the longer transit times and distance. Border crossings create delays, uncertainty and documentation issues. Location decisions are more complex. In addition to the factors discussed in Chapter Four, special decisions need to be made on which country to keep inventory. Shrinkage will also be greater because of longer transit times and distance. Larger safety stock is needed due to the increased uncertainties of international sourcing.

International Inventory Issues
Longer performance cycle
More in-transit inventory
Border crossings
More complex locational decisions
More shrinkage
More safety stock

Finally, there are some transportation issues when we look at inventory management in an international environment. Regarding the **characteristics of transport modes**, the speed of different transport modes will affect the level of in-transit inventory. Shrinkage will be affected by the different levels of damage from transit. Reliability of delivery time varies depending on the mode of transport. The **handling characteristics** affect international shipments differently that domestic moves. Some products are easier to move internationally, others are more difficult or costly. For example, live animals are not often shipped internationally because of health concerns. Also, if the cargo needs to go through customs, the ability to inspect the cargo becomes important. Cargo that may be inspected has an easier time clearing customs.

Selling terms and contracts of sale affect inventory and play a special role in international inventory management. Given that there are massive amounts of in-transit inventory in the world at any one moment, shippers may ask, where is my inventory? Companies should know where their inventory is, and this may not be known. If the selling terms are for EXW, then everything enroute is in the buyer's inventory. If the terms call for transfer of ownership somewhere in between, such as CIF or FOB, in-transit inventory management will be very difficult to calculate.

Free trade zones, discussed in Chapter Nine, are a key resource in international inventory management. Their main function is to provide a place to hold inventory without the cost or hassle of clearing customs. FTZs alleviate many of the customs-related problems associated with inventory management in a global operation.

Precipice

Who would have guessed? A 'thriller' novel about logistics. Daniel Pollock wrote *Precipice*, published by the Council of Logistics Management. The book is about a woman whose father dies suddenly, and she must take over his chain of stores. She soon runs into a series of logistics problems that turn out to be sabotage. Reviewers have mostly applauded it as a midway solution between logistics and suspense. They say it's not a John Grisham novel, but then again, it's not a textbook titled *International Logistics* (ouch, that hurts –the author). Pollock follows in the steps of renowned economist Paul Erdman who wrote a series of bestsellers, such as *The Palace*, making corporate finance sexy.

PRINCIPALS OF INVENTORY MANAGEMENT

Inventory management in an enterprise should be based on an **inventory policy**, instead of just reacting to changes in demand or supply. This policy can be both at the tactical level and the strategic level. It provides the criteria to decide such questions as where to place inventory, when to replenish stock, and how much to allocate to sub-units.

Capacity constraints are an important planning consideration. This refers to such things as the warehouse space, the ability of the information system to keep up with inventory changes, and so on. The inventory policy needs to consider, for example, what is the maximum amount of inventory that could be held if needed. The information system is considered a constraint because an inventory policy may create expectations that cannot be accommodated. For example, can the information system provide real-time status reports on inventory on hand? And to what level of accuracy?

Customer expectations are increasing, and they especially want to find whatever it is they are looking for when they walk into a store. When companies order from their suppliers, they want their order complete and on time. The ultimate goal is called the **perfect order**, in which all orders are complete, with no missing items. Failing this, there is what is known as **immaculate recovery**, which is when a mistake is fixed before the customer is affected.

The basic goal of inventory management is to provide the designated level of customer service at the lowest cost. The designated level of customer service is usually measured in terms of **availability**. The **service level** this is a target designated by management for the level of orders filled. There are a few ways to define service level, such as the amount of orders filled from available stock, the percentage of each order that is filled from available stock, or the time it takes to reorder stock. The important thing to note is that any service level can be achieved. The question is whether the service level can be achieved at a reasonable cost. Service level can be measured in three ways:[6]

- **Stockout frequency.** The probability that a stockout will occur.
- **Fill rate.** Measures the magnitude or impact of stockouts. For example, if a customer orders 100 units and only 70 are available, the fill rate is 70%.
- **Orders shipped complete.** The measure of how many times an order was filled on time and complete.

Stockouts hurt in two ways; they usually mean lost sales, and they are bad for the company's reputation. Inventory problems can translate into a competitive disadvantage if availability is worse *relative to the competition*. The possibility of a stockout is called **inventory risk**. This risk can be **wide** or **deep**. Retailer's inventory is wide because they stock many SKUs, but not very deep. A wholesaler's risk is deep if it stocks a large quantity of an SKU.

The consequences of a stockout vary depending on the product. There is the possibility that if one item is not available, there are **substitutes**. Some products have easy substitutes, such as dishwashing soap. Other products have almost no substitute, such as a specialized machine part. Substitutes greatly reduce the chance of a stockout. If there is one acceptable substitute, orders filled goes from 70% to about 90%, and if there are two acceptable substitutes, the average goes to 97%.[7]

Bowersox and Closs describe three steps in inventory strategy:[8]

1. **Classify products/markets** Different products have different characteristics and their inventory needs differ. There is no single way to classify products, but among the most common are:

   Fine line or ABC classification- products are grouped based on their characteristics.

   Market performance measures- products may be groups based on sales volume, profitability, inventory value, usage rate, or the nature of the item. Pareto's Law, also known as the 80/20 Rule, states that 80% of sales are from 20% of the products. Another common measure is 'A' products that are fast movers/high volume, 'B' is moderate, 'C' slow.

2. **Make strategy for each class**- the reason products are broken down into groups is that some are more important than others. Priorities should be assigned to each group so that the most attention is given to those products with the greatest importance. The inventory policy determines which group is given priority. Within each group may be a different service object and management procedure. For example, the 'A' products that are most important may be reviewed daily to insure there are adequate stocks, while the less important 'C' products are only reviewed monthly.

3. **Develop policies for each class**- the final step is to develop operational procedures that address each product classification.

Just In Time (JIT) inventory policies were introduced a few decades ago, although there is still a mistaken belief that this is appropriate for every company. A company tries to reduce its inventory by having supplies arrive more often, in smaller batches. This can be logically extended to the point where a shipment arrives exactly when it is needed, so there is no inventory needed at all. This procedure has been developed into an entire management philosophy and a science because it implies changes in the overall operations. It requires tighter control on operations because any delay or mistake will have much greater effect than if inventory was held as a buffer. This means a greater emphasis on the quality of operations management.

JIT can be done in a global environment although it is more complex and risky. Some international operations are well organized and the risks of transport and border crossing are low enough to make this possible. In requires careful management of the transportation and customs issues. Carriers and some customs agencies are changing their procedures in recognition of this fact. However, since the 9-11 attacks, new safeguards at borders have made an international JIT program more difficult.

E-commerce has not eliminated the need for inventory. In some industries, the opposite seems to have occurred. The Wall Street Journal reported at the height of the dot-com boom that e-commerce retailers were building more warehouses[9]. They noted how some of the top Internet-based retailers were building more warehouses to hold the inventory that would otherwise be sitting in a retail store. However, their primary example was Webvan, a company that sought to provide home grocery delivery. After ordering $1 billion of warehouse space, the company went bankrupt. Was this just a result of the dot-com boom, or will an Internet economy change the need for inventory? It is not clear at this time. Recall that Ginter and La Londe found that inventory levels have actually been increasing, which they believe to be a result of the demand for more variety.

Inventory trends are also affected by the nature of **organizational design and corporate policy**. Centralized warehouses decrease the need for inventory, to be discussed shortly. Larger companies would have more centralized inventory management. For example, chassis pooling is a common practice in which carriers share each others chassis, resulting in more efficient utilization of these assets. However, this is only done in the US because chassis are carrier controlled. In other countries, there are more chassis and they are often left sitting unused.

Inventory policy will include a plan for how to allocate inventory among the subunits. In other words, each subsidiary needs some inventory, and the headquarters needs to decide how to divide its inventory among them. Two of the most common methods are fair share allocation and DRP.[10] **The fair share allocation** is used to allow each subsidiary their 'fair share' depending on its needs. For example, if there is a scarce resource, such as a popular new product, each of the subsidiaries will be trying to get as much of it as possible. The **DRP** method, as introduced in Chapter Four, is a more sophisticated, computerized approach that seeks an optimal solution based on consumer demand and product supply.

Some of the inventory controls are **proactive**, which means that the replenishment is done in anticipation of it being needed. Sometimes inventory controls are **reactive**, which means that nothing is done until the inventory levels are consumed to a certain point. An enterprise would like to always be proactive, but that is not always feasible.

When to use proactive inventory controls
Highly profitable markets
Dependent demand
Economies of scale
Supply uncertainty
Source capacity limitations
Seasonal supply buildup

When to use reactive inventory controls
Cycle time uncertainty
Demand uncertainty
Destination capacity limitations

INVENTORY ECONOMICS

Inventory has some important economic implications. We know that inventory is costly, yet accountants realize there is a problem identifying the true value of inventory.[11] The benefits of inventory, such as the safety they provide against stockouts, are hard to quantify. Another major issue of inventory economics is **carrying costs**.

Carrying costs (also known as holding costs) can vary depending on the conditions of storage and the nature of the product. They also vary internationally. Some regions have cheaper or more expensive storage costs. The risk of pilferage varies greatly. Warm regions have a greater problem with spoilage. For example, food can be left in a container and last a long time in a temperate or cold environment. In the tropics, even durable goods may be harmed by sitting in a container when the temperature gets extremely high. In the US, it is common to use 25% as a general estimate of carrying costs.

Inventory Carrying Costs[12]
Capital costs

Inventory Service Costs	**Storage costs**	**Inventory Risk costs**
Insurance	Plant warehouses	Obsolescence
Taxes	Public warehouses	Damage
	Rented warehouses	Pilferage
	Company-owned warehouses	Relocation

There is a basic **tradeoff between transportation costs and inventory costs** (Inventory costs are also known as carrying costs of inventory).

If inventory is to be kept at a minimum, this requires more frequent deliveries with smaller lot sizes. This increases the transportation costs because 1. Deliveries are more frequent and 2. There is less of a volume discount. It may also increase costs because the amount of the purchase is smaller.

$$X_2 = X_1 \sqrt{\frac{n_1}{n_2}}$$

where n_1 = number of existing facilities
n_2 = number of future facilities
X_1 = total inventory in existing facilities
X_2 = total inventory in future facilities.

Centralization of warehouses (or other locations for holding inventory) affects the need for inventory. If there is one warehouse supplying all stores, there is a relatively small need for each SKU. Yet if there are multiple warehouses, each one of them needs to have at least one and probably more of each item. Centralization may not be possible in some cases. For example, the warehouses may be in each country, holding stock unique to that market.

Effect of Centralized Warehouse on Inventory[13]
Example: Five warehouses hold 1,000,000 items. If they were to combine into one warehouse, only about 447,000 items is needed.

Effects of Warehouse Centralization

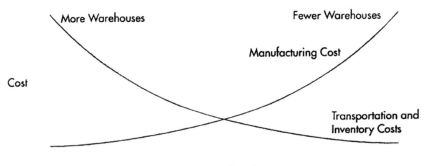

There are two methods for calculating inventory value, actual cost and market value. **Actual cost** is the amount paid for the products. This measure can be skewed if the amount paid does not reflect its actual value. If the products come from an internal transfer, the transfer price (also subject to different values) would be used. **Market value** is an estimate of its value if it were sold at that time, which may be somewhat subjective.

When you have the actual cost of the goods, there are three ways to estimate the cost of goods sold. LIFO, FIFO and Average. **Last in, first out (LIFO)** when one uses the cost of the most recent entries into inventory. In most situations with a positive inflation rate, this increases the value of inventory because the last-in inventory is usually the most expensive. It also reflects the most current costs, which makes this a more accurate measure of current costs. **First in, first out (FIFO)** is the opposite of LIFO, in which the inventory is valued beginning with the oldest items. This may reduce the overall value of the inventory given inflation. **Average** is an average of what was paid for the entire inventory.

Example of LIFO, FIFO and Average

A machinery seller has three generators in stock, A, B, and C. Generator A has been sitting around the longest, and the seller paid $500 for it. B cost $700 and C, costing $1000, is the most recent acquisition. Somebody buys a generator, and now the seller must calculate his cost of goods sold (CGS):

LIFO - $1000

FIFO - $500

Average - $733

INVENTORY CONTROL METHODS

There are some operational methods used that will be briefly considered. Because international logistics looks at the issue from a higher perspective, these methods are only introduced briefly.

Inventory levels need to be reviewed occasionally to see if there is a need for replenishment. This can be done in a few ways. **Perpetual review** is when stock levels are constantly watched. **Periodic review** is when stock levels are checked at periodic times, such as weekly or monthly. The **reactive** method is when stock levels are not checked, but is reordered when they run out.

At what point in the performance cycle should one order a replenishment? The order is made at the re-order point, when inventory is reduced to the level with just enough time for new stock to arrive just when the safety stock level is reached. If there were any delay in the order, the safety stock would then begin to get used.

$$EOQ = \sqrt{\frac{2C_oD}{C_iU}}$$

where EOQ = economic order quantity (EOQ)
  C_o = cost per order
  C_i = annual inventory carrying cost
  D = annual sales volume, units
  U = cost per unit

Example:

$$EOQ = \sqrt{\frac{2 \times 10 \times 500}{.11 \times 3.00}}$$

How much should be ordered? The **lot size** is the amount of product that arrives in one shipment. We already mentioned the tradeoff between transportation costs and inventory costs. This is exactly where that tradeoff comes into play. There is an ideal lot size called the **economic order quantity (EOQ)** that minimizes overall transport costs and inventory costs. The EOQ calculation can be altered to fit real-world variables. Some of the more common are volume transportation rates, quantity discounts, and special lot sizes. **Volume transportation rates** are when carriers offer reduced rates when large volumes are moved. **Quantity discounts** is a similar concept, but is when larger lot sizes are used. Finally, the EOQ can be adjusted for **special lot sizes**. This may be used when the production schedule requires that a certain amount of product is needed, and this lot may be exceptionally large or small.

The alternative to an EOQ method is the **fixed-order-interval system**, in which orders are made at fixed intervals, but the amount of the order varies. Under EOQ, the timing of orders varies, but the lot size, being an ideal lot size, stays the same. The fixed interval system is used in three situations. First, some companies do not automatically check inventory levels, which make the EOQ method troublesome. Second, some vendors offer discounts if orders are made at certain times, such as when their inventory is too high. Third, some companies use their own trucks to pick up supplies, so the benefits of the EOQ are not relevant, and thus different lot sizes may be used.[14]

Economic Order Quantity

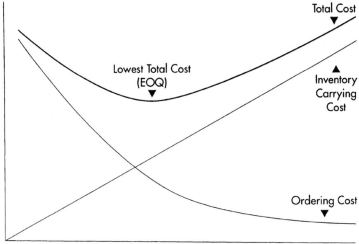

Annual demand volume	3,000 radios
Unit value	$20
Inventory carrying cost	20%
Ordering cost	$20

The EOQ makes the following assumptions:

Satisfies all demand

Continuous, constant and known demand

Constant and known replenishment performance-cycle time

Constant price of product that is independent of order quantity or time

Infinite planning horizon

No interaction between different items of inventory

No inventory in transit

No limit on capital available

If Only Horses Had Wings

by Elizabeth Kaye McCall

Hollywood horse trainer Rex Peterson was working north of Los Angeles on the Sharon Stone film *Simpatico* when a frantic call arrived one Thursday from the Baltimore production set of *Runaway Bride*, where he'd recently finished a job. A new horseback scene was being added to the screenplay and star Julia Roberts was insistent that Peterson's trick horse *Hightower* be on the set Monday morning.

The nuances of transporting horses are a specialty field itself. From stringent import/export criteria mandated by the U.S. Department of Agriculture (USDA) to considerations like Live Animal Regulations outlined by International Air Transport Association (IATA) and The Animal Transportation Association's (AATA) new Registered Equine Attendant Program, the animal's psyche also becomes an element that can impact even a short trip from home.

"You're dealing with a two-year-old child at best", remarked Rex Peterson, one of the film industry's top horse trainers. "Make that a 1,200 pound two-year-old programmed by Nature to flee, not fight, when threatened" and the logistical complexities begin to emerge. Granted, the scenario is somewhat different than that depicted in a 1943 article Post-War Pegasus, which ran in horse-racing's weekly magazine, The Blood-Horse. "It requires no great imagination to predict that after the war we shall travel by air to many race meetings almost as a matter of course, and the next stop will be the transport of horses by air", journalist Meyrick Good wrote.

The journey either way would probably entail much less disturbance to the horse, or loss of form, than an ocean journey, and the benefits that would accrue from increased international turf rivalry would be immense. Good even surmised, if compelled to go above the stratosphere, horses could be fitted with oxygen masks, and would no more object to them than to the muzzles that they have to wear on occasions to prevent them eating their bedding. Fifty-plus years later, many of Good's projections hold true.

Air travel is the dominant mode of international horse transport and significantly changed racing and competitive events. However, within countries (including the US) and even throughout the European continent, 90 percent of the horses are shipped by truck, due to the high cost of air transport. Factors uniquely associated with horse racing make it a logistical category of its own.

Mersant International Ltd., a multi-faceted corporation founded in 1977 as a customs broker and freight forwarder has achieved worldwide recognition as specialists in the shipment of horses. Headquartered in New York, Mersant handles freight ranging from fine jewelry, wearing apparel, and aircraft parts, to horses and exotic zoo animals. An entire department of its operations is devoted exclusively to horse transportation. The official agent for the prestigious Breeders Cup Races and the US agent for the Dubai World Cup, the world's richest horse race, Mersant also maintains offices in Kentucky and California.

Photo courtesy of Port of Oakland.

Joseph N. Santarelli, Sr., one of Mersant's founders, owners, and its current Treasurer, explained what brought the closely-held family company from relative obscurity to prominence. "There are no short cuts to success. It requires a steadfast company policy of pleasing the customer", remarked Santarelli, a licensed customs broker and AATA Board Member who serves as the US Chair of AATA's Equine Committee. Santarelli, who is also a member of the American Horse Council's Health and Regulatory Committee, added, "I don't believe anything is impossible to do, provided you do enough homework."

A case in point involved a prized horse that Mersant imported in the early 1980s from then-Communist Russia, a job riddled with problems. "The horse could not be directly imported to the United States due to a health matter and veterinary care needed to treat it was unavailable in Russia. I ended up finding I was able to bring this horse to Finland", recalled Santarelli, who enlisted the help of a client in Finland to find a boarding location and vet to treat the horse. Subsequently, he had to convince the owner of the merits of a plan that involved flying the animal to Finland on a Russian cargo plane. "He was a very valuable horse", explained Santarelli (and the only horse on the plane). After treatment in Finland the horse was able to meet quarantine requirements and was successfully imported to the US.

"Those bloodlines may not have gotten here if we hadn't gone through those exhaustive measures", said Santarelli. "We did this with full knowledge of all the ministries on how we were doing it, the best way to do it, and the legal way to do it. We did not do it illegally." He quickly added, "When you fool around with things like this you really could introduce certain diseases to countries where they don't now exist, just by lying and trying to trick the system." The New York native elaborated, Hoof and Mouth Disease is something America does not want. African Horse Sickness, if it hit here and was active and endemic, could kill the entire herd in the United States within a matter of weeks.

Horses arriving in the United States are subject to a USDA Entry Quarantine of up to 60 days, although normal quarantines run 24-72 hours. (Special quarantine procedures apply to race horses arriving for events.) Incoming horses must also have an International Health Certificate endorsed by the Ministry of Agriculture of Country of Origin and test negative upon arrival to four diseases: Dourine, Glanders, Equine Prioplasmosis (equi and caballi), and Equine Infectious Anemia (EIA). Any horse testing positive is refused entry and immediately required to return to its country of origin at the owner's expense, or else the USDA will euthanize the animal. Although the USDA will perform a 'courtesy' blood test prior to shipping, the horse must again test negative upon arrival. Logistical considerations for horse exports also involve USDA and U.S. Customs requirements.

Perhaps the biggest wild card in transporting horses by air is the issue of 'qualified attendants'. While specified by IATA's Live Animal Regulations<one attendant is required per pallet of horses up to a certain ratio, the criteria for 'qualified attendant' has been subject to loose interpretation. "Right now there are no international standards per se. Everything is voluntary," commented Santarelli, who spent five years developing AATA's first-ever Equine Attendant Registration Program, which launched in 1998. "It is suggested and highly recommended that an AATA groom be used." He stressed, "It is not required yet."

Unlike lions, zebras, and other animals shipped fully crated, horses are led into a stall-type system that carries three horses (optimum) per pallet position. In most cases, parts of the stall are not completely closed. "Imagine being on a plane when a horse has decided it doesn't want to be there, said Santarelli. "It would make most people cringe and run." The point is underscored in the following excerpts from a February 1999 AATA request made to IATA's Live Animal & Perishables Board for changes to the Live Animal Regulation text: "Unqualified attendants pose potential risks to the carrier, to the aircraft, the airline personnel, and to the ground handling agents of the airline in event of an emergency. Unscrupulous companies wishing to save minimal expenses of a professional groom have sent dangerously inexperienced personnel who travel for free airfare."

"College students are often recruited to pose as grooms in such situations", Santarelli noted, and blithely assured "the horse will be fine." Meanwhile such surreptitious transporters cut their costs by dodging qualified handler costs and items like a standard return ticket home. "What happens to the kid when something happens with that horse?" asked Santarelli, "and where's the humanity for the horse?"

Even privately-arranged domestic van (truck) transportation for horses is not the domain of amateurs. Hollywood horse training pro Rex Peterson, who maintains his own equipment for vanning, utilizes different methods to haul his trick Quarter Horse stallion cross-country (24-hour rest stops) from his practices for Hightower, a gelding. But the pleading 2:00 a.m. call from *Runaway Brides'* production office brought Peterson and horse their most unusual journey to date.

"When she called I thought they were pulling my leg," recalled Peterson. "I said *Hightower* is way too tired." Not to mention his own condition, both were on the *Simpatico* set just hours before. Already Friday, the clock was ticking. But ironically, the late night before left their *Simpatico* work finished. Come Monday in Maryland, Julia Roberts rode Hightower again for *Runaway Bride*. How he got there and home is already Hollywood legend.

---

[11] *Delaney and Wilson, Ibid.*

[2] *Delaney and Wilson, Ibid.*

[3] *"An Historical Analysis of Inventory Levels: An Exploratory Study", James L. Ginter and Bernard La Londe, The Supply Chain Management Research Group, November 2001.*

[4] *Bowersox and Closs, Ibid, p. 30.*

[5] *Bowersox and Closs, Ibid, pp. 247-9.*

[6] *Bowersox and Closs, Ibid, pp. 68-70.*

[7] *"Customer Service Strategy and Management", Douglas M. Lambert, in Robeson and Copacino, Ibid, pp. 82-83.*

[8] *Bowersox and Closs, Ibid, pp. 298-302.*

[9] *"Virtual reality: Web firms go on warehouse building boom", September 8, 1999, p. B1.*

[10] *Bowersox and Closs, Ibid, pp. 289-91.*

[11] *"The Development of an Inventory Costing Methodology", Douglas M. Lambert, National Council of Physical Distribution, Chicago, 1976, p. 3.*

[12] *"Logistics Cost, Productivity, and Performance Analysis", Douglas Lambert, in Robeson and Copacino, Ibid, p. 280.*

[13] *"The Management of Business Logistics", John J. Coyle, Edward J. Bardi, and C. John Langley, 6th ed, West Publishing Company, Minneapolis/St. Paul, MN, 1996, p. 212.*

[14] *"Contemporary Logistics", 7th, James C. Johnson and Donald F. Wood, Prentice Hall, Upper Saddle River, NJ, 1999, p. 326.*

CHAPTER 16
INFORMATION SYSTEMS

In logistics and transportation, just as in other industries, the information system plays a critical role in the overall success of the operation. As we enter the 21ˢᵗ century, information, dot coms and e-commerce have become a near obsession in the business world. Technology and business models are changing rapidly, yet there are aspects of information management that remain the same. The challenge is to identify what is changing and what remains constant, to give credit to the benefits of information systems while eliminating the hype. This chapter discusses information system in general as well as logistics information systems. Much of what we need to understand applies to all industries. There are some aspects of information system unique to the logistics and transportation industry, and these issues will be addressed toward the end of the chapter.

There are technical differences between the Internet and the World Wide Web, but they do not affect the issues discussed in this chapter. Therefore we will only use the abbreviated form of the Internet, the Net.

THE NATURE OF INFORMATION AND INFORMATION SYSTEMS

The first step in understanding an IS is to note that it is not the same as computers or technology. An IS is any system that manages the collection, manipulation and distribution of data. A paper filing system is an information system, and sometimes it can be a better solution than an expensive, high tech computer. In other words, do not mistake information management with the tools. Computers, the Net, and telecommunication are only the tools. Because they are just tools, they will only work as the person using them. One industry leader note, "If you do not first get the people and the process right, investments in technology will not be a solution."[1]

Information is arguably the most important factor in the modern business environment. Those factors that traditionally have determined success, such as the ability to manufacture, or a good geographical location, are not as important as they used to be. Yet high quality information and its management, especially about customers, is the distinguishing quality of high performance organizations. Besides the advantage information gives the company, the role of IS is important because quality information is often expensive to acquire. For example, UPS has spent $1 billion a year for the past decade on their IS.[2]

An Example of the Information Systems Applications for Logistics

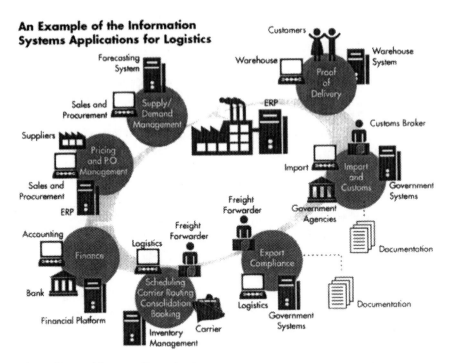

Source: Courtesy of Capslan

It often seems that the Net is going to change every aspect of our lives, yet in other ways, it has had little effect. There are **two opposing ideas of what effect of the Net will have on supply chains and logistics**. One view suggests that increased efficiency will create rewards for all parties in a supply chain. The opposite view is that the Net will overwhelm supply chains, spot markets will make it impossible to plan, and instability will dominate.[3]

The good news is that we have seen stability, even during recessions, in most of the economy. "Web-enables supply chains may be the future but in the near term a manufacturer lives or dies on the quality of its goods and its relationships with suppliers and customers."[4] Does this mean the Net is not as useful as we thought it would be? The answer is mixed. In this chapter we see that information systems are used to fulfill our needs, and people and companies have proven very adaptable at learning how to use these tools.

There is also a human element that cannot, and many say should not, be eliminated. Our jobs provide social and cultural meaning, which can be seen from the fact that one of the CLM's publications was not on logistics so much as organizational culture.[5] In another example, the air cargo booking industry, dominated by GF-X, has only captured 10% of the market (2002). Most people still want the personal touch of calling the carrier even if it was a routine administrative procedure. Information does not just exist by itself, but is strongly influenced by cultural, personal and structural influences.

There are four **industry groups** discussed in this chapter. First, logistics and transportation firms are the ones that move cargo. These firms generally make good use of information systems, but managing information systems is secondary to their main job. Second, the information systems industry is composed of the firms that offer software, computer hardware, and related services. The third group is those organizations specializing in logistics and transportation information. These include, for example, GF-X in the air cargo industry, and the tariff publishers in the ocean shipping industry. Finally, there are the shippers. These companies make decisions about which carrier to use based, in part, on the IS capabilities of the carriers. In other words, they are the beneficiary of the carriers' efforts in IS. However, these customers have their own IS, which means the challenge is find carriers that can work with their systems.

Four Industry Groups
Logistics companies
IS companies
Logistics IS companies
Shippers

E-logistics simply refers to the use of electronic information systems to conduct logistics. Since virtually everyone is using some form of computerization and telecommunication, it does not seem very useful to suggest that anyone is doing anything special by conducting e-logistics. What that term implies is that their information system is more advanced and sophisticated than others.

E-commerce refers to the use of the Net as a marketplace. In other words, the Net is another channel of distribution. From a logistics standpoint, it makes relatively little difference whether the order is received over the Net versus the traditional methods of ordering. Electronic data interchange (EDI) has been used for many years now for B2B procurement. This is essentially the same, although the technology of EDI versus the Net is different.

That is not to say that e-commerce makes no difference in logistics. There are some significant differences in the consumer market as a result of the Net. Individuals are shopping online in increasing numbers, which means the logistics of supplying the retail sector is changing. Online retail sales are expected to top $100 billion by 2003, which is about 6% of U.S. domestic retails spending.[6] That shifts the logistics from supplying stores to shipping directly to customers. This implies package delivery to individuals instead of larger shipments to stores. As for international logistics, the effect is relatively small since individual consumers rarely shop internationally.

The answers you get depend on the questions you ask. Information systems depend first and foremost on what people are looking for. Information is "data that has been endowed with relevance and purpose".[7] Customers expect some level of information as part of the overall service provided. Shippers want to know from their carriers, where is my package? When will it arrive? Warehousers are asked about the levels of inventory available. The point is whether the information is available. Information systems play a very practical role in reducing inventory levels and savings in labor. If a supplier knows how much product its customer is going to need in the next week, it knows exactly how much inventory to keep. As we learned in Chapter Fifteen, a major part of inventory is to cover up for a lack of information.

It is clear that industry leaders are those that manage their companies well. But how much of a difference does the IS make? The best-run companies have a high-quality IS, but it would be false to say that a good IS will make up for an otherwise poorly managed companies. Information management is a support function, not the core operation of a firm (unless, of course, it was an information company).

Strategic and tactical planning is based on information. As we will discuss later in this chapter, some information is used in operational, day to day activities, while another key role of information is to see long term trends. However, information systems cannot tell a manager what to do regarding future plans; that is a judgment decision requiring the insight of a person. There are six basic building blocks of an information system:[8]

1. Input   4. Technology
2. Output   5. Data base
3. Models   6. Controls

Some of these six blocks are self-explanatory, but a couple deserves explanation. Of the six, information **technology** is the glue that binds the other blocks together. **Inputs** and **outputs** usually refer to data coming in or going out (in this case, data and information mean the same thing). When a technician goes into a system to make changes, this is not considered input. When information is fed into the computer to be processed or stored, that would be an input. Output is also data, not to be confused with the way that output is used. Managers interpret information from the output to make decisions. A management information system (MIS) cannot generally make management-level decisions.

Logistics Database

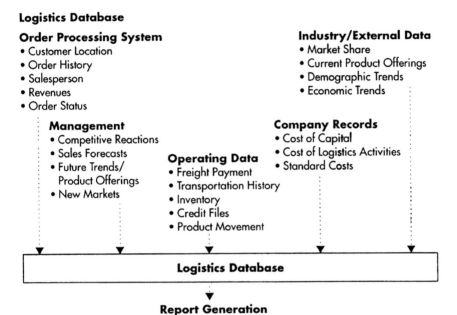

Order Processing System
- Customer Location
- Order History
- Salesperson
- Revenues
- Order Status

Industry/External Data
- Market Share
- Current Product Offerings
- Demographic Trends
- Economic Trends

Management
- Competitive Reactions
- Sales Forecasts
- Future Trends/ Product Offerings
- New Markets

Operating Data
- Freight Payment
- Transportation History
- Inventory
- Credit Files
- Product Movement

Company Records
- Cost of Capital
- Cost of Logistics Activities
- Standard Costs

Logistics Database

Report Generation
- Order Performance
- Shipment Performance
- Damages and Returns
- Product Tracking and Forecasting
- Performance and Cost Reports for Logistics

Source "Fundamentals of Logistical Management", Douglas M Lambert, James R Stock, Lisa M Ellram, Irwin Mc Graw-Hill, New York, 1998, copyright the Mc Graw-Hill Companies

A **model** can be thought of as the structure of the system. Imagine an enterprise headquarters and its field offices. Should the information system go directly from headquarters to the offices and back, so the headquarters act as a hub? Or should all the offices be linked with each other for direct communication? In one example, the North American headquarters of a major shipping company had a mainframe computer serving the needs of that region. The company redesigned the system, eliminated the North American mainframe, and used only one computer located in Asia. This was intended to insure that all the data was consistent throughout the different regions. These are two very simple examples of the 'model' aspect of an information system.

Controls are the ways of managing the information system. They can be a very important and very tricky part of the system. For example, we have heard the expression 'garbage in, garbage out'. This refers to the tendency for data inputs to be inappropriate, so the output is also inappropriate. A control system can be used to check to make sure the inputs are correct, and thus improve the output. For example, imagine a hardware company that wants to store information about its inventory in a database, but products are often entered with the wrong name. They received 100 hammers, but it was entered under the name of 'saws', and now the system is telling us that there are 100 saws in inventory. A control system needs to be in place to check for accuracy. In this case, a control system might be a supervisor who double-checks the database against receipts, or a bar-code entry that would not permit hammers from being entered as saws.

Data is the basic building block of an IS, and the form of data can dramatically affect the quality of the overall system. In fact, one of the main goals of the IS is to acquire data and make it useful. *Keeping Score* describes the varying levels of what they refer to as 'content', which essentially means data. We can see from their typology that there is a hierarchy in which data can be simple and of limited use, which they refer to as 'raw content', to highly useful and sophisticated data, which they refer to as 'syndicated content'.

Content Management Requirements for E-Commerce[9]

Content types	Content Definitions
Raw Content	Unstructured, inconsistent, and inaccurate information from original sources
Clean Content	Raw content that has had inaccuracies and inconsistencies removed
Rationalized Content	Clean content that has been structured to reflect a specific environment or content
Normalized Content	Rationalized content that can be used for comparing alternatives, making decisions, and taking action
Syndicated Content	Normalized content available to those who need and can use it.

Telecommunications is another major aspect of the information systems. Globalization is in part the result of the ability of people around the world to communicate with each other. Once given the chance to communicate, it is inevitable that business was conducted, and that along with affordable transportation created the global economy. Telecommunications is not the same as an IS. The ability to convey messages is one of the many functions of an IS.

Because it has become so easy to transmit information, one of the biggest problems of information management has been to stop or control the flood of information that businesses and individuals receive on a daily basis. Each US Postal Service employee moves 223,000 pieces of mail per year. AOL employees facilitate 13.8 million email and instant messages, and 43 million 'hits' on the Net.[10] Yet the efficiency and effectiveness of individual messages is declining.[11]

The key to understand why and how telecommunications improves a firm's performance is to understand the nature of communications in general. The Institute for the Future[12] found that Americans who work send and receive more than 190 messages a day, a figure that is increasing strongly.[13] According to Michael J. Critelli, CEO of Pitney Bowes, "Although we have increased our ability to generate more messages, we have not significantly increased our capacity to receive and respond to them." We often believe, erroneously, that more information is always better. What quality information systems seek is not more information, but better information.

The cutting edge of IS is the ability to understand the recipient. The key to effective communication is for senders to understand the recipient.[14] If the IS is not receiving useful information, a possible solution may be for the recipient to help the senders understand his needs better. One problem with the current telecommunications systems is that it is too cheap to send messages, and thus they become undervalued. Costly messages such as a letter sent by courier means that the message must be important.[15] That is why, even in the era of email, we find ourselves making personal phone calls, and important documents are often sent hardcopy by express mail at great expense.

Critelli goes on to note that "All messaging systems will be affected by the degree to which they can gain more information about recipients, including their needs for goods and services and their message receipt preferences. Privacy laws will increase the complexity of any effort to gain information about potential recipients..."[16]

Tariff Publishing

Tariff publishing presents an interesting story in the management of information and the effects of government regulations. Recall from previous chapters that in the ocean shipping industry, common carriers were required to file tariffs. This was intended to identify if the conference system was resulting in anti-competitive behavior, and to enforce the conference rules on its own members. Tens of thousands of tariffs were being transmitted to the government (the Federal Maritime Commission, to be specific), yet how was the public to make sense of all this information? Besides maritime shipping, many industries file tariffs with the government.

This creates a huge volume of data (the tariffs) that are in a form not very useful to those doing market research. Therefore, companies, known as tariff publishers, take this information and present it in a manner that could be accessed by anyone who would pay a fee. Note that all these companies were doing was taking public information and making it useful, using EDI technology.

As a result of deregulation, tariff publishers are faced with finding new business or being eliminated. Now that ocean carries are offering service contracts, most of these contracts are in hard-copy. Every time they negotiate a new contract, there is a lot of manual labor, and information that would be valuable in a negotiation with customers is being lost. Large shippers have over 2,000 contracts in effect at any time, and shippers may have over 20 contracts.[17]

Therefore, the tariff publishers are trying to shift their operations from simple rates to service contracts. This is not as easy as it may seem. A tariff is simply a commodity and a rate. Yet a service contract is more like a book. The publishers will survive or not depending on their ability to adjust to this new environment.

Technology that allows you to make changes easily also increases the instability of the system. Recall what life was like before computers (there is surely a lot of readers too young to remember such an age), when administrative tasks were done on paper. The demands made on office workers were in proportion to what they could handle. Increased capacity usually means increased demands. If, for example, the traffic manager can look at the supply and demand for transport and make last minute changes, which means forecasts become invalid because everyone is making last minute decisions.

Logistics depends on the integration of people and organizations, yet we can see that one of the greatest challenges will be to learn how to use the tools currently available more effectively. Traditionally, the solution was for more powerful computers, faster communications, and larger databases.

These are no longer a solution. What the logistics industry needs is better information, better data, and better communications.

INFORMATION SYSTEMS AND THE FIRM

An MIS serves three basic functions: data collection and storage, data communication, and data processing.[18] Ideally, the MIS is used throughout the organization and is completely integrated. In other words, every department has the same general system or a compatible one. There is also in some enterprises a **logistics information system** (LIS). An LIS has been defined by Dr. Richard Dawe as "An interacting structure of people, equipment, and controls involved in the logistics process, which creates both an internal and an external information flow capable of providing an acceptable base for logistics decisions."[19]

Three Functions of the Mis
Data Collection and Storage
Data Communication
Data Processing

The LIS should be part of the overall MIS, not a system that is separate from the enterprise's main information system. This would make sense because the whole point in information systems and logistics is that every part of the enterprise works in harmony.

The MIS is typically designed for a given enterprise, not a generic system that can be placed in any company. Different enterprises have their own goals, organization, structure, and so on. Logistics, however, is fundamentally concerned with integrating the operations of different enterprises. Imagine the supply chain, with several enterprises all with very good information systems, all well designed to serve the needs of their own organization. However, there is nothing that says those systems are compatible with each other.

In Chapter Four we discussed benchmarks. There are benchmarks for what is considered good use of information technology, yet it varies by industry. High tech, aerospace, and motor vehicles are on the cutting edge, while heavy equipment and construction is on the laggard end.[20] Part of the difference may be the demands of the industry. The automobile industry is highly competitive and requires sophisticated scheduling.

Other industries have competitive pressures that are not readily solved by information technology. According to one study, the transportation sector in the US ranked 14th in the amount spent on emerging technology, an average of $2,367 per employee, well below the financial services and food and beverage industries. In terms of general spending on technology, which included such things as fax machines, computers, printers etc, transportation ranked 9th at $9,359 per employee.[21]

One of the newer demands made of the best ISs is **personalization**, which is "applying knowledge about customers...and products to present appropriate products, cross-selling opportunities, and up-selling opportunities."[22] Traditionally, computers were good at repetitive tasks. Now they are being called upon for single-case tasks using complicated decision processes. Personalization is based on rules processing, in which the rules are established and the computer applies the proper rule to the proper situation. Pattern recognition is used for high-volume cases such as B2B.

One of the fundamental issues regarding the distribution of information is **confidentiality**. We already discussed supply chains and how parties in a chain protect their information. This also affects an IS. Covisint is an e-marketplace funded by the three big American automakers (GM, Ford and Daimler Chrysler), in which these and other users can buy and sell auto parts, a market in the hundreds of billions of dollars. Yet many of the vendors are concerned about their intellectual property being available in such a system, and they have good reason to be concerned. The top purchasing executive for one of these companies was caught taking confidential information from his old employer to a new job. Furthermore, the number of suppliers to the major automakers has been reduced from 50,000 to 30,000, leading one observer to ask whether the supply chain has become a noose.[23]

Another area of concern is the relationship between companies and the government. U.S. Customs officials are struggling to sort out what information can be disclosed and what needs to be kept private. The US Customs Service's Automatic Export System (AES) collects Exdec data, some of which is made available to the public. The AES is a very comprehensive system and has provided enormous amounts of data on everything from competitor actions to global trade patterns (see the section in this chapter on databases).

Companies have requested that carriers release information about who is shipping what in order to combat the 'gray market' and counterfeiting.

Carriers do not want to release this because it is confidential business information, and could aid competitors. Sometimes the information is of immediate relevance. Yet, there are cases where even if the information about a shipment is released later, it would reveal patterns of shipments.[24] This is the sort of information that can also assist thieves and terrorists.

Within the US government, the US Customs and Treasury Departments were at one point disputing who controls the electronic interface with shippers. There should be one gateway for shippers. It would be inefficient to require shippers or carriers to file separate paperwork with different government agencies, especially if it is the same information. Not only would this be less efficient, but it would result in a lower quality of data. By having one agency collect the data, there is less chance of mistakes, and less chance the person submitting the data could change the information for his own benefit.

Geophysical Postioning Systems

A satellite tool important for transportation is **geophysical positioning systems** (GPS). This is a military tool that has been adapted to commercial use. The US military launched satellites around the world, that all emit signals down to earth. A GPS receiver, now about the size of a cellular phone, receives those signals and calculates where on earth it is located. GPS receivers can be accurate to within a couple meters. Differential GPS uses a land-based station to compensate for the satellite transmissions, thus allowing for extremely precise readings needed for airplanes to land or for ships to enter tight harbor channels. This is known as DGPS.

GPS is used in many parts of the logistics industry. Packages may be tracked with a GPS receiver that then transmits back its location. DGPS is being used in airports so aircraft can approach from any direction, thus making much more efficient use of the airspace.[25]

LOGISTICS INFORMATION SYSTEMS

This section identifies what is special about the IS of the logistics industry. The first step is to understand what is unique about the industry, and from that we can determine what the special information needs are. Logistics is about moving things, which means the information is usually related to physical goods. Logistics is also about coordinating with the different parties of a supply chain, and the parties are geographically separated.

Technology can also increase the capacity of what is normally considered fixed assets. How can it do this? By reducing redundant movements, rationalizing equipment, operating 24-hour schedules, or mechanization. FedEx once used a computer simulation of their Tennessee airport hub to look at the nightly plane movements. By finding the most efficient pattern of landing, loading, offloading and parking, they were able to save millions of dollars.

What exactly are the logistics information systems being called upon to do? An **Enterprise Resource Planning** (ERP) is an information system the handles all the information needs of an enterprise. The term MIS is used to refer to the system that provided information to the managers to help them make decisions. Manufacturers companies would have an MRP system, discussed in Chapter Four. As you can see, there are a lot of terms being used that often just refer to the same thing, but it is important to recognize them since they are being used in business. Large and diversified firms may make a clear distinction between the ERP for their manufacturing divisions, and the MIS that controls the overall firm.

The ERP is the main example of a firm-wide logistics IS. Besides ERPs, there are many other specialized applications. Supply chain **event management** refers to a program that combs through a database and looks for potential problems. The transportation industry has many IS applications. It would be futile to try to list out all such applications, but this chapter includes some representative examples.

The Net and other telecommunications advances have created a new marketplace on which logistics products and services are exchanged. Note that in this case, the logistics services are essentially the same, but the way shippers access these services is different. As of year 2000, there were about 50 marketplaces or trade sites for logistics-related services, which are expected to be consolidated to about 10 or 15.[26]

A trading exchange has been defined as "an online Net-based marketplace that facilitates the transfer of goods, money, and information."[27] These exchanges may he independent firms, such as GF-X, which are independent of any of the service-providers. Some exchanges are only for a specific industry, known as an independent vertical exchange, or it can be open to a wide range of products or services, a horizontal exchange. A consortium exchange refers to those in which a group of the traders themselves are running the exchange, and a private trading exchange is one in which only the traders are allowed.

Earlier in this chapter we noted two competing views of what the Net will do to the economy, and it seems that instead of overwhelming supply chains, we are seeing stability. Why is this? Although the trading exchanges offer some valuable services, not every company wants the risk of dealing with a spot market. The increase in service contracts means that fewer companies are using the spot market for buying transportation services.[28] In other words, many companies prefer stability and other benefits of long-term relationships over short term advantages of trading exchanges.

Types Of Trading Exchanges[29]
Independent vertical exchange
Independent horizontal exchange
Consortium Trading exchange
Private trading Exchange

International logistics entail some special IS-related issues. One of them is **language**. Computers use programming languages in English, which is fairly universal. But the user interface varies from country to country. A database of products in English would not work well in a Spanish speaking country. **Legal issues** are also involved. Privacy laws, for example, are different in each country. In one country, one can keep a database of customers with information that would be illegal in another country for reason of privacy.

Supply chains pose a special challenge for logistics and the IS. As already mentioned, supply chains are something of a myth in that companies coordinate with their suppliers and customers, but there is little if any management of logistics beyond that. Therefore, it should come as no surprise that the IS is virtually always not a 'supply chain information system' but the company's IS that is applied to a supply chain.

Why is this an important distinction? Imagine a company deciding how to design a new MIS. They do not know who will be their suppliers. That is a very different decision. Yet eventually their purchasing department will want to coordinate electronically with their suppliers' information systems. Suppliers often change, so they cannot base their MIS on any one supplier. Therefore the IS is made unique to that firm, which further reduces their ability to coordinate along the extended supply chain.

Communication Protocols

If computers are going to talk with each other, they need to speak the same language. EDI is the "computer-to-computer exchange of business formats". Data can be transmitted with EDI, but it can be costly to set up, especially for smaller companies. If the Net is used, value-added networks (VANs) are needed. A VAN is essentially a 3rd party that acts as a hub for receiving and routing data. They normally change a setup fee of about $50 and per-transaction charges ranging from 7 to 11 cents, depending on the volume of transactions.

The current protocols for communications across the internet or email are known as AS-1 or AS-2. AS-1 uses Simple Mail Transfer Protocol, STMP, also known as email. AS-2 uses Hypertext Transfer Protocol (HTTP), which we see in the front of every www address for websites. HTML is well known as the language for websites, handling text and images. XML (eXtensible Markup Language) is a language more powerful and flexible manner, and has also evolved into ebXML, which is specifically for e-commerce.

The Data Interchange Standards Association is the leading organization for bringing industry leaders together to agree on standards. These include a lot of specialized organizations, but the Accredited Standards Committee (ASCX12) is charges with developing "uniform standards for interindustry electronic exchange of business transactions-electronic data interchange". There is also the United Nations Electronic Data Interchange for Administration, Commerce and Transport, UN/EDIFACT, for developing standards internationally.

Source: DISA and "EDI for Everyone", Ann Saccomano, JOC, January 6-12, 2003, pp. 20-1.

An IS is designed to save the firm money. This occurs in a number of ways, mostly from reducing inventory and transaction costs. Yet as Bauer et al notes, "...most of the supposed savings tended to be pushed around the supply chain network rather than actually being eliminated."[30] The most efficient and powerful firms are able, through their effective use of information, to push costs onto their suppliers or customers. But is this a bad thing? Recall that the goal of a company is to improve its own profitability, not that of the entire industry.

The trend toward **alliances and partnerships** has created concern for the ownership of the IS. When companies enter an alliance, they presumably adjust their internal systems to be compatible with the other members. This often entails no change to the company's core IS, but an add-on. Companies should be wary of making changes to their system based on partnerships, which often do not last. Jack Welch, CEO of General Electric warns, "Never let a dot com get between you, your customer and your suppliers."[31] When doing a 3PL arrangement, negotiate the software applications rights so that if the arrangement does not work out, at least you are not also losing the software as well.

Richard Hallal of Logistics Development developed the chart in which IS providers can be arrayed based on their industry knowledge and information technology competence. The 3PL "fronts" are 3PL providers that also offer IS services. In other words, their main service is logistics, but they have added the information component to their offerings. Ports and Software based players refer to pure IS companies (ports here refer to information portals, not transportation ports). In the high/high quadrant we have application service providers, companies that combine logistics and IS.

Databases

There are some commercial databases available that are widely used for analysis and business decision support. By far the largest is PIERS, owned by the Journal of Commerce. They capture information filed to the US Customs Service, as well as other sources such as the US Census Bureau, and offer one of the most comprehensive sources of data on international trade.

Manalytics International is a consulting company in San Francisco that uses PIERS for their projects. For example, a company may want to know what is going to happen to demand for their product. Using PIERS, other databases and their own research, they can look at trends in international shipping.

PIERS also offers data down to the level of the individual shipper, so one can see what a given company shipped, the volumes and the times. Why would any company want to divulge this information? It is legally required for importers and exporters to file this information. However, shippers can withhold their name. Some companies do this, while others hide behind the name of an intermediary. Even so, a good researcher can figure out who the shippers are. For example, if "a company" is shipping a specific type of chemical from a city where only one chemical manufacturer operates, the answer is quite easy.

There is a wide variety of sources of data, all of which have their strengths and weaknesses. The United Nation, the OECD, and the European Union all offer important datasets of trade statistics. The important part of the job is to know where to look for the data, an how to interpret it. One must understand how the data was collected, and what the level of aggregation is. For example, some datasets offer the amount of trade in tonnage. While this is useful to in some industries, what sense is it to look at garments by tonnage? PIERS offers data as containerized and non-containerized, and according to shipper.

Internet Based Providers

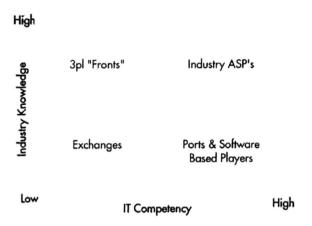

Source. *Logistics Development, Corp*

MEASUREMENT OF BUSINESS PROCESSES

In *Keeping Score*, the authors dealt with the measurement of logistics processes. One of the most important roles of the IS is to provide accurate and useful measures of the firms activities. For this reason, the role of measurement deserves close attention.

The authors of *Keeping Score* note that most logistics managers are focusing their measurement internally – on the performance of warehousing, transportation, and other logistics activities – not externally, on those activities that *collectively* satisfy their customers.[32] A measurement program should be an integral part of the firm's strategy, and the IS. The first step in a measurement program is to identify what processes one wants to measure. This can only be determined from looking at the strategy, and from that determining what the firm's critical success factors are.

Measures Captured on a Regular Basis within the Company

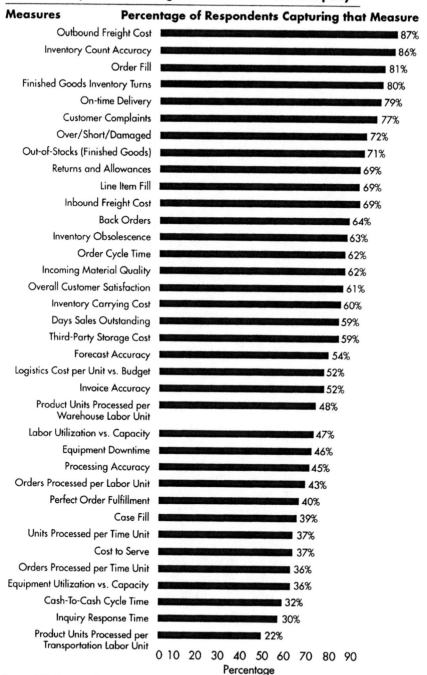

Measures	Percentage of Respondents Capturing that Measure
Outbound Freight Cost	87%
Inventory Count Accuracy	86%
Order Fill	81%
Finished Goods Inventory Turns	80%
On-time Delivery	79%
Customer Complaints	77%
Over/Short/Damaged	72%
Out-of-Stocks (Finished Goods)	71%
Returns and Allowances	69%
Line Item Fill	69%
Inbound Freight Cost	69%
Back Orders	64%
Inventory Obsolescence	63%
Order Cycle Time	62%
Incoming Material Quality	62%
Overall Customer Satisfaction	61%
Inventory Carrying Cost	60%
Days Sales Outstanding	59%
Third-Party Storage Cost	59%
Forecast Accuracy	54%
Logistics Cost per Unit vs. Budget	52%
Invoice Accuracy	52%
Product Units Processed per Warehouse Labor Unit	48%
Labor Utilization vs. Capacity	47%
Equipment Downtime	46%
Processing Accuracy	45%
Orders Processed per Labor Unit	43%
Perfect Order Fulfillment	40%
Case Fill	39%
Units Processed per Time Unit	37%
Cost to Serve	37%
Orders Processed per Time Unit	36%
Equipment Utilization vs. Capacity	36%
Cash-To-Cash Cycle Time	32%
Inquiry Response Time	30%
Product Units Processed per Transportation Labor Unit	22%

0 10 20 30 40 50 60 70 80 90
Percentage

Source CSC/University of Tennessee Logistics Survey 1998

There are barriers to measurement programs. Collecting information, which will be discussed in the next section, is expensive. In order to justify the expense, the information needs to be useful, but its usefulness needs to be proved to everyone involved. Collecting the information typically involves many people, and if they are not committed to the process, the inputs will be poor, and thus the overall measurement program may be jeopardized.

Barriers To Measurement[33]
Measurement is difficult.
The links between measurement and strategy are often unclear.
Functions and processes are complex and often misaligned.
People being measured may be resistant to sharing information.
There is a significant lack of consensus on definitions of terms.

Data, once collected, needs to be made useful to the consumers. Many measurement programs have fallen short because once the information was acquired, it was presented in a form not useful to the rest of the company. Data should thus have the following characteristics:

- **Timely.** An ocean shipping company may want to know the status of one of its ships, which take three weeks to cross an ocean. An air cargo company wants to know the status of one of its planes that take three hours to cross the ocean. 'Timely' does not mean 'quick', but rather in a time that is appropriate for that situation.
- **Relevant.** Given the flood of data available to most enterprises, sorting through it and picking only that which is relevant can be difficult.
- **Accurate**-The quality of data is at least as important as the quantity. Inaccurate data can be more harmful than none at all.
- **Appropriate form** For example, some MIS' may need input to be in text form, while others in numeric form. Another common issue is the software used; 'appropriate form' may mean that the data is in compatible software.
- **Available** Data may exist, but in a location or form that prevents it from being used. Access to data may be controlled for security reasons. A major issue in logistics is that companies have information that would help their suppliers or customers coordinate operations, but do not want to release that information.

A Partlow chart is an example of how information systems can be simple yet effective. Refrigerated cargo is a highly lucrative trade, but how to insure that the temperature remains uniform so the contents do not spoil? The Partlow chart is placed inside refrigerated containers and records the temperature. We can see in this example a line at roughly -15 degrees. If there are any spikes in the line, that indicates a failure of the refrigeration at some point.

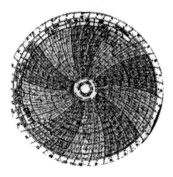

Characteristics of Good Measures[34]

A good measure:	Description:
Is quantitative	The measure can be expressed as an objective value.
Is easy to understand	The measure conveys at a glance what it is measuring, and how it is derived.
Encourages appropriate behavior	The measure is balanced to reward productive behavior and discourage "game playing".
Is visible	The effects of the measure are readily apparent to all involved in the process being measured.
Is defined and mutually understood	The measure has been defined by and/or agreed to by all key process participants
Encompasses both outputs and inputs	The measure integrates factors from all aspects of the process measured.
Measures only what is important	The measure focuses on a key performance indicator that is of real value to managing the process.
Is multidimensional	The measure is properly balanced between utilization, productivity, and performance, and shows the trade-offs.
Uses economies of effort	The benefits of the measure outweigh the costs of collection and analysis.
Facilitates trust	The measure validates the participation among the various parties.

Critical Lessons In Measuring Logistics Processes[35]

1. Make sure logistics measures are in synch with strategy.
2. Truly understand customer needs.
3. Know your costs in providing customer services.
4. Take a "process" view of logistics.
5. Focus only on key measures.
6. Stop ineffective measurement activities.
7. Use information technology.

Textile Central

Illegal transshipment of textiles has been an ongoing problem for those countries that restrict their importation. For that reason, the US Customs Service created the New York Strategic Trade Center, a 'war room' that monitors trade statistics and trends looking for signs of illegal transshipment. It is easy to see that of the $60 billion in textile imports to the US there are many violations, but identifying a company or a specific shipment is the hard part. Customs agents can refer to the Center for any textile-related questions, though the public is not allowed to query them directly.[36]

COMPETITIVE INTELLIGENCE AND INTERNATIONAL LOGISTICS

By Craig Weicker

For over 100 years, Barney International Corporation has produced and marketed fresh fruit and other food products internationally with annual revenue of $3 billion dollars. Production and exports taking place throughout Central America, South America, Caribbean, Africa, Philippines, Indonesia, China, and others for all major markets around the world. Close to 3 million 40-pound boxes of bananas are shipped and trucked weekly, transported by large refrigerated, "reefer" ocean vessels in containers, pallets or break bulk.

The bnana is a highly perishable and fragile product, so handling and time-to-market issues are critical factors for ensuring a quality product. From the time of harvest from the plant in the isolated fields to processing, to source port, to market port, to truck to retailer could be a matter of days or 5 weeks depending on distance traveled between source and destination. Margins are low and competition intense. Logistics costs represent greater than 30% of every dollar in revenue for the company. Under these parameters and costs, the International Logistics department function is a vital role for this company.

Competitive Intelligence

No company operates in a vacuum. In addition to operating professionally to generate profits and adequate shareholder return, any business entity must compete against others for every consumer dollar. Therefore, Senior Management is concerned about the competitive environment in which it operates. The President of General Motors European operations, Louis Hughes, stated in a discussion concerning the market and competitors, "You can fall very deep and very fast if you don't pay attention." At Barney, there has always been a dedicated office reporting directly to the President of the banana operations providing some sort of Competitive Intelligence information.

Competitive Intelligence (CI) is timely and meaningful information about the competition worldwide that will have an impact on the strategic decision making process and outcomes. Collecting CI entails knowing how to pick up and interpret weak, ambiguous signals early enough to make a difference for the future of the company. CI differs from market research in many ways. It is more strategic and comprehensive in nature, with emphasis on the analysis of the data and the strategic implications of the information. It is also more comprehensive in that it also focuses on other functions and issues such as production, R&D, logistics, marketing, future decisions and scenarios and planning. Membership in the trade association, Society of Competitive Intelligence (SCIP), is currently increasing at a rapid rate. Growth of CI is being influenced by increased competition, a complex global and operating environment, and improvements in technology and communication.

www.scip.org

CI and International Logistics at Barney

The structure of the CI department consists of a manager and analyst studying production volumes and capacity, future production scenarios and estimates, source countries, by competitor and for total industry, shipping vessels, market share, the regulatory environment, non-core product information and activity, partnerships, and customer analysis. The regulatory environment plays a significant role on the operations of X and can change dramatically, quickly and without warning.

For example, the European Union created a new banana policy in the early 90's to create import quotas and favorable status to their former colonies in the Caribbean and Africa. Following this change, Barney's market share in Europe fell from 40% to 20%. To make matters worse, this was Barney's most profitable market. The United States International Trade Office representative and the WTO have been fighting the EU to change their policies, and implementing European Import duties into the US for various products as retaliation. The CI department provides critical competitive information to help fight the quota system and also future scenarios to help Senior Management planning in the event of any future changes in policies.

The information comes from all departments and employees within Barney, government operations and representatives, especially Trade and Customs departments, and the Net, as well as other sources. Technology has created more information and improved the tools for communication so it is necessary to set up a system and organization to manage and process all the data.

The core of the CI department is a custom built relational database tracking all vessels, products, for all companies, all source and market countries, ports, etc. The reefer ships are the most significant assets for each company in this industry, so an understanding of how the competition uses these assets (own verses charter, rotations, port calls, etc...) is vital. Daily vessel operating costs exceed $10,000 and unloading, loading and other port costs can run in the several hundred thousand dollars. Data was collected and analyzed and distributed in weekly, quarterly, annual and other ad hoc reports. With more than 40,000 employees worldwide, it was a constant job to advertise the CI department services to all divisions and management and also have them participate in the information gathering.

The department recently created an intranet site accessible to senior management worldwide via password. The site distributes reports and allows users to obtain database files for business intelligence reporting purposes thereby putting the analysis in the hands of the end user. It also links to hot daily news information from sources all over the world, often translating from source language into English using Net translation tools.

The International Logistics department would make short and long term plans for future production volumes, shipping decisions, port operations, scheduling of vessels and trucks and market destinations. Significant industry changes would affect those short and long-term plans.

For example, if a hurricane hit Guatemala and the CI department estimates a 40% loss of production capacity for one competitor, analysis would be done and communicated to International Logistics department. Questions such as, what markets will be affected and when? Will they be able to secure other volume elsewhere and how will this affect their logistics/shipping operations? What destination ports, markets and customers might be affected? Based on these types of questions and scenarios, the strategic planning of senior management, in particular the international logistics department would be able to make better decisions.

The options to monitor and analyze the competitive landscape are endless in CI. More and more companies are creating in-house departments or are relying on outside companies specializing in CI showing their commitment to this function and its relevance to business operations, especially International Logistics.

---

[1] *Delaney and Wilson, Ibid.*

[2] *O'Reilly, Ibid, p. 106.*

[3] *Delaney and Wilson, Ibid, p. 14.*

[4] *Delaney and Wilson, Ibid, p. 15.*

[5] *Gablel and Pilnick, Ibid.*

[6] *Forrester Research, Inc.*

[7] *"The Coming New Organization", Peter F. Drucker, Harvard Business Review, January/February 1988, p. 8.*

[8] *"Information Systems", John G. Burch and Gary Grudnitski, 5th ed, John Wiley and Sons, New York, pp. 40, 42.*

[9] *"eBusiness", Bauer et al, p. 44.*

[10] *"The Post Office and the Digital Switch", Thomas J. Duesterber, in "Mail@the Millenium", Edward L. Hudgins, editor, Cato Institute, Washington DC, 2000, p. 140.*

[11] *"The Future of Messaging", Michael J. Critelli, in "Mail@the Millenium", Edward L. Hudgins, editor, Cato Institute, Washington DC, 2000, p. 127.*

[12] *An organization funded by Pitney Bowes, maker of postage handling equipment.*

[13] *Critelli, Ibid, p. 127.*

[14] *Critelli, Ibid, p. 126.*

[15] *Critelli, Ibid, p. 129.*

[16] *Critelli, Ibid, p. 133.*

[17] *"Beyond Tariff Publishing", Helen Atkinson, JoC Week, February 25-March 3, 2002, p. 34.*

[18] *Ronald H. Ballou, "Business Logistics Management", Prentice-Hall, Englewood Cliffs, NJ, 1973, p. 58.*

[19] *Dawe, Ibid, p. 10.*

[20] *"E-Business: The Strategic Impact on Supply Chain and Logistics", Michael J. Bauer, Council of Logistics Management, Oak Brook, Il, Charles C. Poiser, Computer Sciences Corporation, Lawrence Lapide, John Bermudey, AMR Research 2001, p. 9.*

[21] *"Worldwide IT Trends and Benchmark Report", 1999, as cited in JOC, October 1, 1999, p. 6.*

[22] *"eBusiness", Bauer et al, p. 39.*

[23] *Delaney and Wilson, Ibid, p. 16.*

[24] *JOC, March 5, 1999, p. 11A.*

[25] *"Local, global positioning", The Economist, September 26, 1998, p. 84.*

[26] *"eBusiness", Bauer et al, p. 13.*

[27] *"eBusiness", Bauer et al, p. 55.*

[28] *Delaney and Wilson, Ibid, p. 17.*

[29] *"eBusiness", Bauer et al, pp. 60-3.*

[30] *"eBusiness", Bauer et al, p. 82.*

[31] *CBS "Sixty Minutes" interview, October 22, 2000.*

[32] *Keebler et al, Ibid, p. 3.*

[33] *Keebler et al, Ibid, p. 73.*

[34] *Keebler et al, Ibid, p. 8.*

[35] *Keebler et al, Ibid, pp. 5-7.*

[36] *JOC, September 17, 1999, p. 4.*

CHAPTER 17
PUBLIC LOGISTICS

Not all logistics is conducted by private firms. Governments have been active in running major logistical operations. When public agencies are managing logistics, many of the concepts we have studied are the same, but others are different. This chapter reviews 'industries' in which the government or public agencies are managing logistics and transportation. The primary difference between public and private logistics is the need to fulfill obligations to society and not just the stockholders. We also see the influence of political factors. This chapter includes sections on the transportation of hazardous waste and humanitarian relief. These are operations that, although sometimes managed by private firms, are influenced by public pressures and thus deserve special attention.

The distinction between public and private is becoming increasingly vague. Private companies often have pubic obligations in the same manner as government agencies. The discussion on corruption noted that the definition was expanded from 'government' to 'trusted authority' because many private entities have obligations much like government officials. There are an increasing number of companies that are government owned but operate as private companies. Their management, and thus logistics, would be a mix of private and public.

PUBLIC AND PRIVATE INFRASTRUCTURE

Throughout this book we have discussed many types of logistics services, some of which are provided by private companies and some by public agencies. Why is it that some functions are performed by the government and others not? The characteristics of some industries require that they be done by a public agency, while others may safely be relegated to the private sector. The Postmaster General of the U.S. notes that "privatization merely defines ownership".[1] Yet ownership plays a fundamental role in the nature of the incentives by which an organization operates. The main issues are efficiency, ability to raise capital, safety, fairness, and corruption.

Efficiency means the ability to provide the services in a cost effective manner. The Port of Mumbai was used as an example of an extremely inefficient port. Government provided services are generally inefficient because there is little incentive to do better. James Q. Wilson is a noted scholar on organizations, and while he does not like to criticize public organizations by referring to them as "inefficient", he emphasizes how they have other priorities that private companies do not have.[2]

The **ability to raise capital** becomes an issue in the poorer countries. Levi notes that "the history of state revenue production is the history of the evolution of the state".[3] In rich countries we take it for granted that a government can raise taxes to pay for expensive things like bridges and roads. The poorest governments often lack the administrative ability to tax society. It is often assumed that the state has the ability to raise an infinite amount of capital. If the amount is limited, privatization can allow the state to use its limited capital on what it can do best.

Fairness refers to the need to provide some services to all member so society. In a later part of this chapter we discuss the need to provide postal services to everyone, even if it means losing money providing the service to some areas. Roads are another example where not everywhere in a country may be able to pay for them, but there is a social obligation to provide accessibility. The universal service obligation (USO) is described as "an obligation imposed on the provider to ensure that anyone in its service area has access to an affordable, minimum level of a standard quality service bundle".[4]

Roads and rail transport is particularly subject to the special needs of the poorest sectors. International shipping and air transport would not likely involve the extreme poor, but they would need to move around their area for work. This means that privatizing roads and rail systems would need to take their needs into consideration. The USO may be dealt with in either of two ways. Either the concessionaire must provide universal service, even to users that are money-losers, or the government pays for the money-losing sectors.

Corruption, which was already discussed in Chapter Fourteen, applies to government functions and is symptomatic of an organization that does not have profit incentives. One of the common reasons for privatizing government functions is to reduce corruption. Yet the privatization process also creates a prime opportunity for further corruption, as the government can sell the assets to someone for an unfairly low price. Rose-Ackerman notes, "privatization can reduce corruption by removing certain assets from state control and converting discretionary official actions into private, market-driven choices. However, the process of transferring assets to private ownership is fraught with corrupt opportunities".[5]

Three factors increase the chances of corruption during the privatization process. [6] There is no clear way to determine the value of the asset being privatized. The operators can provide selective information to the public and to those who will buy it. Finally, there is the possibility of structuring the privatization to give the privatized firm monopoly powers.

Many aspects of transport infrastructure are what economists refer to as **private goods**. This means they meet two conditions, the excludability conditions and the rivalry condition. Ocean ports, airports, busy roads, and some others meet these two conditions. Others, such as quiet rural roads, do not meet the excludability condition. Most of these goods are actually **club goods**, in that the operator can exclude users, but the optimal size of the club is relatively large. Thus they tend to be natural monopolies. **Public goods** are those that cannot be withheld from anyone, such as a city street. Since the benefits of a city street cannot be denied to anyone, no company could feasible collect revenue from the users.

Economic theory suggests that private goods should be provided by the private sector, and the government should only provide public goods. This theory is subject to broad interpretation and disagreement, as we see vast differences in the amount of government activity in different countries. However, privatization is being carried out in most countries, rich and poor, as it becomes better understood how to regulate industries. In other words, privatization does not seem to be the choice of capitalist countries at the expense of society.

Many industrial sectors have traditionally been dominated if not entirely run by government. Most train companies have been the one and only State company. Ports have almost all been publicly run. Roads are also provided by the government. Given our new understandings of regulation and economics, the challenge is to disaggregate an industry horizontally and vertically to separate the private goods that can be privatized from the public goods that should remain under public control.

For example, the traditional State train company may be divided into tracks, rolling stock (trains), and terminals. This is an example of vertical disaggregation. The train company may also be divided into regions, which is horizontal disaggregation.

Privatization does not mean selling everything off. There are a variety of methods depending on the need of the government to retain control. **Divestiture** is the extreme case where all assets are sold off. **Greenfield** projects are ones in which a new asset is built by private parties. This is intended for raising capital that the government would not be able to do. **Operate and maintain projects** (O&M) involve a private party managing government owned assets. Finally, **concessions** are where a private party manages an asset and pays the government a portion of the revenues.

When offering a contract, there is the decision as to whether to offer multiple contracts and divide the market, or offer one contract and create a private monopoly. There are economic reasons in which an industry needs to stay under one entity. Transport infrastructure is characterized by indivisibility, joint production, and the inability to store transport services. Indivisibility refers to the ability to divide the service depending on the users and the markets. For example, a railroad track system cannot be divided, whereas trucking services can. Joint production refers to the fact that demand for one sector, such as the port, drives demand for other parts, such as the terminals.

Exclusive contracts should be used in three situations.[7] First cross-subsidies among different users are denied. For example, fairness may dictate that the same rate is charged to all users. In this situation, one cannot charge lower fees to the poorest users and make it up from the richest users. Second the initial risk in the sector or the country is high. Third, the service the concessionaire will provide is a natural monopoly.

When assets are in the private sector, the government's role changes but are not eliminated. There is now the challenge of acting as economic regulators instead of operators. One of the reasons for privatizing assets was that government is inefficient as operators. Yet why would they be any better supervising the industry? "Governments are recognizing that one of the main reasons for poor regulatory governance is that the civil servants recruited to staff these agencies do not have the necessary technical skills to transform them into effective economic regulators."[8]

When it was noted that private companies operate more efficiently, that depends on the nature of the incentives. **Static efficiency** refers to minimizing cost and allocating resources to where there is the greatest yield to society. **Dynamic efficiency** is to stimulate the right amount of innovation and investment to meet long term needs. When the private company owns the assets, as in a divestiture, this is not a problem. When concessions are offered, then there is the concern that capital costs are included. For example, if a company gains the concession to manage a toll rode for a short term, they may not invest any money into its long term maintenance.

If a private concessionaire did need to pay for all capital costs, this may result in an unreasonably expensive situation. Many assets, such as the larger bridges, the Chunnel between Britain and France, and many highways could not be paid for under the traditional system of marginal cost, in which users are charged for the full costs of capital and variable costs. A second best alternative has been long run marginal costs (LRMC), which is short range marginal cost (SRMC) plus the marginal cost of capacity (MCC). In other words, the concessionaire needs to cover the short run costs plus the MCC, which varies depending on how fully utilized are the assets. When roads are under utilized, there is almost no cost to an additional truck. When the road is close to capacity, then the MCC becomes great.

Where is privatization heading? Governments are still learning from experience. There have been some disasters, in which public assets were placed into private hands without understanding the incentive system. The resulting service was poor or the assets were neglected. This does not necessarily mean that anyone had bad intentions, but these issues are highly complex. Recently, there have been many success stories in which competition is introduced through well controlled privatization and society benefited immensely.

Airports Council International's
Key Challenges Facing The World's Airports
Airports Council International is the Geneva-based organization that represents the airport and the airport industry. Their members included the great majority of the airports in the world. They consider the following issues to be the most important facing the industry at this time. It is interesting to note how airports, the vast majority of which are public or semi-public, must address social and economic needs together.

Coping with the Giants
In the not too distant future, airports in all regions may have to handle aircraft capable of carrying 600 passengers. The major manufacturers have developed plans for these giants. ACI is ensuring that all the factors arising out of the operation of these aircraft are taken into consideration, including increased wingspan and wheelbase. gross weight, the effect of wake vortices and the design implications for terminal facilities.

Technological Changes Leading to Increased Capacity
Technological progress in navigational aids will increase capacity on the ground and in the air. ICAO's decision to adopt a satellite-based Communications, Navigation and Surveillance/Air Traffic Management System (CNS/ATM) is only the first step.

Airport Environmental Capacity Constraints
Present indications suggest that the phase-out of Chapter 2 aircraft by 2002 will not alleviate the noise problem as effectively as was once thought. As air traffic increases, overall aircraft noise levels around airports are expected to rise some time after 2002. ACI's goal is to ensure that traffic growth can be maintained within the environmental capacity limitations imposed upon airports by governments. These limits may be set lower than the current capacity of airports. Future traffic growth will only be possible if noise and emissions are reduced at the source. The industry cannot wait! More stringent noise and emissions certification standards have to be achieved in order to encourage manufacturers and airlines to produce and operate quieter and cleaner aircraft.

Expanding the Travel Market
The transition of air transport from an elite to a mass transport mode will be the most important development affecting regulatory practices in the coming years. This could mark the end of special treatment of air transport by some governments. Increased exposure to competition may well cause a conflict of interests between airlines and airports. While airlines are endeavoring to capture markets and dominate hubs, it is in the interests of airports to seek to prevent such concentration and domination. ACI supports air transport liberalization, provided that it is accompanied by appropriate government action designed to maintain competition, airport autonomy and consumer choice.

Airports Need to Make Money!
The changing business environment of the aviation industry will affect the management and operation of airports. Airports need to make money. They provide an important service, which must be commercially viable, just like all the other services which make up air travel. Airlines have a tendency to blame airport charges for their financial problems. However, airport charges worldwide have been a very stable component of the operating costs of airlines. On average, they have remained at around 4 per cent of the costs of airlines over the past 10 years.

Speeding Up the Journey

Travelers often remember delays more than super efficient airports. The slow and inefficient clearance of passengers, baggage and cargo by government inspection services at many international airports causes bottlenecks and delays. These delays may deter travelers. With a view to helping airports speed up passenger, baggage and cargo processing, ACI recommends standards and procedures and helps to implement them. It is also taking a close look at the development of new techniques and methods and participates in field missions to assist in facilitation matters. New technology, such as advance passenger information (API), machine readable travel documents (MRTDs) and biometric identification systems will assist airports in alleviating facilitation problems, but only if their implementation is coordinated by all the agencies concerned.

Security - Strengthening the Weakest Link

Aviation security is complicated by the fact that neither governments nor the industry control the agenda. The initiative tends to lie with those who carry out acts of unlawful interference with civil aviation. The introduction of checked baggage screening will probably be the next major step in the field of airport security. In order to preserve airport capacity, airports have to play a leading role in the planning, coordination and installation of new screening systems. ACI is also tracking developments aimed at strengthening measures to prevent freight, courier and mail from being used to introduce explosive devices onto aircraft.

Enhancing Airport Safety Everywhere

The traveller expects and deserves the same level of safety at all airports around the world. The gap between developing and industrialized countries is not likely to diminish in the near future. However, through ACI, the world's airports are doing their utmost to contain this split by promoting and expanding the activities of the ACI Fund for Developing Nations' Airports and by enhancing assistance between rich and poor airports. Among its activities, the ACI Fund offers airport staff the opportunity to attend recognized training courses.

MILITARY LOGISTICS

Success on the battlefield is determined mostly by logistics. This may seem odd considering the emphasis on high tech weapons, tactics, and training. Yet in the war novel *Red Storm Rising*, a Russian General captured the idea well when he said, "The tactics...no, amateurs discuss tactics. Professional soldiers study logistics."[9] An individual battle may be decided by tactics and equipment, but the overall war effort is determined by the ability to get the equipment to the battlefield.

Logistics becomes an integral part of tactics during a conflict, in which one can attack enemy personnel, or their transport capabilities. The Austro Prussian War of 1866 was seen as the beginning where logistics was the essence of battle. An army could get on a train and move 200 miles in an afternoon. However, the Generals relied on train watchers to keep track of the trains and telegraph back to HQ. The other side would simply pull down the telegraph lines, so the Generals were operating blindly.[10]

World War II was won by the Allies who not only had more resources at their disposal, but had better transportation assets. Many military authorities believe the most important aspect of that war was fought not in Europe, but on the North Atlantic, for control of shipping lanes. Civilian sailors on cargo ships had a higher fatality rate than any unit in the military.

The scope of logistics of that war can be seen in some interesting statistics. The US built 100 aircraft carriers, 285,000 aircraft, and pipelines under the English Channel for the fighting in France. Bombers were sent to the South Pacific to runways that were only completed while the plane was still in the air. Medicines and blood products stayed refrigerated from the US to the remotest parts of the world. New roads connected China with the Bay of Bengal and Alaska with the lower 48 states.[11] Never before had such an extensive logistics system been created.

More recently, the Gulf War in 1991 was also won in part by logistics. It is true that the coalition forces against Iraq had more firepower. But the fact that the war was won in one week had more to do with the coalition's ability to cut off Iraq's overland routes. This was done because the coalition could move faster and with more control. It is no surprise that the head of logistics, Lieutenant General William Pagonis, became a celebrity among logisticians for his role in that conflict.

Military organizations include the armed forces of the national government in each country. There are also military units at other levels of government, such as state guards, police organizations and so forth. There are also non-government military such as militias and guerrilla organizations. For purposes of this section, we will assume that the military is the legitimate, national military force.

There are some fundamental differences between military logistics and business logistics, yet there is a lot of debate exactly what those differences are. Businesses are trying to act more like the military, and military organizations are adopting the best practices of the private sector.

Logistics normally is not exciting reading, but one exception is Lt. General Pagonis' book, *Moving Mountains*, the story of logistics during the Gulf War. He notes in the first paragraph "Running logistics for the Gulf War has been compared to transporting the entire population of Alaska, along with their personal belongings, to the other side of the world, on short notice. Supporting the US forces included 1.3 billion gallons of fuel, 122 million meals (over a one year period), and enough mail to fill 28 football fields six feet deep."[12] Pagonis sees only two real differences between military logistics and business logistics:[13]

- **Profit versus lives.** Instead of making money, military logistics is about human lives. He stated that the goal was to save lives, but one could also make a case for the fact that the goal of military logistics is just the opposite. This is obviously a very controversial field but outside the scope of logistics.
- **Average age/responsibility of personnel.** The average age of military personnel is relatively young. Military personnel control far more resources, however one wants to measure it, than civilian personnel. A new lieutenant in the US Army, for example, is usually around 25 years old yet is responsible for about $1 million worth of equipment and (politically sensitive) weapons, plus a dozen or more personnel.

Pagonis does not believe that **patriotism** separates the military from business, and that the loyalty of workers versus soldiers is any different. Many civilian workers are fiercely loyal to their employer. He did not state this in his book, but it can also be noted that not everyone in the military is particularly loyal; some are simply doing a job.

There is a common myth that soldiers follow orders without questions. In fact, professional armies expect their soldiers to make decisions and act independently. Pagonis noted how soldiers have an interesting way of encountering 'radio difficulties' when they receive a dumb order. For-profit companies sometimes expect their employees to do their work without question, while others empower workers to make their own decisions.

The military tends to have a much **higher turnover**, as soldiers are transferred from one unit to another. In the business world this would be a big problem since training new employees is expensive and time consuming. Yet in the military everyone has a common base of knowledge as soldiers. They all go through the same training and professional schools, and every unit follows somewhat standardized policies. His observation is based on the US military. Other countries such as Great Britain have a regimental system, which means that everyone in the unit stays together for their entire career and the unit travels on missions together.

Finally, Pagonis puts much emphasis on a **kingpin**, one individual responsible for overseeing the entire logistics function. He took this policy to Sears where he became the Vice President for logistics. Many companies are bringing all the logistics control into a centralized office or under one person. Yet companies are also contracting out major parts of their operations. There is no clear overall trend, but some companies are decentralizing their overall structure, while others are pushing for more centralization.

Businesses often have a **war room**, a room in the headquarters that acts as a nerve center for special situations that need quick and coordinated responses. Ironically, the author has seen some military organizations that do not have a 'war room', but a 'situation room'. In complex, social/political/military conflicts, the military did not want to give the impression that they were only going in to fight a battle, but instead to manage the situation.

Inventory management is fundamentally different in the military organizations. Instead of minimizing inventory and receiving it only when needed, the military needs to keep large stocks on hand. In fact, military inventory is vast compared to civilian needs. When a conflict breaks out, there is generally not enough time to order materials.

One example of how countries have managed their military inventory is Australia's agreement with the US Air Force in which the two countries co-own spare parts for a certain plane (the F-11). The Australians draw from the stockpile what they need, and the US Air Force charges them for the administrative costs, which are less than if Australia maintained their own inventory.[14]

The US military has several large ships filled with supplies strategically positioned around the world just in case they are needed. During the battle over the Falklands/Malvinas Islands, Great Britain chartered nine 25,000-ton tankers from BP to refuel warships enroute. They were positioned at various places in the middle of the sea. BP and Ministry of Defense had agreed to the outfitting of some commercial tankers so they could refuel the 100-ship task force at sea. "From a logistics standpoint, the Falklands' location was probably about the worst location for the British to have a fight. The challenges of supply were probably greater than the challenge of battle."[15]

Henry Kissinger estimated that for the US to launch a full-scale attack in the Middle East, about 5-6000 miles away, would take ten days.[16] In one study, it was estimated that an army could move at the following rates:[17]

1816–1918   250 miles/day
1919–1945   375 miles/day
1945–   500 miles/day

Bueno de Mesquita argues that a state's power declines as the distance to the battlefield increases. He cites four reasons: organizational and command difficulties, long distances hurt military morale, long-distance conflicts encourage domestic dissent, and fighting far from home is more debilitating for soldiers and equipment.[18]

POSTAL SERVICE

Of all the logistics performed by government, mail distribution is one of the largest and most complex, with the possible exception of military operations. Europe's postal services cost $72 billion, or 1.4% of EU's GDP, employing 1.4 million people.[19] The US Postal Service (USPS) has revenues of $63 billion. Postal services are undergoing privatization, which is one of the most dynamic and far-reaching events in the world of logistics. Multi-billion dollar postal services are being privatized, and government postal services are being transformed beyond recognition.

Mail delivery has traditionally been done by the federal government based on the assumption that an effective mail service is a basic necessity for a well run society. As a public good, it was kept under public ownership. Unlike similar services such as small package delivery, mail agencies usually have a policy where the service must be provided according to the USO introduced earlier in this chapter.

USO standards vary from country to country, but the developed countries have very liberal definitions of their service area. Even residents in remote rural locations get mail service at or near their residence. This remote service is not cost effective and is therefore subsidized by those who live in urban areas. Although intended for individuals, the USO applies to all postal customers, which means individuals, households, businesses, or any other entity. The U.N. Declaration of Human Rights states that "every person has the right to look for, receive and send without limitation of frontiers, information and ideas by any means of expression whatever."

Weighing Up the Mail

Domestic Mail Revenues of Universal-Service Providers

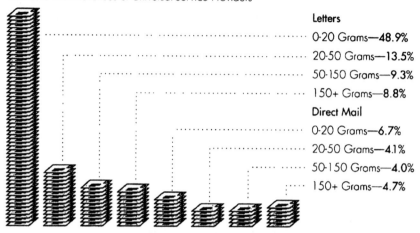

Letters

0-20 Grams—**48.9%**

20-50 Grams—**13.5%**

50-150 Grams—**9.3%**

150+ Grams—**8.8%**

Direct Mail

0-20 Grams—**6.7%**

20-50 Grams—**4.1%**

50-150 Grams—**4.0%**

150+ Grams—**4.7%**

Postal services were usually self-financed, but have a monopoly in the shipping of letters or other items. That is why package delivery companies usually have a large cardboard envelope to put even the smallest of items. That large envelope is just a little larger than what the law defines as a 'letter'. As the small-package industry grew, it became apparent that the government monopoly may not be necessary. Currently, the USPS is the largest government postal service in the world, and competes with the package delivery companies. Most of the postal services in Europe are being privatized, so there is now a heated debate about the future of the USPS.

The USPS holds the monopoly on delivery of first and third class mail, and since 1934 they alone can use the home and business mailbox. Revenues for first and third class mail exceed sales of Coca Cola, McDonalds, Wells Fargo and Microsoft combined, $43 billion.[20] Only nine American companies had higher sales and none had more employees.[21] The USPS does not pay for parking tickets, license plate fees or taxes, yet UPS and FedEx do pay these charges. In one analysis, the costs of being a monopoly came to $9 billion, and the benefits of economies of scale came to $6 billion, so there was a net cost to society of $3 billion.[22] Another study found the cost to society of the USPS estimated at $12 billion.[23]

The question is what would happen if the USPS were open to competition? This is one area that has been subject to extensive analysis by economists and others. One thing is certain, many jobs are at stake. Labor is about 80% of USPS costs, and these jobs pay much higher than the private firms.[24] Therefore, labor costs are the primary variable that will affect how the USPS competes with other carriers. In a competitive environment, the USPS could provide universal service, but not necessarily with the same number of deliveries per week.

The USPS has reacted to these pressures in a variety of ways. The Postal Reorganization Act of 1970 to make the USPS more business-like[25]. The USPS has been attempting to enter the Internet market by offering online services. It has been trying to move into markets that were traditionally only for private companies. This raises the question of whether a government agency should be allowed to compete with the private sector in non-monopoly sectors.

International mail requires some special regulatory and management issues. In 1874 a conference in Berne adopted the Universal Postal Convention and established the **Universal Postal Union** (UPU). States agreed on charges and procedures for handling international mail. When a letter is sent internationally, the sending country collects the money. Foreign mail is to be treated in the same manner as domestic mail. There are strong norms on the free movement of mail, and only in rare cases have governments failed to respect this. Mail is considered property of the sender until it is delivered to the final recipient. UPU believes that postal services should only be held liable for items that are registered or insured.[26]

A postal service keeps all revenue for outgoing international mail, but does not receive any money for incoming mail. It was assumed that there would be a roughly balanced flow of mail between nations. Mail flows did not turn out as expected. Of the international mail, 70% originates in the developed states. International mail for developed states is about 1.5-4% of their total mail load, while for developing states it is 10-20%. For many developing states, incoming mail is much larger than outgoing.[27] Thus the poorer countries are losing money in the reciprocal arrangements with other countries.

State postal services sometimes compete with other states by offering lower international postal rates. This has created an industry known as **remailing**, in which private companies collects mail in one country, brings it to another and mails it from there. This practice has been credited as a boost of competition and free market incentives in an otherwise government monopoly. However, there is a norm against competing for more mail. The UPU states that Postal administrations should not negotiate with different foreign administrations, and not price differently for different routings and not take business from other states by using private carriers

European postal agencies are now actively eliminating state monopolies, and it is in this continent where we see the results of open competition. A 1997 directive kept for states the monopoly on letters of less than 350 grams (12 ounces), which only opened for competition 3% of letter volume and 5% of incumbent operators' revenues.[28] EU's Council of Ministers decided in 2000 to privatize or end the state monopoly on many industries, including gas, electricity, and postal services.[29] Britain's Post became a private firm in 2001, with the government as only a shareholder.)[30]

Germany's mail service, Deutsche Post, is now changing from a traditional postal service to a publicly owned distribution company competing directly with private sector businesses. Germany's mail services will be completely deregulated by 2003, forcing Deutsche Post to become competitive if it is to survive. It is doing this by diversifying its services, mostly through acquisition. This Post now offers logistics services with revenues at over $20 billion. Another national postal service, TNT Post Group of the Netherlands, has also been expanding into non-mail services.[31]

The privatization debate is not just about whether or not to eliminate the government monopoly, but in what ways. For example, reducing the state monopoly in Europe to 100 grams would mean 20% of their business in which they previously had a monopoly is now open for competition. If the limit on what other carriers can carry is reduced to 50 grams, 27% of the European posts' business would be subject to competition. This does not mean the incumbents would lose 27% of their business, but that they would need to defend it against competition.[32]

What of the USO in Europe? The cost of USO varies from 5 to 14%, according to an EU study. Countries with remote areas have higher costs.[33] Rural customers may have to pay up to four times as much.[34] Some say the USO is not so critical. Most of the concern is for private letters, which make up only 8% of volume, and Christmas cards are half of this.[35] Sweden liberalized their postal service in 1994, and while the USO is protected under law, Sweden Post shed a quarter of their workers.[36]

Postal services and other carriers are not the only ones interested in the fate of this industry. Large shippers are also concerned. The Federation of European Direct Marketers (FEDMA) would seem to want liberalization. In fact, they are concerned that rapid opening would allow the Posts to crush competitors and still maintain their position. They want a controlled opening.[37]

THE INTERNATIONAL TRADE IN TOXIC WASTE

Just as smuggling is a logistics function with important social effects, the trade of toxic waste is a major environmental problem that revolves around the movement of the goods. According to the Organization for Economic Cooperation and Development (OECD), a cargo of hazardous waste "crossed a national frontier more than once every five minutes, 24 hours a day, 365 days a year." About 2.2 million tons of toxic waste crosses borders each year.[18] International transportation has been cited as one of the primary reasons for the creation of this trade. Controlling it depends in part on the effective identification of cargo that has been discussed throughout this book.

Why is this a concern? This waste requires careful treatment, and irresponsible handling can cause environmental nightmares. One way that regulations can be evaded is to ship the waste to a country that does not enforce proper standards. Theoretically, any country can provide safe disposal, but supervising this in many countries is practically impossible. For this reason, the global community has been seeking to prevent the international movement of hazardous waste.

The first problem is the large amount of toxic waste being produced, and the cost of disposing of it properly. The 1976 U.S. Resource Conservation and Recovery Act defines "hazardous" material that which may "pose a substantial present or potential hazard to human health or the environment when improperly treated, stored, transported, disposed of, or otherwise managed." This definition includes a wide variety of wastes, some of which were not considered particularly hazardous before the Act passed. The Act marked the beginning of a new era in the control of pollutants. It began a process of regulation known as "cradle to grave".

The U.S. produces about 500 million tons of toxic waste a year[39], which is one ton of hazardous waste for every woman, man and child in the country.[40] In 1978, it cost about $2.50 to bury a ton of paint sludge, while burning it would cost $50. After the passage of the RCRA, the prices increased, to $200 for burial and $2,000 for burning.[41] The cost and regulatory environment is roughly similar in other countries, in that government controls are strict and costs of handling the material are high. Hazardous waste disposal hits small companies the hardest.[42] Smaller quantities of waste are easier to dispose of illegally than large quantities. Yet the cost of disposal in Africa was estimated to be around $40 and as low as $2.50.[43]

These high costs of disposal soon opened a market in some countries to offer cheap disposal. The reason they could offer cheap disposal was obvious. The governments were either not interested in protecting their people from the toxic effects of the waste, or they were not able to regulate the industry properly. In some cases, the local laws required proper safeguards, but the rulers ignored these. A California firm signed a multi-million dollar contract with the Kingdom of Royaume de Tetiti, only to find out that the country consisted of a few uninhabited islands and the 'king' was a con artist wanted by the authorities in Singapore and Thailand.[44]

Opposition to the international trade of toxic waste has been developing in many countries as well as at the international level. The 1972 declaration adopted at the United Nations Conference on the Human Environment in Stockholm stated that each nation is responsible for ensuring that activities within the country do not harm the environment of other countries, known as Principle 21.[45]

The issue came to a head with the 1989 **Basel Convention** on the Transboundary Movement of Hazardous Wastes and Their Disposal. All African and most Third World delegates wanted a ban on the trade, while industrial nations for the most part refused. The convention did manage to create a requirement that the waste exporter notify the importer before any shipment may proceed. US opposition doomed one proposal that a country may not ship waste to a country with standards lower than their own.[46] The US is the only major country not to join this Convention, though 120 other nations are members.[47]

In the early 80's, it was thought that the US was producing about 40 million tons of hazardous waste annually. The EPA did a study and found that it was more like 250 million tons. Later research indicated it could be as high as 500 million tons. Yet, remarkably, nobody really knows.[48] About 70 million batteries were discarded in the US, consisting of 70 million gallons of sulfuric acid and more than a billion pounds of lead. About 10-20% of them are dumped illegally.[49] In one scheme, burnable toxic waste was put in a tank of diesel or fuel oil, and the other 90% of the tank was the regular product. This tainted fuel was then driven from the US to Ontario where consumers burned it unknowingly.[50]

In 1980 only 12 notices of intent to export hazardous waste were issued by the EPA to foreign countries, by 1988, this had jumped to 522.[51] Controlling this practice is an ongoing challenge. The US government has also been the customer in some illegal dumping operations. A legal front company bids for the contract to dispose of government waste, but then turns it over to another organization that disposes of it illegally. Note that it is perfectly legal to ship and sell overseas products, such as DDT, that are illegal in the US.

One major effect of Basel was a shift from "waste" to "recycling", which was not covered by the Convention. Observers noted a major increase of material to recycling facilities, and some of this was hazardous materials avoiding the ban.[52] In 1995 the Bases Convention was amended banning the export of toxic wastes from OECD to non-OECD nations. This opened an entirely new chapter in the international negotiations as the recycling industry seemed to become caught up in a debate. Recycling is a $160 billion a year industry employing 1.5 million people around the world. Basel's definition of waste was very vague, and although there have been attempts to clarify terms, problems remain. As one industry leader notes, "scrap is not waste".[53] Only a small percentage of the global scrap shipments are hazardous, but they provide a cover for illegal shipments.

The US/Mexico border has been a hot spot for the trafficking in waste. A barrel of waste can cost as much as $1000 to dispose of legally, giving a strong incentive to find alternatives. The officials at the border have other issues, weapons, drugs and illegal immigrants in particular, to worry about. Waste coming from the US into Mexico is relatively low on their priority list. NAFTA requires that waste be shipped back into the US. There is now a tracking system of hazardous waste crossing the border, using a database known as HAZTRAKS[54], though it is difficult to track illegal and clandestine shipments.

The recipient countries have also been reacting to the situation. Guinea-Bissau at one point was planning to earn $120 million a year by offering themselves as a disposal site, more than its total annual budget, but later changed its mind.[55] When the U.S. State Department learned that a Colorado company planned to ship waste to Africa, it stepped in to warn that this could cause anti-American protest overseas.[56]

One of the most famous stories of trafficking in waste was what has come to be known as the Khian Sea Odyssey. In 1987, The City of Philadelphia struck a deal with a local contractor to dump 200,000 tons of ash from the city's garbage incinerator. The ship drifted for 27 months, from one port to the next, often never seeing land before it was refused entry. It changed its name twice and changed the nation of registry from Liberia to Honduras.

Eventually 10 nations refused the Khian Sea entry. The Amalgamated Shipping Company, owner of the ship, found itself stuck in the middle. At one point it showed up back at New York, where the Coast Guard refused to let it enter the port. The ship then took off, and the Coast Guard refused to try to keep it there by force. Haiti allowed it to dump, but after part of the load was offloaded, the deal was terminated, and the ship was forced to leave. By November 1988, the ship, now under new ownership, appeared in Singapore, without its cargo. Why did they not dump the cargo at sea? Opening the hatch is dangerous while at sea, and would lose insurance protection. It was speculated that India or Pakistan are likely places for it to be offloaded.[57]

THE LOGISTICS OF HUMANITARIAN RELIEF[58]

As commander of the 353rd Civil Affairs Unit, U.S. Army Captain David Elmo had to be prepared to handle a wide variety of logistical challenges. The 353rd is a New York City-based reserve unit assigned to deal with civilian populations in Army operations. It is involved in disaster relief projects worldwide. These are projects that are organized and conducted on a contingency basis because they are non-routine events that cannot be forecasted accurately. In a non-routine environment, a logistical system cannot be tailored to specific, detailed tasks, but instead must be able to handle a plethora of possible outcomes or contingencies.

For example, during the Persian Gulf War, the Iraqi Army drove Kurdish refugees into the northern mountains of Iraq, where they were left stranded and hungry. Elmo's group, which was comprised of approximately 200 members, was assigned to provide logistical support for their rescue. With barely a week to study the situation and prepare his team, Elmo was sent to Incirlik, Turkey, where he directed a vast amount of food, water, fuel, and assorted supplies to isolated locations nearly 500 miles away in a combat zone.

When a population is at risk of famine, logistics is the most important element of a disaster relief project. Famines occur not because there is not enough food in the world, but because the food is not exactly where it is needed. Disaster relief is a highly specialized form of logistics for tasks that are non-routine and cannot be anticipated with any significant degree of specificity. Relief operations often are described as "paramilitary" because they are special events that take place in emergency environments. In contrast to disaster relief, industrial logistics are designed for more routine actions, such as repetitive manufacturing.

Humanitarian relief operations are an increasingly busy industry, as natural and man-made disasters occur almost constantly. The U.S. Department of Foreign Disaster Assistance maintains a "disaster history list" of 1,622 events that occurred between 1900 and 1981. CARE[59], an Atlanta-based relief and development organization, was involved in more than 25 emergency relief projects in a recent 12-month period alone. This dramatic trend is the result of increased human population and its concurrent stress on the natural environment.

Characteristics of Disaster Relief Logistics.

Captain Elmo's first task for the Kurdish relief project, named Operation Provide Comfort by the U.S. Department of Defense, was to organize his team and prepare for the movement of supplies, even though he did not know exactly what the recipients needed. Preparing for an effort such as Operation Provide Comfort varies greatly from a manufacturing environment, in which just-in-time (JIT) systems require detailed planning and close coordination that comes only with experience. When every operation is unique, planning and preparation are very different. Plans and schedules must be much more flexible and open to last-minute changes.

First, strategic planning is done to prepare for emergency projects. Planners identify assets that would be available if needed, and assess strengths and weaknesses of a plan based on likely scenarios. The next step is the actual project planning- also known as tactical planning, which is done when disaster strikes and the relief work is activated. In an industrial logistical system, the time horizons differentiate strategic from tactical planning: strategic planning is long term, while tactical planning is short term. In a disaster environment, strategic plans prepare the organization for what must be done in an emergency, and tactical plans are developed when that emergency is realized. These two distinct activities are not necessarily separated by time.

The commodities handled in a disaster relief project usually are much more varied than those handled in repetitive process systems. Unlike a warehouse that handles packages in a certain range of sizes and shapes, the commodities required by disaster relief can include, for example, a 1,000-ton shipload of grain, refrigerated medicines, mail, water, and even people. Because every disaster is unique, specific needs vary. In Operation Provide Comfort, an air drop of food and water initially proved difficult when pallets of water bottles fell apart as the parachutes opened up. Loose water bottles fell onto populated areas, creating a potentially lethal hazard. The problem was resolved only after personnel experimented with different rigging systems.

Transportation and distribution also play critical roles in disaster relief. Incirlik, Turkey, served as Operation Provide Comfort's base because it was the closest location that provided modern utilities necessary to direct the operation. Because Incirlik was nearly 500 miles from the refugees, the operation required the use of not only trucks and trains, but anything else that could possibly move the relief supplies quickly into rugged terrain. The operation had at its disposal the most sophisticated transportation means available, including helicopters and airplanes designed to parachute-drop supplies.

"The first challenge was developing a point of entry for supplies and equipment- things like improving roads, or getting a bulldozer up a mountain to clear a helicopter landing zone," says Elmo. "We had to repair the infrastructure to allow more direct, sustainable methods of getting food and supplies to the refugees, and ultimately that was based on a truck transportation system, once we had repaired the roads."

Personnel involved in disaster logistics contrast with other logisticians in many ways. Military units have sophisticated logistical expertise designed for an emergency environment. Their training is "mission oriented", which means they must identify what needs to be done, then carry out the actions to accomplish the task. Within the U.S. Army are civil affairs units-such as Captain Elmo's-designed to handle the civilian populace during wartime and oversee disaster relief.

In addition to the military, non-governmental organizations (NGOs) such as CARE and the United Nations conduct major relief efforts. Most people from development agencies, however, have backgrounds in public policy or Third World development, and professional logisticians are rare. This shortage of logistics professionals affects the efficiency of distribution efforts, a common criticism of relief operations. There is also a distinction between

NGOs such as CARE and OXFAM (an Oxford, England-based relief organization), whose primary purpose is development; and disaster-specific agencies such as the International Committee of the Red Cross. The development-oriented agencies are very familiar with a region, while the disaster-oriented agencies often know more about disaster relief and logistics.

The organizational structure of relief institutions reflects their emphasis on public policy issues over logistical issues. CARE, for example, is organized geographically, not functionally. The personnel are familiar primarily with their region of the world, and must become adept at all tasks of development and relief. Distribution and transportation requirements differ dramatically around the world, so emphasis is placed on knowing the regions.

Military units are at the opposite extreme; they are functionally organized. This organizational structure includes transportation units which are trained to move anything in any environment. The 353rd Civil Affairs Unit is designed for any kind of mission that entails civilians. It has at its disposal transportation professionals who can be assigned for a specific operation.

The quality of a relief project is affected significantly by logistics, and the humanitarian concerns of the operation are as much a priority as cost efficiencies. Aid is often considered a form of foreign intervention that can hinder a development as much as it can help. For example, during a food shortage, the most "efficient" method of food distribution may be to provide centralized facilities, at which people can come and receive their rations. However, famine victims are weak, travel can be hazardous to their health and safety, and such an arrangement would cause people to leave their homes and become refugees clustered around the food depots. Decentralization of food distribution to the end user, therefore, is of utmost humanitarian importance, even though it can be much more expensive.

Information Systems

Information systems are arguably the single most important factor in determining the success of an emergency logistical operation. In the context of relief operations, an information system is comprised of several elements:

The information that must be communicated and processed
The methods of communication
The reporting procedures
The hardware

In an industrial environment, the information department is synonymous with the computer system. In disaster relief, the non-computerized aspects of the information flow play a much more important role. A manufacturing-oriented logistical system has the advantages of experience and inertia to limit the amount of information that must be transferred. It does not take an excessive amount of work to coordinate a delivery truck that arrives every day. In emergency situations, however, most events are unique, requiring more coordination and the transfer of much more information. Operation Provide Comfort team members worked with large volumes of extremely time sensitive data, but without the benefits of an information system that would provide easy answers.

How, then, does a disaster logistics system handle disorganized information rapidly? Real time communications, such as radios and telephones, are the single most important method of reacting quickly for effective coordination. Written reports and other forms of delayed information usually cannot accommodate rapidly changing information needs.

International disaster relief projects sometimes suffer from translation problems because the workers often come from different countries and speak different languages. Another form of real time communication is the computer and modem, but relief agencies are not well funded, and many information systems are not sophisticated enough to use this technology effectively.

When Special Forces units assigned to assist Operation Provide Comfort first arrived in northern Iraq to assess the needs of the Kurdish refugees, they used satellite phones to contact their commanders, who were stationed in Rome, Italy. The commanders, of course, were not directly involved in dropping relief supplies in the area. Consequently, supplies provided were not necessarily what the Kurds needed, and the drops often landed either too far from refugees and workers, or literally dropped right on top of them. Once direct communication links were established, the operation became much more efficient because it was tailored to the needs of the refugees. The volume of information remained high, because delivery locations were changing quickly and transportation modes switched from air drops to helicopter flights, and finally to overland truck delivery.

Sophisticated computer systems are used primarily by larger institutions. For example, the United Nations headquarters in New York administers the United Nations International Emergency Network (UNIENET), which supports UN agencies with databases containing contact lists and equipment stockpiles. The World Food Program (WFP) administers the International Food Aid Information System (INTERFAIS) from its Rome, Italy, headquarters. Its reports include detailed descriptions of obscure ports and logistical facilities in underdeveloped parts of the world. Although WFP is the single largest provider of food aid in the world, INTERFAIS also tracks food aid from every other type of organization from which data is available.

While the UN's computer systems are useful for the headquarters office, they currently provide little useable information to operations units at the local level. To address this problem, Captain Mark Wolfenberger of the U.S. Army Civil Affairs Command in New York developed the Disaster Assistance Logistics Information System (DALIS), a pc-based database program that tracks supply inventories. The program, which was introduced during Operation Provide Comfort, is designed expressly for local operations by matching the needs of a location with the supplies available in the region. Such a task has become a major challenge because of the multitude of agencies involved in any given relief operation, and the inter-organizational coordination difficulties they pose. A major drawback of DALIS is that it operates independently of the programs previously described. If telecommunications technology could be applied to DALIS, operational units could be provided with all the information available at institutional headquarters, and headquarters would have a more realistic idea of what is needed in the field, thereby enhancing efficiency.

Information systems have suffered from a lack of investment in part because of a contradiction of priorities. Relief agencies are judged by the percentage of their funds used on relief supplies, and those agencies that spend the least on overhead are often considered the most efficient. Thus, agencies are reluctant to spend money on a sophisticated information system that would actually improve their efficiencies in the long run. Hundreds of institutions world-wide play a role in emergency relief operations, and the lack of inter-organizational coordination is becoming a major impediment as institutional rivalries takes priority over logistical efficiencies. A universal information system would benefit a large number of agencies, but no agency thus far has been willing to accept the leadership responsibility and development costs.

Conclusion

After several months of facing the challenges of rugged terrain, long distances, and political instability, the job of safeguarding the Kurdish refugees was over, and the 353rd Civil Affairs Unit packed up and returned to the U.S. In a private company, when you are successful, you continue to do your job. In disaster relief, the best sign of success is that you have worked yourself out of a job. Elmo's team members returned home as experts on northern Iraq and its logistical demands. What would their next challenge be? Because they never know, all they can do is prepare for every contingency. Information and communications are critically important to the success of logistical support in any emergency situation. Efficient transportation methods such as planes and helicopters, and technological advances such as bar coded pallets, make the job easier. However, the most vital tools of all are the professionals who can deploy sophisticated logistical assets and execute complex operations anywhere, anytime.

[1] "A View from the Postal Service", William J. Henderson, in "Mail@the Millennium", Edward L. Hudgins, editor, Cato Institute, Washington DC, 2000, p. 19.

[2] "Bureaucracy", James Q. Wilson, Basic Books, New York, 1989.

[3] "Of Rule and Revenue", Margaret Levi, University of California Press, Berkeley, 1988, p. 1.

[4] Estache and de Rus, ibid, p. 41.

[5] "Corruption and Government: Causes, Consequences, and Reform", Susan Rose-Ackerman, Cambridge University Press, New York, 1999, p. 35.

[6] Rose-Ackerman, ibid, p. 36.

[7] "Concessions for Infrastructure: A Guide to Their Design and Award", Michel Kerf, R. David Gray, Timothy Irwin, Céline Lévesque, and Robert R. Taylor, The World Bank, Technical Paper No. 399, Washington D.C, 1998.

[8] Estache and de Rus, ibid, p. 3.

[9] Wood et al, ibid, p. 1.

[10] "War: Ends and Means", Paul Seabury and Angelo Codevillo, Basic Books, New York, 1989, p. 111.

[11] "International Logistics", Donald F. Wood, Anthony Barone, Paul Murphy, Daniel L. Wardlow, Chapman and Hall, NY, 1995, p. 1.

[12] "Moving Mountains", Lt General William Pagonis, with Jeffrey Cruikshank, Harvard Business School Press, Boston, 1992, pp. 1-2.

[13] Pagonis, Ibid, pp. 210-220.

[14] "From one air force to another", Robert Mottley, American Shipper, November 1998, p. 22.

[15] "International Logistics", ibid, p. 14.

[16] "White House Years", Henry Kissinger, Little/Brown, Boston, MA, 1979.

[17] "The War Trap", Bruce Bueno de Mesquita, Yale University Press, New Haven, CT, 1981, p.104.

[18] Bueno de Mesquita, Ibid, p. 41.

[19] "Europe's Last Post", The Economist, May 13, 2000, p. 65.

[20] "The Postal Service's Market Grab", Michael A. Schuyler, in Hudgins, ibid, p. 25.

[21] Schuyler, Ibid, p. 33.

[22] "Consequences of Competition", Robert H. Cohen, in Hudgins, ibid, p. 109.

[23] "Postal Service Problems: the Need to Free the Mails", Peter Ferrara, in "The Last Monopoly: Privatizing the Postal Service for the Information Age", Edward L. Hudgins, editor, Cato Institute, Washington DC, 1996, p.p. 23-32.

[24] "The Coming Revolution in Mail Delivery", Edward L. Hudgins, in Hudgins, Ibid, p. 4.

25 *"The Postal Service's Market Grab"*, Michael A. Schuyler, in Hudgins, Ibid, p. 25.

26 *Zacher with Sutton, Ibid, p. 192.*

27 *"1989 Statistique des Services Postaux"*, UPU, Berne, 1990.

28 *"Europe's Last Post", The Economist, May 13, 2000, p. 65.*

29 *"Europe's Last Post", Ibid, p. 65.*

30 *"Europe's Last Post", Ibid, p. 66.*

31 *"Going postal", American Shipper, February 1999, p. 20.*

32 *"Europe's Last Post", Ibid, p. 68.*

33 *"Europe's Last Post", Ibid, p. 65.*

34 *"Europe's Last Post", Ibid, p. 65.*

35 *"Europe's Last Post", Ibid, p. 66.*

36 *"Europe's Last Post", Ibid, p. 65.*

37 *"Europe's Last Post", Ibid, p. 66.*

38 *U.S. Environmental Protection Agency, "Report of Audit E1D37-05—456-80855", "EPA's Program to Control Exports of Hazardous Waste, March 31, 1988.*

39 *Unreleased draft of "National Survey of Hazardous Waste Treatment Storage, Disposal and Recycling Facilities: Final Report", submitted to the Environmental Protection Agency by Research Triangle Institute, Durham, NC, as cited in "Global Dumping Ground", Bill Moyers and the Center for Investigative Reporting (CIR), Seven Locks Press, Cabin John, MD, 1990, p. 6.*

40 *Moyers and CIR, ibid, p. 6.*

41 *The Economist, April 8, 1989, p. 24.*

42 *Moyers and CIR, ibid, p. 55.*

43 *Clapp, ibid, p. 23, and "The Environmental Politics of the International Waste Trade", Laura Strohm, Journal of Environmental Development 2, no. 2, 1993, p. 133.*

44 *"To Tonga with Love", Andrew Porterfield, California Business, December 1987, p. 68.*

45 *Moyers and CIR, ibid, p. 9.*

46 *Moyers and CIR, ibid, p. 13.*

47 *"UN plan shifts cleanup of toxic waste to exporter", Jack Lucentini, The Journal of Commerce, November 3, 1998, p. 1A.*

48 *Moyers and CIR, ibid, p. 103.*

49 *"A Cleaner Environment: Removing the Barriers to Lead-Acid Batteries", James G. Palmer and Michael L. Sappington, St. Paul, MN, October 1988.*

50 *Moyers and CIR, ibid, p.94.*

51 *San Jose Mercury News, August 25, 1988.*

52 *Clapp, ibid, p. 3.*

53 *Scott Horne, quoted in Clapp, ibid, p. 87.*

54 *Clapp, ibid, p. 115.*

55 *"The Global Poison Trade", Newsweek, November 7, 1988, p. 66.*

56 *Moyers and CIR, ibid, p. 10.*

57 *Moyers and CIR, ibid, pp. 17-32.*

58 *Reprinted with permission of the Institute of Industrial Engineers, 3577 Parkway Lane, Suite 200, Norcross, GA 30092, 770-449-0461*

59 *The acronym stands for Cooperative for Assistance and Relief Everywhere, Inc.*

Selected Bibliography

Advisory Commission on Conferences in Ocean Shipping, "Report of the Advisory Commission on Conferences in Ocean Shipping", Washington DC, April 1992.

Albaum, Gerald, Jesper Strandskov, Edward Duerr, and Laurence Dowd, "International Marketing and Export Management", Addison-Wesley, Reading, MA, 1994.

Atkins, Captain Warren H, "Modern Marine Terminal Operations and Management", The Port of Oakland, Oakland, CA, 1983.

Ballou, Ronald H, "Business Logistics Management", 4th edition, Prentice Hall, Upper Saddle River, NJ, 1999.

Boeing Corporation, "World Air Cargo Forecast", annual.

Bowersox, Donald J, and David J. Closs, "Logistical Management: The Integrated Supply Chain Process", McGraw-Hill, New York, 1996.

Bowersox, Donald J., Patricia J. Daugherty, Cornelia L. Droge, Dale S. Rogers, and Daniel L. Wardlow, "Leading Edge Logistics- Competitive Positioning for the 1990's", Council of Logistics Management, Oak Brook, IL, 1989.

Bowersox, Donald J, David J. Closs, and Theodore P. Stank, "21st century Logistics: Making Supply Chain Integration a Reality", Council of Logistics Management, Oak Brook, IL, 1999.

Boyer, Kenneth D, "Principles of Transportation Economics", Addison-Wesley, New York, 1997.

Braithwaite, John, and Peter Drahos, "Global Business Regulation", Cambridge University Press, New York, 2000.

Branch, Alan E, "Elements of Shipping", 7th edition, Chapman and Hall, New York, 1996.

Carroll, Brian J, "Lean Performance ERP Project Management: Implementing the Virtual Supply Chain", St. Lucie Press, Boca Raton, FL, 2002.

Carroll, Glenn R, and Michael T. Hannan, "The Demography of Corporation and Industries", Princeton University Press, Ithaca, NY, 2000.

Cass, S, "Port Privatization: Process, Players and Progress", IIR Publications, London, 1996.

Clapp, Jennifer, "Toxic Exports: The Transfer of Hazardous Wastes from Rich to Poor Countries", Cornell University Press, Ithaca, 2001.

Cooper, James, Michael Browne and Melvyn Peters, "European Logistics: Markets, Management and Strategy", 2nd edition, Blackwell, Malden, MA, 1994.

Coyle, John J, Edward J. Bardi and C. John Langley, "The Management of Business Logistics: A Supply Chain Perspective", 7th edition, Thomson South-Western, Mason, OH, 2003.

Coyle, John J, Edward J. Bardi and Robert A. Novack, "Transportation", 5th edition, West Publishing, St. Paul, MN, 2000.

Czinkota, Michael R, and Ilkka A. Ronkainen, "International Marketing", 5th edition, Dryden, New York, 1998.

De Ruse, Antonio Estache Gines, editor, "Privatization and Regulation of Transport Infrastructure: Guidelines for Policymakers and Regulators", World Bank Institute, Washington DC, 2000.

Deming, W. Edwards, "Out of the Crisis", Massachusetts of Technology, Cambridge, MA, 1986.

Dempsey, Paul S, "Airport Planning and Development Handbook: A Global Survey", McGraw-Hill, New York, 2000.

Dicken, Peter, "Global Shift: Transforming the World Economy", 3rd edition, Guilford, New York, 1998.

Doganis, Rigas, "Flying Off Course: The Economics of International Airlines", 2nd edition, Routledge, New York, 1985.

Dornier, Philippe-Pierre, Ricardo Ernst, Michel Fender, and Panos Kouvelis, "Global Operations and Logistics: Text and Cases", John Wiley and Sons, New York, 1998.

Dawe, Richard, "The Impact of Information Technology on Material Logistics in the 1990s", Transportation and Distribution and Ernest & Young, 1993.

Estache, Antonio, and de Rus, Gines, "Privatization and Regulation of Transport Infrastructure: Guidelines for Policymakers and Regulators", The World Bank, Washington DC, 2000.

Ethier, Wilfred J, "Modern International Economics", 3rd edition, W.W. Norton, New York, 1995.

Faulks, Rex W, "International Transport: An Introduction to Current Practices and Future Trends", CRC Press, Boca Raton, FL, 1999.

Fernie, John, and Leigh Sparks, editors, "Logistics and Retail Management: Insights into Current Practice and Trends from Leading Experts", St. Lucie Press, Boca Raton, FL, 1999.

Fredendall, Lawrence D, and Ed Hill, "Basics of Supply Chain Management", St. Lucie Press, Boca Raton, FL, 2001.

Gablel, Jo Ellen, and Saul Pilnick, "The Shadow Organization in Logistics: The Real World of Culture Change and Supply Chain Efficiency", Council of Logistics Management, Oak Brook, IL, 2001.

Gomes-Ibanez, Jose A, and Clifford Winston, editors, "Essays in Transportation Economics and Policy", Brookings Institution Press, Washington DC, 1999.

Gourdin, Kent N, "Global Logistics Management", Blackwell, Malden, MA, 2000.

Gross, K. Hawkeye, "Drug Smuggling: The Forbidden Book", Paladin Press, Boulder, CO, 1992.

Harmon, Roy L, "Reinventing the Warehouse: World Class Distribution Logistics", Free Press, New York, 1993.

Heizer, Jay, and Barry Render, "Operations Management", 6th edition, Prentice Hall, Upper Saddle River, NJ, 2001.

Hill, Charles W, "International Business; Competing in the Global Marketplace", 4th edition, McGraw-Hill Irwin, New York, 2003.

Hinkelman, Edward G, "Dictionary of International Trade: Handbook of the Global Trade Community", third edition, World Trade Press, Novato, CA, 1999.

Hudgins, Edward L, "Mail @ the Millenium: Will the Postal Service Go Private?", Cato Institute, Washington DC, 2000.

Journal of Business Logistics (quarterly)

Journal of Commerce (weekly)

Keebler, James S, Karl B. Manrodt, David A. Durtsche, and, D. Michael Ledyard, "Keeping Score: Measuring the Business Value of Logistics in the Supply Chain", Council of Logistics Management, Oak Brook, Il, 1999.

Kopicki, Ronald J, editor, "Best Policies and Practices for Supply Chain Development in Emerging Markets", MIT Press, Boston, MA, 2000.

Krugman, Paul, "Geography and Trade", MIT Press, Cambridge, MA, 1991.

Kyle, David, and Rey Koslowski, "Global Human Smuggling: Comparative Perspectives", John Hopkins University Press, Baltimore, 2001.

Lakshmanan, T. R, Uma Subramanian, William P. Anderson, and Frannie A. Leautier, "Integration of Transport and Trade Facilitation: Selected Regional Case Studies", The World Bank, Washington DC, 2001.

Lambert, Douglas M, James R. Stock, and Lisa M. Ellram, "Fundamentals of Logistics Management", Irwin McGraw-Hill, Boston, 1998.

Langenwalter, Gary A, "Enterprise Resources Planning and Beyond: Integrating Your Entire Organization", St. Lucie Press, Boca Raton, FL, 2000.

Locke, Dick, "Global Supply Management: A Guide to International Purchasing", Irwin, Chicago, 1996.

Lynch, Clifford A, "Logistics Outsourcing: A Management Guide", Council of Logistics Management, Oak Brook, Il, 2000.

McCarthy, Patrick, "Transportation Economics: Theory and Practice, A Case Study Approach", Blackwell, Malden, MA, 2000.

Mendenhall, Mark, Punnett, Betty Jane and Ricks, David, "Global Management", Blackwell, Cambridge, MA, 1995.

Mueller, Gerhard O. W, and Freda Adler, "Outlaws of the Ocean: The Complete Book of Contemporary Crime on the High Seas", Hearst Marine Books, New York, 1985.

O'Laughlin, Kevin A, James Cooper, and Eric Cabocel, "Reconfiguring European Logistics Systems", Council of Logistics Management, Oak Brook, IL, 1993.

Pagonis, William, "Moving Mountains: Lessons in Leadership and Logistics from the Gulf War", Harvard Business School Press, Boston, 1992.

Peng, Mike W, "Behind the Success and Failure of U.S. Export Intermediaries", Quorum Books, Westport, CT, 1998.

Peterson, Barbara Sturken, and James Glab, "Rapid Descent: Deregulation and the Shakeout in the Airlines", Simon and Schuster, New York, 1994.

Pollock, Daniel, "Precipice", Council of Logistics Management, Oak Brook, IL, 1997.

"Reuse and Recycling-Reverse Logistics Opportunities", Council of Logistics Management, Oak Brook, IL, 1993.

Robeson, James F, and William C. Copacino, "The Logistics Handbook", The Free Press, New York, 1994.

Sampson, Anthony, "Empires of the Sky: The Politics, Contests and Cartels of World Airlines", Random House, New York, 1984.

Schaffer, Richard, Beverly Earle, and Filiberto Augusti, "International Business Law and Its Environment", 3rd edition, West Publishing Company, New York, 1996.

Schoenbaum, Thomas J, "Admiralty and Maritime Law", West Group, St. Paul, MN, 2001.

Shapiro, Alan C, "Multinational Financial Management", 6th edition, Allyn and Bacon, Boston, 1999.

Sletmo, Gunnar, and Ernest W. Williams Jr, "Liner Conferences in the Container Age", Macmillan, New York, 1981.

Southern, R. Neil, "Transportation and Logistics Basics", Continental Traffic Publishing Company, Memphis, TN, 1997.

Stock, James R, and Douglas M. Lambert, "Strategic Logistics Management", 4th edition, Irwin McGraw-Hill, Boston, 2001.

Stern, Louis W, Adel I. El-Ansary, and Anne T. Coughlan, "Marketing Channels", 5th edition, Prentice Hall, Upper Saddle River, NJ, 1996,

Stindt, Fred A, "Matson's Century of Ships", Modesto, CA, 1982.

Stopford, Martin, "Maritime Economics", 2nd edition, Routledge, London, 1997.

"Strategic Planning for Logistics", The Ohio State University for Council of Logistics Management, Oak Brook, IL, 1992.

Stutz, Frederick P, and de Souza, Anthony R, "The World Economy: Resources, Location, Trade, and Development", 3rd edition, Prentice Hall, Upper Saddle River, NJ, 1998.

Thord, Roland, "The Future of Transportation and Communication: Visions and Perspectives from Europe, Japan and the USA", Springer-Verlag, Berlin, 1993.

Tyworth, John E, Joseph L. Cavinato and C. John Langley, Jr, "Traffic Management", Waveland Press, Prospect Heights, Il, 1987.

United States, "Customs Law Handbook", Gould Publications, Binghamton, NY, 1989.

Vollmann, Thomas E, William L. Berry, and D. Clay Whybark, "Manufacturing Planning and Control Systems", 3rd edition, Irwin, New York, 1992.

Waters, Donald, editor, "Global Logistics and Distribution Planning: Strategies for Management", 3rd edition, CRC Press, Boca Raton, FL, 1999.

Wells, Alexander T, "Air Transportation: A Management Perspective", 3rd edition, Wadsworth, Belmont, CA, 1994.

Wood, Donald F, Anthony Barone, Paul Murphy, Daniel L. Wardlow, "International Logistics", Chapman & Hall, New York, 1995.

Wood, Donald F, and James C. Johnson, "Contemporary Logistics", 7th edition, Prentice Hall, Upper Saddle River, New Jersey, 1999.

World Trade Magazine, monthly.

Zacher, Mark W, with Brent A. Sutton, "Governing Global Networks: International Regimes for Transportation and Communications", Cambridge University Press, New York, 1996.

NAME INDEX

Abell, Derek F, 8

Adler, Freda, 272

Agusti, Filiberto, 42n, 268n, 288n, 304n

Albaum, Gerald, 91n, 124n, 268n, 288n

Anderson, William P, 21n, 91n, 235n

Armstrong, Richard, 332

Atkinson, Helen, 383n

Baasch, Henrik, 148n

Bagley, Constance E, 288n

Baldwin, Tom, 124n, 304n

Ballou, Ronald H, 383n

Bangsberg, P.T, 235n

Bardi, Edward J, 124n, 147n, 360n

Barnett, Chris, 124n, 326n

Barone, Anthony, 408n

Bauer, Michael J, 383n

Berhardt, Eric, 198

Bermudez, John, 383n

Bernstorff, Andreas, 172n

Berry, William L, 81-4, 91n

Betancor, Ofelia, 235n

Blasi, Joseph R, 326n

Bonney, Joseph, 235n

Bowersox , Donald J, 6.7,19, 21n, 26, 41n, 58, 59n, 124n, 341n, 344, 348, 360n

Brennan, Terry, 124n

Burch, John G, 383n

Calantoni, 58

Christopher, Martin, 73, 91n

Clapp, Jennifer, 172n, 409n

Closs, David J. 6, 7, 19, 21n, 26, 41n, 59n, 124n, 341n, 344, 348, 360n

Cohen, Robert H, 408n

Codevillo, Angelo, 408n

Cooper, M. Bixby, 41n

Cooper, Martha C, 26

Copacino, William C, 11, 59n, 91n

Coyle, John J, 124n, 147n, 360n

Critelli, Michael J, 367, 383n

Cruikshank, Jeffrey, 21n, 91n, 408n

Curry, James J, 59n

Dalenberg, Douglas R, 235n, 304n

Daley, James M, 235n, 304n

Dalhart, Glenn A, 341n

Damas, Philip, 235n, 341n

Dauer, Angela Calise, 326n

Daugherty, Patricia J, 235n

Dawe, Richard L, 21n, 66, 67, 91n, 369, 383n

De Ruse, Ginés, 235n, 408n

de Souza Anthony R, 91n, 124n

Delaney, Robert V, 21n, 49, 59n, 341n, 360n, 383n

Dicken, Peter R, 68, 69

Dornier, Philippe-Pierre, 59n

Dowd, Laurence, 91n, 124n, 268n, 288n

Drucker, Peter F, 3, 21n, 383n

Duerr, Edwin, 91n, 124n, 268n, 288n

Duesterber, Thomas J, 383n

SUBJECT INDEX

Russia, 80, 136, 137, 144, 298, 320

Sacramento International Airport, 198-202

Safety, 128, 137

Sandwich wars, 192

Sanitary Certificate, 276, 280

Scheduled receipts, 84

Security, 127, 184, 24-1, Chp. 13, 271, 312-5

SED, see Export Declaration

Selling terms, see Terms of Sale

Shipbreaking, 152

Shipbuilding, 152, 158

Shipper, 17, 111
    Large vs Small, 100
    Similarly situated shipper, 113

Shipper's Association (or Council), 104, 336

Shipping
    Inland, 144-5
    Maritime, see Maritime
    shipping

Siemens, 10, 331

Similarly situated shipper, see Shipper

Simple Mail Transfer Protocol (STMP), 374

Sinotrans, 332

Singapore, 38, 109, 110, 144, 211, 215, 402

Site, versus Situation, 95

Smuggling, 283, 242

Source reduction, 90

Sourcing, 9, 23, 25, 26

Spain, 16, 80

Specificity, 71

Speed, 97

Spending on logistics, 11, 12, 13, 15, 16

Staggers Rail Act of 1980, 143

Stanford Research Institute, 97

Starboard, 156

Stern, 156

Stevedores, 128

Stock Keeping Unit (SKU), 57, 343, 348

Storage, see Inventory

Subsidies, 241

Substantial transformation, 253

Sudan, 127, 224

Supply chain, Chp. 3, 373
    Also see Channels
    Defined, 6, 43
    Integration, 44
    Leadership, 51
    Management, 5, 6
    Myth, 46
    Orientation, 46
    Power, 50, 58
    Velocity, 51
    Waste, 45-8

Surveyor, 301

Switzerland, 15

Taiwan, 35, 77

Tajikistan, 343

Tankers, 103-4, 158, 166-8

Tare, 181

Tariffs, 36, 230, 317

Tariff engineering, 27

Tariff publishing, 368

Tariff shift rule, 253